The Ergonomics of Working Postures

Models, Methods and Cases

The Ergonomics of Working Postures

Models, Methods and Cases

Edited by
Nigel Corlett and John Wilson
University of Nottingham, England

and
Ilija Manenica
University of Split, Yugoslavia

The Proceedings of the First International
Occupational Ergonomics Symposium
Zadar, Yugoslavia, 15–17 April 1985

Taylor & Francis
London and Philadelphia
1986

UK	Taylor & Francis Ltd, 4 John St., London WC1N 2ET
USA	Taylor & Francis Inc., 242 Cherry St., Philadelphia, PA 19106–1906

British Library Cataloguing in Publication Data

International Occupational Ergonomics Symposium
   (1st : 1985 : Zadar)
   The ergonomics of working postures: models,
   methods and cases.
   1. Human Engineering
   I. Title   II. Wilson, John   III. Corlett,
   Nigel   IV. Manenica, Ilija
   620.8'2   TA166

   ISBN 0-85066-338-5

Library of Congress Cataloging in Publication Data

International Occupational Ergonomics Symposium
   (1st : 1985 : Zadar, Croatia)
   The ergonomics of working postures.

   Includes bibliographies and index.
   1. Human engineering—Congresses. 2. Human mechanics—Congresses.
   3. Posture—Congresses.
   I. Wilson, John, 1951– . II. Corlet, E. N. (Esmond Nigel). III.
   Manenica, Ilija. IV. Title. [DNLM: 1. Biomechanics—Congresses. 2.
   Human Engineering—Congresses. 4. Posture—Congresses.
   W3 IN8416 1st 1985e/WE 103 1575 1985e]
   T59.7.I574   1985   620.8'2   86-14459

   ISBN 0-85066-338-5

Cover design by Russell Beach
Typeset by Alresford Typesetting & Design,
Alresford, Hants.
Printed in Great Britain by Taylor & Francis (Printers) Ltd,
Basingstoke, Hants.

Preface

One of the earliest fields of interest in ergonomics which was able to provide predictive models was that of dynamic work. Oxygen consumption, heart rate and physical activity were brought together to provide models, capable of taking into account certain individual differences, which could be used to assess the workloads to which people were exposed and to compare different jobs in terms of their physiological cost.

The problems of static work, however, although recognized, did not yield so readily to modelling and prediction. Particular problem areas, such as seating, lifting and some aspects of posture, were explored extensively, but the understanding which tied these together was sparse.

Recent years have seen new techniques and new methods of analysis develop. EMG records are being investigated in more sophisticated ways; longitudinal studies of the effects of posture are more common, helped by more precise means for the recording and analysis of postures themselves. The increased sophistication of biomechanical analysis and of instrumentation in general now permits insights not previously available, whilst the ubiquity and power of the computer has opened up the potential for wider analyses of data.

The prevalence of posturally related diseases, and their relationships with working life, have also become more evident to the general public. There is a desire that the incidence of such diseases should be reduced and this has increased the need, and the opportunities, to pursue the development of a better understanding of posture.

These factors, coming together, denoted a re-awakened interest in this quite old area of ergonomics, and encouraged us to believe that it was appropriate to bring together those engaged in research into the various aspects of posture in order to assess the current research position, and to see how the various fields of postural research were linking together.

This belief was translated into action by setting up an international symposium on the Ergonomics of Working Postures, at Zadar, Yugoslavia, in April 1985. This symposium was seen as the first in an occasional series which would explore fairly specific areas of occupational ergonomics with the purpose of drawing together

v

major current trends. The intent is that the grouping of information achieved in this way will help researchers to re-assess the current state of play, and enable both new research approaches and new ways of applying the research to be developed.

Whether these objectives have been achieved, or are even approached, must be for the reader to decide. The symposium itself, which those who attended considered to have been very accurately described by its title, provided a fruitful exchange of information and mutual understanding. It is now the challenge for both the attenders at the symposium and the readers of this book to advance the study of posture so that we can move from the position of adopting simple, single-factor solutions to this highly complex problem to the stage of having a workable model of the relationships between muscle and joint actions, external forces with their directions and frequencies, and the bodily responses which lead to fatigue, adaptation and disease. The role of the ergonomist must be, primarily, a preventive one, and the effects of our inadequate knowledge of the responses to, and consequences of, working postures are too vast in their extent and effects to permit us to be satisfied with anything less than a practical ability to predict what is safe, acceptable and efficient with regard to this important aspect of working life.

Acknowledgements

The editors acknowledge with gratitude all the support and assistance they have received, from friends and colleagues in the UK and Yugoslavia, during the running of the Symposium and the production of this book. Those who have helped are far too numerous to mention individually, but special thanks must go to David Grist, our editor at Taylor and Francis, for his encouragement and efforts, to Moira Tracy for her work on the index, to our contributors for their patience with the editing process, and especially to Ilse Brown for her forbearance with two of the editors, and the effort she has put into organizing both Symposium and book.

E. N. Corlett
J. R. Wilson
I. Manenica

Contents

SECTION 1

POSTURAL RISK FACTORS AND DISEASE

A continuing question in the study of diseases is the identification of their causes, and where a disease is believed to have an occupational origin the concern to identify the cause is increased. This heightened interest arises because of the possible consequences to other as yet unaffected workers, the likely attentions of regulatory bodies in the outcome and the legal consequences of identifying an occupational cause. This first section presents five studies where the links between musculo-skeletal diseases and work behaviours have been investigated.

Interest must naturally focus on techniques as well as results, for the lack of long-term and large-scale epidemiological studies requires investigators to seek ways of linking causes and effects by using many factors which, by supporting each other, will reinforce the probability of a causal link. The use of a questionnaire to make the link between work activities and their consequences has some self-evident weaknesses, but can provide important initial data from which more focused investigations can follow. The unexpected point from the use of this procedure by Buckle, Stubbs and Baty is the appearance of an apparently strong dose–response relationship for foot disorders, these appearing when more than about 30% of the working time is spent on the feet. It may be that we should be using questionnaires more extensively than heretofore, perhaps as a regular and more standardized method for the exploration of the effects of work, prior to more detailed studies.

Pioneering work in this respect, using standardized questionnaires for investigating musculo-skeletal problems, has been supported by the Nordic Council of Ministers. A working group has produced four forms to date, a general one and three additional ones which go into more detail concerning problems experienced in the neck, the shoulder and the back, respectively. These are translated into Norwegian, Swedish, Danish and Finnish. They are also available in English from B. Ydreborg, at the Department of Occupational Medicine, Medical Center Hospital, S-701 85 Örebro, Sweden. A data base is being compiled from the users of these forms and, having the benefit of a widely agreed set of questionnaires designed for self-administration without expert assistance, it would seem good sense for these to be used even more widely. A computer analysis service is available for completed forms from the above department. This initiative could provide the data for a major investigation of the utility and validity of the method in practical applications.

Other papers in this section show that even low levels of load which are repeated can lead to considerable levels of sickness absence, and that these levels can be estimated from field studies, even though field samples are biased by having only the survivors at the more experienced end of the distribution. Static muscle load stands out as important, particularly when combined with a moderate or large extension or flexion of joints. An important qualification to be put on this state-ment comes from the paper by Kilbom, Persson and Jonsson, where they demon-strate that the important individual differences affecting musculo-skeletal complaints between operators doing the same work lie in the numbers of changes in posture adopted, and the more or less relaxed postures taken up by the different people in their samples. The investigation and analysis of these individual differ-ences require cost-effective techniques and these authors describe their method. Other possible procedures are presented in a later section.

But what is to be done about prevention of these disorders? The identification of threshold limit values to protect, say, 95% of the population seems an unlikely direction for progress, since there is, as yet, no means of specifying a load factor at a joint which would incorporate the effects of the static load, the contribution of repeatability and the forces imposed on the system by the work demands. Indivi-dual differences in people and in working techniques are other aspects which add to the difficulty. Tempting though it may be, therefore, to specify, say a percent-age maximum voluntary contraction, repetition frequency and angles of departure of a joint from some 'standard' position, the results are likely to be of little use. Even if the attempt were to be made, the danger of assuming that anything within the boundaries was safe would put many people at risk with a much reduced opportunity of getting their situation investigated.

Furtherance of knowledge about the effects of work activities and postures on health would be more rapidly advanced if there was a greater commonality of methods. Whether or not a standardization of methods is necessary is debatable, as yet the current methods available are not, in themselves, flexible enough to constrain investigators to a limited range of techniques. Yet it could still be useful if researchers, when they seek a method to gather the data they require, considered how those data could also be assessed against the findings of others with concerns beyond the range of their particular study. All research is limited by cost, but the selection of techniques with an eye to the use of their results as contributions to a pool of knowledge, leading towards the specification of the influence on disease of a limited number of factors, could lead to a fuller understanding of the problem in a relatively short time. There is a premium, in science, on doing original work rather than following a lead, but perhaps this paradigm can be followed too far in the pursuit of understanding.

At this intermediate state of knowledge it would seem desirable if company medical services were knowledgeable and up to date about causes, and preventive actions, in the field of work-induced musculo-skeletal injuries. Undoubtedly the best of them are, but the prevalence of such diseases, and the recent focus in industrial health on repetitive strain injuries (by whatever name), suggests that too little is appreciated. It is a problem which increases with automation and computers, due both to the more sedentary activities and the rapid responses possible from equipment. The links between behaviour and disease must be sought with some vigour, but meantime the approaches which enable potentially dangerous work practices to be identified and examined, such as are described in the following section, should be more widely used. They are imperfect, but will improve with more extensive application and development.

Chapter 1
Muscle Load and Illness Associated with Constrained Body Postures

R. H. Westgaard, M. Wærsted and T. Jansen

Institute of Work Physiology, Gydas vei 8, Oslo 3, Norway

and A. Aarås

Standard Telefon og Kabelfabrik A/S, Østre Aker vei 33, 0581 Oslo 5, Norway

1. Introduction

It is commonplace to learn about a worker contracting an injury in the back, shoulder, arm or other structure of the body while occupied with his normal duties at work. The incident may be a chance occurrence, even though related to activities at the workplace, or it may indicate that activities at the workplace constitute a general health risk for those working there. A further complication in the assessment of a possible work-related health risk is that similar injuries may happen as a result of activities outside work, or leisure-time activities may contribute to the development of injuries at the workplace.

The challenge then is to collect and analyse available data in such a way that it is clear whether conditions at work—on average—constitute a health risk. If the analysis is limited to injuries at a specific workplace, the result will be of limited general interest. However, if the measurements also include an assessment of workload, it will be possible to assemble results from different projects relating different levels of workload with the apparent health consequences of such loads, and thereby obtain a general picture of the effect of varying levels of work strain. If more information is available concerning the people who develop specific injuries, other compounding factors may also be evaluated.

Such thoughts are well known in epidemiological research. Work of this general character, trying to establish guidelines for safe limits for manual handling of heavy objects and concentrating on health risks such as back injuries, has been in progress for a long time (Davis 1977, 1981, Snook 1978). However, during the last few years low-level but prolonged static muscle load has appeared as a major risk factor in the development of load-related illnesses. This paper presents data evaluating the risk of different levels of static load on the trapezius (shoulder) muscles, in terms of the development of musculo-skeletal illnesses located in the shoulder and neck.

2. *Methods*

Measurement of muscle load

Muscle load was measured by electromyographic (EMG) recordings, using surface electrodes with a diameter of 6 mm and a centre distance of 20 mm between the electrodes. The EMG signal was quantified by a digital, full-wave rectification of a segment of the signal (50 ms duration), followed by a measurement of the average deviation of the rectified signal from zero signal value. The EMG signal value for longer time periods was calculated as the mean value of two or more intervals of 50 ms. This procedure, where the baseline is calculated as a moving average, has the advantage that small shifts in the baseline level throughout the recording period are compensated for in the analysis. Digital rectification and integration of the EMG signal is also more sensitive at low signal levels than quantification by the r.m.s. (root mean square) procedure.

The calibration of the EMG signal relative to muscle force largely followed the procedure of Jonsson (1978). A calibration recording was performed before the main recording session, and then repeated at the end. If the two calibration values deviated by more than 20% from the mean, the recording was rejected. Most calibrations showed much smaller deviations, but other sources of errors, like the estimation of maximal voluntary force (MVC), must also be taken into account. In the absence of more specific information on the different sources of error, it appears reasonable to retain 20% as a rough estimate of total error in the estimate of muscle force by EMG. Thus, a nominal load of 10% MVC indicates a real load between 8 and 12% MVC, while a nominal load of 5% MVC indicates a real load between 4 and 6% MVC.

The estimation of risk of injury

The estimation of risk of injury is best done by knowing the fraction of workers within a well-defined population who contract a specific injury or illness. In these projects the basis for this estimate has been sick leaves among groups of workers in identified work situations, with similar loads on specific body regions. Medical diagnoses associated with each sick leave have been collected from local health authorities or general practitioners. Sick leaves due to musculo-skeletal illnesses (myalgia, tendinitis, low-back pain, ischiadis, tendovaginitis, etc.) were of particular interest in this study. Such illnesses may be caused by overload on muscles or supporting tissue, whether or not such overload is caused by conditions at work, activities not related to work or other factors. Illnesses like arthritis which are not considered to develop as a consequence of muscle load are not included in the analysis of possible, work-related musculo-skeletal illnesses.

Some of these diagnoses contain further information regarding the nature of the illness. Tendinitis means an inflammation of a tendon. Other diagnoses, like myalgia, simply mean 'pain in a muscle', and the underlying pathological processes are poorly understood (if at all). It is therefore of little value to differentiate between different categories of musculo-skeletal diagnoses, especially since a general practitioner may be mistaken at this level of specificity. However, the common denominator of this class of illnesses is that the patient has experienced a high level of pain, and this is the important feature in our analysis. Sick leave with a musculo-skeletal diagnosis is simply interpreted as an incident where a worker has experienced an episode of pain in a muscle, or surrounding tissue, of sufficient intensity and duration to visit a doctor. The doctor has then agreed that the patient had this medical condition and made a further decision that the patient is unable to continue working. The only other information used in the analysis is the location of the illness on the body, and considerable effort has gone into contacting general practitioners to establish the body location of the illness if this was not clear from the first diagnosis.

When used in this way, the uncertainty in the classification of illnesses should be small. A few sick leaves may have received a 'relevant' diagnosis which is not correct, even by our relatively coarse classification, and the opposite is also possible. In addition, we have been unable to identify diagnoses in the case of about 5% of all sick leaves. The results presented in the following sections probably represent an underestimate of the problem among the different groups of workers, especially since interviews have established that some workers continue to work, and do not see a doctor, while experiencing very high levels of pain.

When estimating 'risk of contracting an injury', it is necessary to take into account the varying time of employment of the different workers. Time of employment in the work situation of interest is also time of exposure to the workload, and an increasing length of exposure to load is likely to increase the chance of contracting a load-related illness. In the statistical evaluation of the material, the workers are therefore divided into groups of at least 20 persons who have been employed a defined length of time. The number of workers who have recorded a relevant sick leave one or more times during their time of employment are then calculated as a percentage of all workers within the same group. In studies where it is of interest to study the development of sick leave with time, following an ergonomic adaptation of the workplace, the sick leave is calculated for each calendar year in terms of percentage of possible worktime.

3. Selection of workers with well-defined, localized workload

The aim of the work presented here is to assess the possible harmful effect of muscle strain in general, rather than of specific work situations. The tolerance to such strain will probably vary from muscle to muscle, as well as from person to person. The particular muscles selected in these studies are the upper trapezius and levator scapula, which is the muscle group lifting the shoulder girdle and is important in stabilizing the scapula when the person is performing arm movements.

In order to be able to use the 'group analysis' approach, it is important to select workers from work situations which demand a specific posture and movement of the arms, and thereby (hopefully) give a well-defined load on the trapezius muscle. Even though some variation in muscle load between workers in the same work situation must be expected, this variation ought to be small relative to differences in load between workers in different work situations.

The work situation in the first project (STK-8B) is extensively documented by Westgaard and Aarås (1984). The work involves production of parts for telephone exchanges, and is very demanding on the muscles in the shoulder. The position of the arms, and thereby the muscle load on the trapezius, varies in a stepwise manner throughout the day. The load on the trapezius is less well documented for this work situation than for the work situations in the three projects which followed, but it is clear that the load must have been very high at long intervals throughout the working day.

Figures 1 and 2 illustrate the work situations studied in the first two follow-up projects. In Figure 1 the work task is again production of parts for telephone exchanges, but the posture of the workers in this project (EB-DF) differs from that of the workers in the first project. In particular, their working height stays the same (and very high) throughout the working day. The third (overall) project is based on sewers making thermal clothing (HH-Fibre) (Figure 2). Again, the working height is the same throughout the working day, but is lower than in the second project, as is evident when comparing the upper-arm position in Figures 1 and 2.

The last project is based on female service workers on two production plat-forms of a North Sea oil field (Statfjord). Their work duties include cleaning and

Figure 1.   Typical working posture in one of the projects. The worker is connecting wiring points on the back panel of a telephone exchange switching unit. Other workers in the same production depart-ment, who were soldering the wiring points, were excluded from the analysis of sick leave.

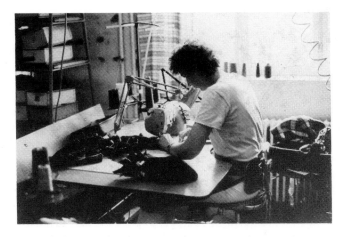

Figure 2.   Typical working posture for a sewer making thermal clothing.

various catering functions, similar to those of a large hotel on land. They are working at a high pace for several hours with few rest pauses. Their nominal working day is 12 hours and they have a work cycle of 14 days on the oil platforms and 21 days at home. Thus, the hours at work are very long, and the work strain very standardized, despite an apparently varied and dynamic work situation.

The workers included in the analysis are all females, and each group consists of 230–540 persons, employed in these particular work situations from a few months to 10 years. The analysis also includes a small control group of 56 female office workers with varied work duties, and with no VDU work. There has not been any measurement of load on the trapezius muscles for this group, but the work duties allowed small breaks to a much larger extent than in the other groups.

4.   Results

Work strain on the upper trapezius muscle

Figure 3 shows four examples of the load on the upper trapezius muscle while cleaning cabins on a North Sea oil platform. The data are collected from four different female workers. For each worker the recording from whichever of the right or the left trapezius showed the highest muscle load was selected. Each point in the four recordings represents mean muscle force an interval of 1 s, as a percentage of MVC. The points are plotted consecutively to show variations in the load throughout the recording period, which lasted from 1 hour 1 min to 1 hour 42 min.

In Figure 3 (*a*)–(*c*) points near 0% MVC are present all the time. There are also short periods of relatively high muscle load, near 50% MVC, at irregular intervals throughout the recording period, but median muscle load is relatively low all the time. Figure 3 (*d*) shows a recording where the muscle load remains larger than zero for periods of a few minutes. This is seen as unmarked patches underneath the band of points which indicate the variation in muscle load. The load on the trapezius muscle must be considered intermittent even for this worker, since the periods with low static load are very short. This work situation can therefore be considered to create a low-intensity, intermittent load on the main shoulder muscles. However, continuous monitoring of activity through heart rate and activity log showed that this work pattern might be sustained for several hours without any breaks, for a total of 8 hours throughout the 12 hour working day.

Figure 4 shows the load on the trapezius muscle of four sewers while making garments. This is similar to Figure 3, except that each worker did two different sewing operations with different sewing machines during the recording period. One sewing operation is represented in all four recordings, the other operations are all different. The muscle load does not change much when changing to a different sewing operation, except for the example in Figure 4 (*c*) where the last operation, using a semi-automatic machine, clearly is less demanding. However, this particular operation only employs 1 of about 80 sewers at any time.

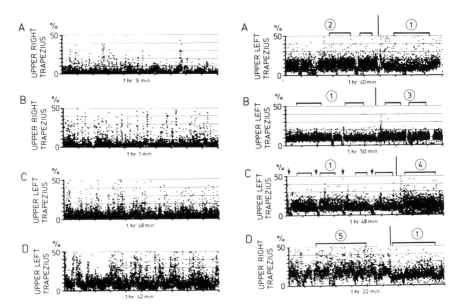

Figure 3.   Muscle load on the trapezius for four different female workers cleaning cabins on an oil platform. The trapezius recording with the highest load from either right or left trapezius is shown. Each point in the four recordings shows mean load in 1 s intervals, as a percentage of maximal voluntary contraction (MVC). Notice the variable length of the recordings, which are shown in full for each worker.

Figure 4.   Muscle load on the trapezius for four different female sewers, making thermal clothing. Each worker performs two sewing operations separated by a vertical bar and marked by a number in a circle. The trapezius recording with the highest load is shown. The figure is otherwise similar to Figure 3.

It is apparent that the band of points which indicate the load on the trapezius is shifted away from zero load for all recordings in Figure 4. The muscles are always working, except for short periods when a set number of garments are finished. The finished garments are moved to a store and a new lot prepared for sewing. This part of the work cycle allows the trapezius muscles to relax. The general work cycle with long periods of static load on the trapezius interrupted by short pauses, is particularly noticeable in Figure 4 (*c*) where the breaks in muscle load are marked by arrows. These breaks are obviously important, but the long periods with static load may nevertheless be considered the dominant feature of the work pattern. The quantification of muscle load was therefore based on periods with static workload, as indicated by horizontal bars on top of each panel. The bars show recording periods used in the quantitative analysis.

The result of the quantitative analysis of workload on the trapezius for the service workers on the oil platforms and the sewers is shown in Figure 5. The figure shows static workload (Jonsson 1978) on the trapezius for nine recordings from nine different service workers (*a*), and for 25 recordings of sewing operations

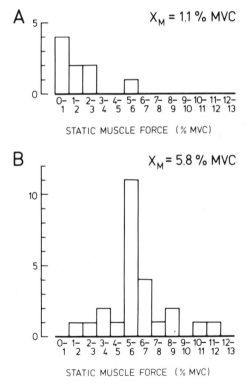

Figure 5. Quantitative analysis of static load on the trapezius for the two groups of workers illustrated in Figures 3 and 4: (a) female cabin cleaners and (b) female sewers. Twenty-five work operations by 15 sewers are included in the lower histogram.

by 15 sewers (*b*). Median static workloads on these muscles are 1·1% MVC for the service workers and 5·8% MVC for the sewers. Most of the latter group work with a static muscle load varying between 3 and 9% MVC. Thus, there is a very clear, statistically significant difference in workload on the trapezius between the two groups of female workers.

The workload in the third work situation, assembling parts for telephone exchanges, has not been analysed quantitatively to the same extent as the above work situations, but inspection of individual recordings has shown that the load pattern on the trapezius for this group of workers is similar to that of the sewers, except that the level of static load is higher. The median static load is likely to be between 8 and 10% MVC. There are also fewer breaks in the static-load pattern for these workers.

Thus, these three groups of workers represent work situations with distinctly different workloads on the shoulder muscles. In particular, there is little overlap in the load on the trapezius between the groups even when taking the variation within each group into account. This situation therefore represents a good basis for comparison of the development of musculo-skeletal illnesses in the shoulders and neck for the different groups of workers.

The development of illnesses in shoulders and neck for different groups of workers

Figure 6 shows the fraction of workers with sick leave due to musculo-skeletal illnesses in the shoulder or neck, as a function of time of employment, for the three follow-up experimental groups with a relatively stable load throughout the working day. (Identification of body location of the different musculo-skeletal ill-

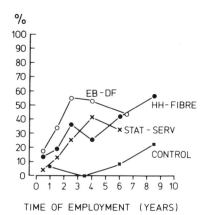

Figure 6.   Percentage of workers with one or several sick leaves due to musculo-skeletal illness located at the shoulder and neck, as a function of time of employment. Workers from three different work situations, with high static load (EB-DF), moderate static load (HH-Fibre) and intermittent, continuous load (Stat-Serv) on the trapezius muscles, are shown. A control group with varied office work is also included.

nesses was not complete at the time of writing. The results of Figure 6 thus contain some sick leaves not yet fully verified as meeting the criteria for inclusion.) In addition, similar data for the control group with varied office work are shown, except that the data available for this group do not allow any differentiation of musculo-skeletal sick leaves into different body locations. All musculo-skeletal illnesses including low-back and lower-arm injuries are therefore included in the data for the control group. Each of the points for the experimental groups in Figure 6 is based on at least 20 persons, varying from 22 to 160. The control group is much smaller, and the data for this group are based on groups of 9–19 persons.

For each of the experimental groups there is a clear tendency for an increasing fraction of workers to develop musculo-skeletal illnesses located in the shoulder and neck in the early years of employment. (The sick leaves may occur at any time during the period of work. They are usually of about 1 month's duration, varying from 3 weeks to almost 1 year.) This is in marked contrast to the control group where only 1 of 35 females record such illnesses for the first 5 years of work, and only 3 of 21 with 5–10 years' job experience. The difference is statistically significant, even with the small number of people in the control group and despite the inclusion of all musculo-skeletal illnesses, regardless of body location, in this group.

When comparing the experimental groups, the fraction of workers with a musculo-skeletal sick leave due to an injury in the shoulder and neck increases from 4 to 25% during the first 3 years of employment for the female workers with the most intermittent load (service workers), from 13 to 36% for the female workers with a moderate static load (sewers), and from 18 to 55% for the female workers with a high static load (assemblers). These data by themselves indicate a continuous, graded risk of developing musculo-skeletal illnesses with an increasing level of static load on these muscles. Differentiation between the three groups are less clear for workers employed for more than 3 years. For two of the groups there is even a reduction in the fraction of workers with recorded musculo-skeletal illnesses in the shoulder and neck among workers with the longest time of employment. This may, in part, be due to a selection process whereby workers who suffer long periods of pain at work try to find alternative employment, leaving only those who are able to meet the physiological demands of the work situation without too much discomfort. The two groups of workers where this effect is most noticeable also showed the highest overall incidence of musculo-skeletal illnesses. (The service workers had a very high rate of injuries in the low back and the lower arms in addition to shoulder and neck.) The fourth experimental group previously reported (Westgaard and Aarås 1984) had a rate of musculo-skeletal illnesses in the shoulder and neck which was intermediate to the two groups with moderate and high static load.

When correlating the load on the trapezius with the occurrence of musculo-skeletal illnesses in the shoulder and neck, it should be pointed out that a measurement of load on the trapezius is not a general measurement of strain in the shoulder–neck region. However, many patients and workers report symptoms located to the belly and tendons of the trapezius muscle, symptoms which are confirmed in medical examinations. The trapezius is the main lifter of the shoulder girdle, and is important in stabilization of the scapula, which is necessary when making precision movements of the arm. These movements are apparent in Figures 1 and 2, and also for the service workers. The load on the trapezius may therefore be considered a reasonable indicator of load on the shoulder and neck region, although caution in drawing too firm conclusions is warranted.

Effect of attempts to improve working posture

All the projects have included an attempt to reduce the problem with musculo-skeletal illnesses among the workers by initiating measures which should improve working postures. An improvement in posture should, by our hypothesis, cause a reduction in postural load on the most affected muscles, and thereby reduce the occurrence of musculo-skeletal illnesses.

Results from two of the projects have been evaluated and are reported here. The project at STK has already been documented by Westgaard and Aarås (1985). The results, in terms of reduced muscle load, reduced sick leave (Figure 7) and reduced labour turn-over (Figure 8) provide strong evidence of a successful attempt to reduce musculo-skeletal sick leave by introducing more flexible work stands, armrests and other aids which reduce postural load. The economic benefit of this work to the company is the subject of a separate contribution elsewhere in this volume (Chapter 34).

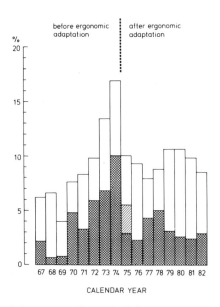

Figure 7. *Long-term sick leave (more than 3 days' duration, percentage of possible working time each year) at STK, Kongvinger, in the years 1967–1982. The hatched part of the columns indicates long-term sick leave due to musculo-skeletal illnesses. Single hatching in 1975 indicates musculo-skeletal sick leave starting in 1974.*

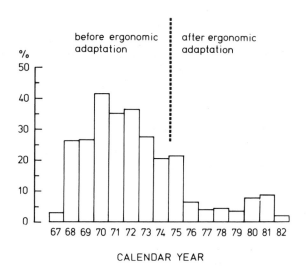

Figure 8.   Labour turn-over (percentage of average number of workers each year) at STK, Kongsvinger, in the years 1967–1982.

Recent, more comprehensive recordings of muscle load in the adapted work situations have shown that the workers at STK are still exposed to a level of static load which appears to be a problem among other groups of workers. Nor is musculo-skeletal-related sick leave eliminated in the adapted work situations. The surprisingly positive effect of ergonomic adaptations at STK may in part be due to a secondary effect of the ergonomic efforts at this factory, which has served as a testing ground for ergonomically designed workstands. This secondary effect is a considerable and continuing education of the workers and plant officials regarding the relationship between workload and the development of musculo-skeletal illnesses. This has without doubt influenced attitudes and thereby, possibly, willingness to accept necessary corrective measures such as short breaks or a change in work duties for workers with symptoms of postural pain. It is difficult to know to what extent this happens, or to quantify the effect of the different factors, but both the ergonomic adaptations, reducing the level of postural load and a possible change in attitudes, allowing necessary breaks in the load pattern, could be of importance for the reduction in sick leave.

Ergonomic adaptations of the workplace were also implemented for the sewers. The main features are illustrated in Figure 9. The table surface was enlarged and given a better shape. The enlarged table surface provided additional storage space for material to be sewn, and also better possibilities for resting arms and elbows. The height of the table was made adjustable by a rack-and-pinion device. A new chair, which was more comfortable and more easily adjustable than the old chair, was supplied. Apart from information provided at the time of introduction of the new workplaces, the ergonomic adaptations were not followed up in any major way with information to the workers regarding the use of the more flexible

Figure 9.   The adapted workplace for the sewers. Note the additional table surface and the crank on the table leg, used to adjust the height of the table. The sewing machine is recessed into the table and the height of the chair is easily adjustable by a gas cylinder.

workstand or the importance of reducing static load on the muscles. Also, it was necessary to maintain a high production rate in order to compete with foreign, low-cost imports.

The recordings of muscle load shown in Figure 4 were made in the adapted work situation. In two instances the previous work situation was recreated by removing the additional work surface, without any effect on muscle load (Figure 10). In one case, where a particular adjustment was made to enable a lowering of the table top and thereby reduce the working height, a clear reduction in muscle load was seen, similar to that demonstrated for work situations at STK. However, there was a distinct impression that the height adjustment facility of the table was little used—some workers did not even know the possibility existed. Thus, the overall effect of the ergonomic improvements in reducing muscle load was probably small. The improvements might also have been difficult to utilize at the high work pace with rapid movements of the arms. Figure 11 shows that there was no effect of the ergonomic adaptations in terms of reduced sick leave for these workers.

The different results in the two projects, one positive and one negative, are interesting. The ergonomic adaptations provided a basis for a reduction in muscle load in both projects, but have been efficient only in the project at STK. However, the workers at STK were working with a static load on the shoulder muscles in the

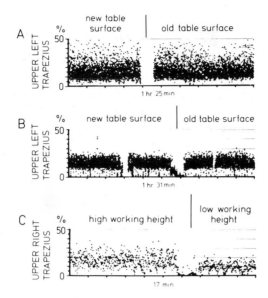

Figure 10.   Effect of ergonomic adaptations of the sewing workstand on the load on the trapezius muscle. The figure is similar to Figures 3 and 4 and shows the effect in terms of muscle load of working with and without the add-on table surface (a, b), and of reducing the height of the table (c).

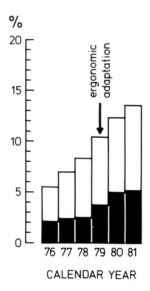

Figure 11.   Effect of ergonomic adaptations on musculo-skeletal and total long-term sick leave for the sewers. The histogram shows total long-term sick leave (full height of columns) and musculo-skeletal sick leave (dark part of columns) as percentages of possible working time during the different calendar years. The ergonomic adaptations were implemented in 1979.

adapted work situations—at a level which probably would be difficult to tolerate for long periods without adequate breaks. ('Breaks' should here be interpreted as pauses in load on local structures, either by interruptions in work or by changing to other work tasks which do not load the particular muscles.) The work situation at STK might have allowed such breaks to a larger extent than was the case for the sewers. This is highly speculative at the moment, but the hypothesis will be examined in further studies.

5. Discussion

The major point to emphasize in this paper is that even a low-level, intermittent but continuous load can provide a risk of developing musculo-skeletal illnesses if the load has to be maintained for a long enough time. Also, the indication of a graded risk of developing musculo-skeletal illnesses, with increasing levels of static muscle load and increasing time of exposure to the load, is interesting. In one of our projects as many as 50% of the workers with high static load and more than 2 years' employment have recorded one or several sick leaves due to musculo-skeletal injuries located in the shoulder and neck. In contrast, there are very few sick leaves of this kind among the control group. The control group is unfortunately not well defined with regard to work duties, and the aim is to improve and expand this material.

These results are based on measurements of load on the shoulder muscles (the trapezius), and it must be emphasized that the tolerance for loads may be different for different muscle groups. Muscles developed to counteract gravity forces have a different muscle fibre composition and a much higher resistance to prolonged load. The shoulder muscles would not be expected to have this kind of tolerance as they have been used intermittently with well-defined rest periods until the last few decades when there has been an increasing demand for high productivity—which for many work situations can be translated into long periods of continuous muscle load. This development would particularly affect the shoulder and neck muscles which are important in the control of position and movement of the head and arms. However, results based on the trapezius muscle may also be used in a general way as an indication of the effect of postural load, until more specific evidence regarding such effects on other body structures is available.

The general impression from our work is that a reduction in the level of postural load in work situations with little flexibility, by means of an ergonomic adaptation of the workplace, may be of major value. However, it is difficult to reduce the load sufficiently, to below a 'harmful limit', by such measures alone. It is therefore advisable to attempt to affect the work pattern also—by both education and motivation of the workers and by organizing the work in such a way that it is no longer necessary to load isolated structures (muscles) in the body throughout the working day.

Chapter 2
Musculo-skeletal Disorders (and Discomfort) and Associated Work Factors

P. W. Buckle, D. A. Stubbs and D. Baty

Ergonomics and Materials Handling Research Unit, Robens Institute,
University of Surrey, Guildford GU2 5XH, U.K.

1.   Introduction

The study reported formed part of a 5 year investigation of musculo-skeletal disorders within female occupational groups. Whilst the majority of this work was concerned with the problem of back pain within the nursing profession (Stubbs and Buckle 1984, Stubbs *et al.* 1983 a, b, 1984), attention was also paid to more general patterns of discomfort and disorder in other female groups.

The magnitude of the problem of musculo-skeletal disorders within the United Kingdom is evident from official statistics (DHSS 1982). These show that for the period 1979-1980 diseases of the musculo-skeletal system and connective tissue rank third for men and fourth for women with respect to spells of certified incapacity due to sickness. The contribution of a number of disorders to this category is detailed in Table 1, where episode and consultation rates are considered (RCGP, OPCS and DHSS 1979). The disorders selected are those which have been the

Table 1.   *Episodes and consultations per 1000 population.*

	Episode rate	Consultation rate
All diseases of the musculo-skeletal system and connective tissue	109·5	211·4
Bursitis	2·8	4·5
Tenosynovitis	4·2	7·0
Synovitis	1·6	2·8
Back pain (BP)		
Disc	6·3	19·1
BP alone	13·3	21·6
BP + sciatica	4·2	9·4
BP + other neuritis	1·2	2·2

Source: RCGP, OPCS and DHSS (1979).

subject of attention by ergonomists in recent years. Most studies have considered specific disorders in relation to workplace design (e.g. Armstrong *et al.* 1982, Herberts *et al.* 1981, Hünting *et al.* 1980) and there is little information relating to general patterns of discomfort or pain within workforces. This investigation reports the experience of regular discomfort or pain at a number of body sites experienced by female employees working within the retail trade in the United Kingdom. The study concentrates on the use of questionnaires in establishing the relationship between work elements, patterns of discomfort and the subjects' own attribution of factors 'causing' this pain or discomfort.

2. Method

The development and piloting of the questionnaire is described in full elsewhere (Stubbs *et al.* 1984) and only a summary is presented here. The questionnaire was designed to be self-administered by female employees in a cross-section of occupational groups. The questionnaire items were derived primarily from those used in previous investigations (Buckle 1983, Corlett and Bishop 1976, Stubbs *et al.* 1983 a) and modifications were made after discussions with interested parties within the organizations involved in the study and following pilot studies. The data examined in this paper were collected using this questionnaire which was distributed to staff in five large stores of a major department store chain and to nine sample sites of a major supermarket chain. In the former, the occupational health sister at each site was responsible for questionnaire distribution. The sister selected the first 100 female employees born in January and February (and where necessary, March). Two days after distribution, return 'reminders' were posted. The sealed replies were collected by the sisters who then forwarded them to the university. The procedure for distribution of the questionnaire within the supermarket chain was as described above with the exception that only ten employees were selected at each of the sites.

The data were subsequently coded for computer storage. Statistical analysis was undertaken on the University of Surrey's Prime network using the *Statistical Package for Social Sciences*, Version M, Release 9.1 (Nie *et al.* 1975). The items in the questionnaire relating to areas of occupational and individual characteristics, health issues and discomfort sites have been detailed in Figure 1. In order to establish the prevalence of regularly experienced pain or discomfort (i.e., more than once per week), subjects were requested to shade in areas, where applicable, on a small blank figure and to attribute, wherever possible, events or tasks that they felt were associated with this discomfort.

Where subjects have been asked to describe the factors associated with their pain or discomfort, the results have been considered only in a descriptive manner and absolute frequencies are not presented. This has been undertaken in order to avoid misinterpretation of data with widely fluctuating, and difficult to control, response rates.

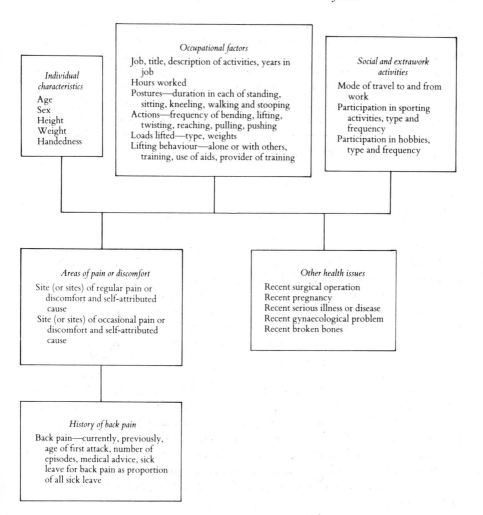

Figure 1.   *Questionnaire variables.*

3.   *Results*

Response rates

The response rates from the two surveys were 68·4% from the department chain store ($n = 342$) and 94% from the supermarket chain ($n = 85$). These response rates would suggest that there exists a high level of interest with respect to such problems amongst these female groups and this is further supported by the wide-ranging ideas put forward by respondents when open-ended items were represented and comments were elicited through the questionnaire. The results which follow are based on the 427 responses to the survey, although in many instances the variable or variables under discussion may represent smaller data sets due to the failure of respondents to complete certain items within the questionnaire.

Prevalence results

Sites of regular discomfort

For those working in department stores who completed this section $(n = 298)$, 64·8% recorded at least one site of regular discomfort, 52·3% at least two sites and 27·9% three or more sites. Figure 2 details the percentage of respondents reporting regular discomfort at each of the body sites indicated. Whilst the feet were considered to be a site of regular discomfort by over one-third of all respondents, these figures may have been biased by the use of discomfort or pain in the feet as an example in the instructions for completion of this section of the questionnaire. It should be noted however that the second example provided, the elbow, does not seem to have generated similar bias. Other immediate areas of concern are those of the upper and lower lumbar spine, the neck and shoulder girdle and, to a lesser extent, the knee.

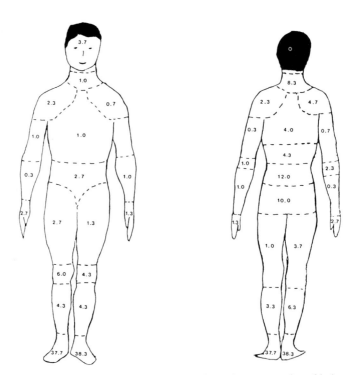

Figure 2.   Prevalence of regularly occurring pain or discomfort at a number of body sites amongst department-store staff. All figures are percentages of those completing this section of the questionnaire $(n = 298)$.

The pattern of discomfort for those working in supermarkets who completed this section ($n = 70$) is similar (see Figure 3), 56% recorded at least one site of regular discomfort, 41·5% at least two sites and 17·1% three or more sites.

Back pain

The importance of the back-pain element within the overall context of musculo-skeletal disorders is sufficiently great to consider this problem in a separate, as well as a joint, context. Accordingly, Table 2 shows the prevalence of this disorder within the two study groups. Lifetime prevalence includes all those with a history of back pain, annual prevalence those reporting back pain in the year preceding the survey and point prevalence is defined as those with back pain at the time of completion of the questionnaire.

The pattern of sick leave, and sick leave associated with back pain, is very similar for both study groups (Table 3). Inspection of this table reveals that approximately one-tenth of all sick leave was attributed to back pain and that this represents approximately half a day per person per year.

The recurrent nature of back pain is well illustrated in the findings reported in

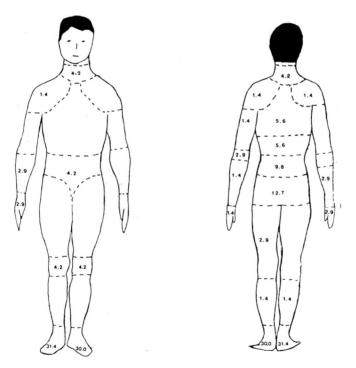

Figure 3.   Prevalence of regularly occurring pain or discomfort at a number of body sites amongst supermarket staff. All figures are percentages of those completing this section of the questionnaire ($n = 70$).

Table 2. *Prevalence of back pain.*

Prevalence (per 1000 at risk)	Department store (n = 342)	Supermarket (n = 85)
Lifetime	715	726
Annual	507	518
Point	192	229

Table 3. *Back pain and sick leave in 1982.*

	Department store (n = 342)	Supermarket (n = 85)
Number reporting sick leave in 1982	202 (59·6%)	51 (60%)
Number reporting sick leave for back pain in 1982	19 (5·6%)	5 (5·9%)
Days lost through sick leave in 1982	1831	456
Days lost through sick leave per person in 1982	5·35	5·36
Days lost through sick leave for back pain in 1982	202	44
Days lost through sick leave for back pain per person in 1982	0·59	0·52
Percentage of all sick leave attributed to back pain	11	9·6
Percentage of those reporting back pain in 1982 who had sick leave because of it	11·1	9·3
Percentage of those having sick leave in 1982 who reported having sick leave for back pain in 1982	9·4	9·8

Table 4. Whilst approximately one-quarter of both study groups reported no history of the disorder, the majority of those who had experienced back pain reported at least five attacks.

Self-attribution of pain or discomfort experienced by department-store staff

Pain and discomfort in the neck and shoulder region were frequently attributed to lifting or carrying of loads or to general fatigue at the end of a working day. Lifting and bending were also cited frequently by those reporting back pain and in some instances specific tasks were mentioned, e.g., stacking shelves. However, static postures, especially sitting for long periods and prolonged standing, were considered to be associated with back pain more frequently than were the dynamic tasks considered above. Problems associated with the upper limbs were most frequently attributed to either load carriage and lifting or previous injuries. Repetitive work tasks were only rarely considered to be contributory. Pain or discomfort in the hips, legs and knees was predominantly considered to be associated with either prolonged standing or general work fatigue as was the case when discomfort in the feet was recorded.

Table 4. Frequency of attacks of back pain.

	Department store		Supermarket	
Number of attacks	Frequency	Percentage of total	Frequency	Percentage of total
0	79	23·1	24	28·2
1	37	10·8	8	9·4
2	36	10·5	8	9·4
3	20	5·8	4	4·7
4	20	5·8	7	8·2
5–10	57	16·7	15	17·6
11–20	16	4·7	5	5·9
More than 20	58	17·0	10	11·8
Missing data	19	5·6	4	4·7
Total	342	100·0	85	100·0

Self-attribution of pain or discomfort experienced by supermarket staff

The pain or discomfort experienced in the neck and shoulder region was frequently attributed to lifting at work although problems with cash-till operation were also mentioned. Work factors at the checkout area were also listed with respect to problems of discomfort in the upper limbs. Abdominal pains associated with lifting were reported by two respondents and lifting was mentioned by many of those who reported regular discomfort or pain in their back. In addition to lifting, prolonged sitting and standing were other work components frequently cited by the respondents as being associated with this area of pain. Whilst pain in the hips, legs and knees was experienced by a smaller percentage of respondents it was often attributed to prolonged sitting or standing and the problem of cramped work-stations (e.g., tills) was again mentioned. The high prevalence of foot pain or discomfort may be due in part to the questionnaire bias mentioned previously. In the majority of instances this discomfort was attributed to prolonged standing.

The remainder of the results section considers three areas of pain or discomfort and investigates the association between these and occupational factors. In total, 37 separate body areas were considered but it would not be practical or useful to examine the findings for all of these within the context of this paper. The first site of pain reported in this section is that of back pain, which was selected because of its undoubted continuing importance as a major contributor to sickness absence and suffering. In addition to back pain, pain or discomfort in two other areas, the shoulder girdle and the feet, have been analysed with respect to associated risk factors.

Back pain

Analyses were undertaken using two groups of back-pain sufferers who were compared with the remaining members of the cohort. The two groups of sufferers comprised those who reported back pain at the time of completion of the questionnaire and those who reported having had at least one attack of back pain in the year previous to the administration of the questionnaire (i.e., 1982). Differences between those with and without the disorder were investigated for each variable using appropriate statistical techniques.

Those reporting back pain at the time of questionnaire completion were found to report spending less of their day walking at work (with back pain = 30·2%, without back pain = 40%, $t = 2·7$, $p < 0·01$) and considered themselves to have to perform frequent (more than ten times per hour) twisting and reaching movements more often than those not suffering from the disorder ($\chi^2 = 10·2$ with 4 df, $p < 0·04$ and $\chi^2 = 10·5$ with 4 df, $p < 0·04$ respectively). A similar analysis for those reporting back pain in 1982 indicated that this group was older than those who did not report pain in the year preceding the survey (with pain, mean = 36·2 years, SD = 13·2; without pain, mean = 33·3 years, SD = 14·4; $t = 2·1$ with 409 df, $p < 0·05$).

Shoulder girdle

An analysis was undertaken to compare the frequency of the hypothesized risk factors in those who did report regular pain or discomfort in this area and those who did not. No significant differences were recorded for any of the variables considered.

Feet

The high prevalence of foot pain or discomfort was investigated further by analysing the exposure to hypothesized risk factors in both the 'sufferers' and the 'non-sufferers'. Significant differences between these groups are reported in Table 5. Inspection of these indicate that sufferers report spending a greater time standing,

Table 5.   *Occupational factors and foot discomfort/pain.*

| | Regular pain or discomfort | | | |
| | Yes | | No | |
	Mean	SD	Mean	SD
Age (years)	32·4	13·3	36·5	14·1**[a]
Hours worked weekly	35·5	7·2	32·6	9·8**
Percentage of working day spent:				
Standing	28·2	27·9	17·2	22·1***
Sitting	14·4	19·6	38·0	35·3***
Kneeling	8·3	18·8	4·1	8·6*
Walking	45·4	26·2	33·5	27·2***

[a]Significance of difference in means (*t*-test): *$p < 0·05$, **$p < 0·01$, ***$p < 0·001$.

walking and kneeling and less time sitting than do non-sufferers.

Prior evaluation of this questionnaire (Baty *et al.* 1985) suggests that ability to estimate time spent during the working day in body postures detailed in the table is poor for standing and walking separately but good when both are combined. This measure of 'time on feet' during the working day has been considered (Figure 4) in relation to the prevalence of those suffering discomfort or pain. A clear dose–response relationship is observed. Of additional interest is the finding that for those spending more than 30% of their working day on their feet the prevalence of discomfort or pain is 48·2%, whilst for those spending less than this the figure is 7·0%.

4.   *Discussion*

This paper has described the pattern of regularly experienced pain or discomfort within a female working population of 427 members of the retail trade. Almost two-thirds (63%) of those who completed the relevant questionnaire section have reported regular pain or discomfort in at least one location and over half have reported two or more locations. It would seem therefore, that within this group the

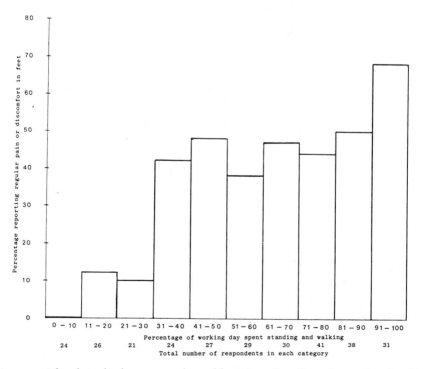

Figure 4.   The relationship between prevalence of foot pain or discomfort and proportion of working day spent 'on feet'.

'eternal triangle' of ergonomics put forward by Van Wely (1970) as being efficiency, comfort and health is sadly distorted. The bodily sites of most concern, as revealed by the prevalence rates, include the feet, the upper and lower lumbar spine, the neck and shoulder girdle and the knee. The prevalence of back pain within the study group may also be compared with results collected from a sample of nurses using the same questionnaire (Stubbs *et al.* 1984). The data were collected at the same time as the reported study. Nurses are considered to be a group of particular concern with respect to the back-pain problem and did show higher lifetime and point prevalence rates than did the study group. However, the annual prevalence rate of the study group was approximately 510 per 1000 at risk, which was only slightly less than that recorded for a nursing group (535 per 1000) (Stubbs *et al.* 1984). This similarity would suggest that there is a very real need to study this disorder further within the retail trade. Similarly, the sick-leave data from the retail-industry sample indicated that one-tenth of all sick leave was attributed to back pain and this figure agreed closely with that reported by the nursing sample. However, these are self-reported data and it would be of interest to know how they compare with official company health records. The confidential nature of the reported study did not, unfortunately, allow such a comparison to be made.

Considering first the risk factors associated with those reporting back pain at the time of the survey, a number of postural demands were identified (e.g., frequent twisting, frequent reaching, less time spent walking). This was not found to be the case when a similar analysis was undertaken for those who reported having experienced back pain in the year preceding the survey. In this analysis the only factor found to differ significantly was that of age. Whilst some evidence (Stubbs *et al.* 1983 a) suggests that for a nursing population different conditions are being assessed in each case, it is still considered surprising that more risk factors common to both groups were not identified.

Self-attribution of factors associated with back pain frequently elicited lifting, bending and prolonged sitting or standing. Such factors are often encountered in the literature (Buckle 1984) from retrospective epidemiological studies. In addition to these, however, specific tasks were often recorded. The problems of identifying such tasks are obvious at a group level in that they may involve only small sets of employees or represent only a small component of the working day. Whilst self-reporting may involve a large bias, it may also be the only rapid method for identifying such tasks for future investigation.

Statistical analysis of the groups who did and did not report pain or discomfort regularly in the neck or shoulders has not found any significant differences with respect to the work variables considered. The analysis was made more difficult by the relatively lower prevalence of these sites of pain. There was, however, some consensus of views when self-attribution of cause was requested. Many subjects concluded that lifting and carrying of loads were responsible for their pain and in some instances, especially among supermarket staff, specific workstations were cited (e.g., cash tills). These findings support the work of others (Ohara *et al.* 1976, Newton 1984) in identifying such workstations as associated with these disorders. The failure to identify factors at a group level may be due in part to the specificity of the factors associated with such workstations and their relatively low frequencies.

Whilst the prevalence data for back pain would indicate a serious area of concern, the extremely high point prevalence of foot discomfort or pain (approximately 340 per 1000 at risk for the total study group) was thought to be of particular relevance for ergonomists. Analysis of hypothesized work factors has indicated the existence of a close relationship between 'time spent on feet' and the prevalence of discomfort. Whilst this finding in itself is perhaps not surprising, and was frequently cited by subjects as being the 'cause', the apparent existence of a dose–response relationship and what could be considered to be an approximate threshold value in terms of exposure (roughly 30% of the working day) merits further study. If the existence of such a relationship is confirmed then it may be possible to provide firm guidance. The reported data do not consider duration of individual spells of 'time spent on feet', only total time as a percentage of the day. As such, the provision of regular rest periods may still provide a simple solution to this problem. Similarly, the lack of concurrent validity when self-reported standing and walking times are compared with those observed, prevents us examining in more detail the risk associated with the predominantly static load when standing still and that of walking, when an increased dynamic muscular component is present.

The likely existence of dose–response relationships, as for foot discomfort, is an encouraging finding in the attempt to utilize simple techniques (e.g., questionnaires) in the identification of tasks associated with sites of regular discomfort. The high prevalence of this particular problem has meant that the relatively small sample size for such an epidemiological approach has not proved a major problem. For less prevalent disorders it may be that case-control studies are to be preferred in the identification of associated occupational factors. However, one of the aims of this study was to consider how a simple questionnaire approach might aid both professional researchers and those with a wider remit in terms of health and safety, especially occupational health personnel, in the rapid identification of areas of concern. More importantly perhaps, the availability of such data may provide a much firmer base when arguing a need for improvements or the involvement of expert advice. Our own evaluation of the concurrent validity of self-reported data with respect to such questionnaires should, however, be borne in mind before such a development is undertaken (Baty *et al.* 1985).

5. Conclusions

The use of a questionnaire in determining factors associated with regularly experienced pain or discomfort within a group has been considered. Where the prevalence of a disorder has been high the identification of associated factors has been encouraging and in one instance a dose–response relationship has been presented. This approach has been usefully supplemented by subjects' own attribution of 'cause' of discomfort or pain. Such an approach is of particular use in identifying the problems associated with workstations used, for example, by only a small part of the work force or for only a small part of the day. The use of this approach may be beneficial to all those responsible for occupational health, especially where data relating to clinically diagnosed disorders are limited or not readily available.

Chapter 3
Cumulative Trauma Disorders of the Hand and Wrist in Industry

B. Silverstein, L. Fine, T. Armstrong, B. Joseph, B. Buchholz and M. Robertson

The University of Michigan, Ann Arbor, MI 48109, U.S.A.

1.   Introduction

The premise of cumulative trauma is that repetitive or sustained microtrauma occurring over time compromises the integrity or functioning of the tissues. There are several ways in which forceful and/or repetitive motions can lead to inflammation of tendons and compression of nerves. Cumulative trauma disorders (CTDs) therefore largely affect the musculo-skeletal system and the peripheral nervous system. The incidence and prevalence of hand and wrist CTDs have not been well documented in United States industry.

The main objective of this cross-sectional investigation was to determine whether there were associations between hand and wrist CTDs and known work attributes of force and repetitiveness, using concurrent primary data sources to estimate health effects and exposure.

The study population consisted of active workers with at least 1 year's seniority on jobs from six different industrial plants in the midwestern and southern United States, and whose job met one of four exposure group criteria. Workers in jobs with low-force–low-repetitive (LOF–LOR) attributes served as an internal comparison population for three other exposure categories: (*a*) high-force–low-repetitive (HIF–LOR); (*b*) low-force–high-repetitive (LOF–HIR); and (*c*) high-force–high-repetitive (HIF–HIR).

2.   Materials and methods

Site and job selection

Criteria for plant selection included:

  1. At least one job for each exposure category with approximately 20 or more workers performing each job.
  2. Stable employment and production patterns in the previous 2 years.

31

3. Plant willingness to release employees for health status evaluation.
4. No active labour–management disputes in progress.

The six plants which participated in the study included electronics assembly, major appliance manufacturing, investment casting of turbine engine blades, apparel sewing, ductile iron foundry and bearing manufacturing.

All jobs with at least 20 workers were identified and reviewed on a first plant visit by the research workers, without any information about the workers' health problems. Excluded from potential selection were jobs introduced within the previous 2 years. Cycle time, production rates and weight of parts handled were estimated at this visit. If the work cycle had a sequence of elements which repeated themselves within the cycle, this was defined as a 'fundamental cycle'. Jobs were also classified into force and repetitiveness categories (Table 1), during the visit.

Table 1.   *Criteria for initial classification of exposure.*

Repetitiveness
   High: < 30 s cycle time or > 50% of cycle time performing the same fundamental cycle
   Low: > 30 s cycle time and < 50% of cycle time performing the same fundamental cycle

Force
   High: estimated force requirements > 4 kg
   Low: estimated force requirements < 2 kg

Job analysis

At least three representative workers in each selected job were videotaped (two cameras) performing the job for at least three cycles. Signal generators displaying date, time, frame number and bilateral electromyographic (EMG r.m.s.) recordings were incorporated into the system's video mixer. EMGs were recorded from each subject's forearm flexor muscles and connected to a pre-amplifier system described by Foulke et al. (1981) (Figure 1). All EMGs were calibrated to known forces before and after the subject was filmed. Wrist posture and grip type were considered in the calibration.

Playback of the videotape in slow motion and stop action revealed a picture of the subject including posture, EMG and time which enabled a more detailed analysis of jobs to be performed. A modified Therblig elemental analysis for the right and left hand was used to identify fundamental cycles and their average number per work cycle. EMG recordings, wrist posture and grip type were abstracted approximately every 20 frames ($\frac{1}{3}$ s). Mean force and standard deviation for the right and left hand were estimated and averaged over subjects performing the same job. In order to characterize force requirements for different types of jobs, a weighting measure was used to take into account extreme variability in force within the cycle. This was referred to as 'adjusted force' [= (variance/mean force) + mean force]. Adjusted force (either right or left hand) of more than 6 kg was defined as *high force* and less than 6 kg was defined as *low force*. The mean adjusted force for the low-force jobs was 3·0 kg and for the high-force jobs was 12·7 kg.

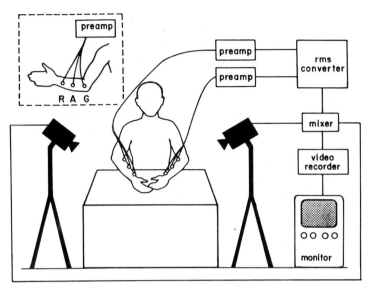

Figure 1.   Schematic of the EMG job-analysis system.

Subject selection and health status evaluation

A random sample of 12–20 active workers per job, stratified by age and sex where possible, was selected from among those with at least 1 year's seniority on the study job. Structured interviews and non-invasive physical examination were used to evaluate the health status of subjects. All health evaluations were conducted in private rooms in the plants during work hours by University of Michigan personnel.

A primary interview, administered by research team interviewers to all subjects, elicited demographic, prior health and work history information. Potential confounder or effect modifier data obtained on primary interview included sex, age, years on the job, prior upper extremity injuries, chronic diseases, reproductive status of females, recreational activities and prior job activities. The remaining questions addressed upper extremity pain or discomfort experienced in the previous 2 years. If the subject had experienced recurring difficulty in one or more parts of the upper extremity, the appropriate secondary interview instrument was used. The secondary interviews were designed to obtain more detail about the subject's complaints including location, duration, onset, aggravating factors and treatment.

All subjects received a standardized non-invasive physical examination from a research team examiner blinded to medical history and exposure. The examination was based on the work of Cyraix (1979), Cailliet (1980) and Waris *et al.* (1979). It included: inspection; active, passive and resisted range of motion testing; palpation; pulses; deep tendon reflexes; and dermatome evaluation. A tendon-related disorder was defined by no pain or limitation on a passive range of motions but increased pain with resisted motion tests. Carpal tunnel syndrome was defined by a positive modified Phalen's and positive Tinel's test. Guyon tunnel syndrome was defined by a positive Tinel's test at the guyon tunnel. A non-specific designation was used if no clear pattern was present on physical examination.

The general criteria for positive CTDs are presented in Table 2. Plant adjusted odds ratios (Mantel–Haenszel) and multiple logistic regression analysis were used to evaluate the association between hand/wrist CTDs and exposure status (Kleinbaum *et al.* 1982).

Table 2.   General criteria for cumulative trauma disorders.

Interview
    1. Symptoms of pain, numbness or tingling
    2. Symptoms lasting > 1 week and/or > 20 times in previous year
    3. No evidence of acute traumatic onset
    4. No related systemic diseases
    5. Onset since working on current job

Physical examination
    1. Characteristic signs of endpoints
    2. Rule out other conditions with referred symptoms

---

3.   Results

Out of 641 eligible workers originally selected from employee rosters, 2% refused to participate, 3·3% were on medical leave of absence, 4·2% did not actually meet selection criteria and 6 were excluded due to active rheumatoid arthritis. The final study population consisted of 574 (89·5%) active workers (287 males and 287 females). The mean age of the study population was 39·3 ± 10·1 years and the mean job seniority was 7·9 ± 5·8 years. Overall, there were no differences between males and females with respect to age or seniority.

However, males and females were not evenly distributed in exposure categories (Table 3). Males tended to predominate in the HIF–LOR category (35·2% of males and 18·1% of females). Females tended to predominate in the LOF–HIR category (34·8% females and 15·0% of males). Within exposure categories, males and females often did not perform the same jobs. There were significant differences in age and seniority on the study jobs by sex–plant–exposure strata ($p < 0.0001$). These three factors were treated as potential confounders in the analysis.

There were no significant differences in health history or recreational activities between sex/exposure groups and these were not considered further.

When those with symptomatic localized osteoarthrosis (LOA) were not considered (11 females and 4 males), there were 105 subjects (18·3%) with hand/wrist CTDs on interview (11·1% of males and 25·4% of females). There were 51 subjects (8·9%) with these disorders on physical examination and interview (4·2% of males and 13·6% of females) (Figure 2).

Table 3.   *Plant combined demographic characteristics by exposure group.*

	Exposure group			
	LOF–LOR	HIF–LOR	LOF–HIR	HIF–HIR
Number				
Males	75 $(26\cdot1\%)^a$	101 $(35\cdot2\%)$	43 $(15\cdot0\%)$	68 $(23\cdot7\%)$
Females	61 $(21\cdot3\%)$	52 $(18\cdot1\%)$	100 $(34\cdot8\%)$	74 $(25\cdot6\%)$
Age				
Males	39·3 (10·4)	40·2 (10·0)	41·3 (9·8)	36·2 (8·7)
Females	39·8 (10·7)	37·6 (7·9)	40·4 (11·4)	38·8 (9·7)
Years on job				
Males	6·6 (4·7)	9·5 (6·6)	8·3 (6·8)	8·6 (6·2)
Females	8·0 (5·8)	5·8 (3·6)	8·0 (5·6)	7·5 (5·4)

[a]Figures in parentheses are standard deviations.

Plant adjusted odds ratios indicated increased risk for hand/wrist CTDs (excluding LOA) in all exposure groups compared with the LOF–LOR group (Table 4). On physical examination and interview, the risk for HIF–HIR males was five times that of the LOF–LOR group ($p < 0.05$). In the sex-combined analysis, both high-force groups had a significantly increased risk of hand/wrist CTDs. The odds ratio was 30·3 ($p < 0.0001$) for the HIF–HIR group. The difference between the combined and sex-specific odds ratios was due to including females from two jobs in the combined analysis that were not included in the female-specific analysis due to no female controls in these two plants. These jobs had a high prevalence of CTDs among females (44·4% in punch press and 30% in hand grind).

The predictors in the logistic regression analysis (Table 5) were quite similar to the odds ratios observed in the stratified analysis. Hand/wrist CTDs were negatively associated with age and years on the job (not statistically significant)

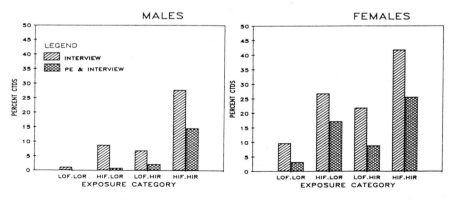

Figure 2.   *Prevalence of hand/wrist CTDs excluding LOA (plants combined).*

Table 4.   *Hand/wrist CTDS (excluding LOA) by exposure group plant adjusted odds ratios.*

	Exposure group		
	HIF–LOR	LOF–HIR	HIF–HIR
Interview			
Males	8·9*	5·0*	10·6***
Females	2·5	2·7	6·4*
Physical examination and interview			
Males	2·7	3·3	4·9*
Females	4·1	2·9	5·2
Combined	4·9*	3·6	30·3****

Chi square (M–H): $*p < 0.05$, $**p < 0.01$, $***p < 0.001$, $****p < 0.0001$.

Table 5.   *Predictors of hand/wrist CTDs (excluding LOA): physical examination and interview multiple logistic regression analysis, n-574. (Plants included in all models.)*

Predictor	Model I		Model II	
	Coefficient (SE)	Odds ratio	Coefficient (SE)	Odds ratio
Sex	1·9644 (0·41998)	7·1	1·5781 (0·41195)	4·8
Age	−0·00875 (0·01788)	1·0	−0·00856 (0·01899)	1·0
Years on job	0·00530 (0·03089)	1·0	−0·00787 (0·03272)	1·0
HIF–LOR			1·6487 (0·80129)	5·2
LOF–HIR			1·1994 (0·79897)	3·3
HIF–HIR			3·3713 (0·81109)	29·1
2 log likelihood	311·98		278·40	

suggesting a selection/survivor bias in the study population. The predicted association between sex and CTDs (odds ratio = 4·8) did not take into account job differences between males and females within exposure categories. When force (0, 1 variable), irrespective of repetitiveness, was entered into model I, the odds ratio for force was 4·4 ($p < 0.0001$). When repetitiveness (0, 1), irrespective of force, was entered into model I, the odds ratio was 2·8 ($p < 0.005$).

We are in the preliminary stages of attempting to estimate the contribution of posture to the difference in prevalence observed between jobs in the same exposure categories. Figures 3–6 present the percentage of worktime in which the dominant wrist is in ulnar deviation and more than 45° of flexion or extension, as well as the hand in any pinch or grip posture. The comparisons are between pairs of jobs within an exposure category having different prevalences of hand/wrist CTDs (excluding LOA) on physical examination and interview (prevalence indicated in legends).

In the LOF–LOR category (Figure 3) the major postural differences which may explain a higher prevalence among wax assemblers are more ulnar deviation and flexion of the wrist. In the HIF–LOR category core press operators used somewhat more ulnar deviation and pinching (Figure 4). In the LOF–HIR category, the Mpoint job required more wrist extension than the drill and tap job (Figure 5). In the HIF–HIR category Ppress required more pinching than the buff job (Figure 6).

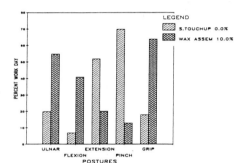

Figure 3.   Postural difference: LOF–LOR.

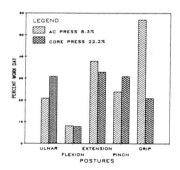

Figure 4.   Postural difference: HIF–LOR.

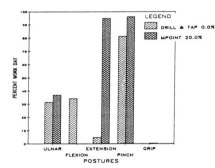

Figure 5.   Postural difference: LOF–HIR.

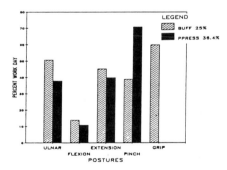

Figure 6.   Postural difference: HIF–HIR.

While the postural data are preliminary, they suggest that stressful postures by themselves (LOF–LOR), or in combination with high force and or high repetitiveness, may increase the risk of hand/wrist CTDs.

4.   Discussion

Force and repetitiveness are positively associated with hand/wrist (excluding LOA) CTDs irrespective of other factors. The combination of high force and high repetitiveness (HIF–HIR) increases the magnitude of association more than either factor alone. Stressful hand/wrist postures may increase the risk of CTDs by themselves or in combinations with high force and/or high repetitiveness. In order to estimate the relationship between these three factors further analysis is required.

The impact of sex as a confounder could not be adequately estimated because males and females within exposure groups were not always performing the same job. Even when they were performing the same job, females tended to be at greater risk for some, but not all, hand/wrist CTDs. It is possible that these observed associations between sex and CTDs were actually a function of increased postural stress due to stature. In order to test this hypothesis, the job of each worker would have to have been videotaped and analysed. This was not done in this investigation.

The findings in this investigation may have underestimated the prevalence of hand/wrist CTDs in several ways. First, subject selection was limited to active workers. Those off the job with CTDs at the time of evaluation (potentially more severe cases) would not have been available for study. Second, the 1 year seniority criterion for subject selection excluded those who might have CTDs and transferred before 1 year, as well as those with CTDs but not on the job for at least 1 year. The finding that hand/wrist CTDs were negatively associated with age and years on the job (Table 5) supports the argument of selection/survival bias in the study population.

Chapter 4
A Pilot Study of Postural Characteristics of Jobs
Associated with
an Elevated Risk of Rotor Cuff Tendinitis

L. Fine, B. Silverstein, T. Armstrong, B. Joseph and B. Buchholz

The University of Michigan, Ann Arbor, MI 48109, U.S.A.

1. Introduction

Disorders of the shoulder which are work related were first described by Ramazzinus in 1717. They generally have been attributed to one of two main factors, high static loads on the muscles of the shoulder girdle on repetitive forward flexion or abduction of the shoulder above midtorso height of approximately greater than 60° (Hagberg 1982, Wright 1945). Rotor cuff tendinitis, while not as common as disorders of the hand and wrist, can be a significant problem for workers in some jobs. In a cross-sectional survey which will be discussed in greater detail in this paper and by Silverstein *et al.* in Chapter 3, we surveyed 574 workers from six industrial sites selected because their jobs met specific exposure criteria. The prevalence of workers with 'positive' questionnaire responses and disorders of the hand and wrist identified through physical examination was 9% compared with 6% for shoulder disorders. The most common type of shoulder disorder was rotor cuff tendinitis, generally supraspinatus. In the four jobs with the highest prevalence rate in this study, 23% of the workers had findings on the questionnaire and physical examination consistent with cumulative trauma disorders of the shoulder.

Using the information from this study we have carried out an aetiological case control study design in order to compare, in a preliminary manner, the postural characteristics of the jobs associated with a high prevalence of shoulder disorders with the jobs associated with a low prevalence of shoulder disorders.

2. Methods

Job selection

The site and job selection criteria have been described by Silverstein *et al.* in Chapter 3. There were 34 jobs studied in six plants. The jobs were ranked in order from the job with highest prevalence of shoulder disorders to those jobs with a zero prevalence. The four jobs with the highest prevalence of positive cases identified by questionnaire and physical examination were classified as the 'high-risk jobs'.

These jobs were from two plants (A and B). Nine other study jobs in these plants had prevalence rates of 5% or less. Among the potential confounding factors of gender, plant, age and exposure categories for hand repetitiveness and force as described by Silverstein *et al.*, we believed that the plant was probably most important. As a result we selected all three available low-risk jobs from plant A, together with the three high-risk jobs. The one low-risk job available from plant B had a high rate of shoulder complaints on the questionnaire responses, suggesting that it might not truly be a low-risk job. Therefore, we selected the remaining low-risk job from the third plant, C, to take with the fourth high-risk job in plant B. The selection process was completed in the absence of knowledge of the postural characteristics of the jobs.

Job analysis

At least one 'representative' worker in each selected job was videotaped from two angles performing his/her job for at least three cycles. Playback of the videotape in slow motion and stop action revealed the subject's posture. A modified Therblig elemental analysis for the right and left hand identified what we called the 'fundamental cycles' and their average number per work cycle. Postures of the shoulder were determined by their location about three axes of rotation in the following order: (i) extension–forward flexion, (ii) abduction–adduction and (iii) external–internal rotation. Extension–forward flexion was divided into five levels: (*a*) < 20° described as 'neutral', (*b*) 21–60°, (*c*) 61–90°, (*d*) > 90° and (*e*) 0–60° of backward extension. Extension–forward flexion were combined into one variable depending on the degree of flexion and abduction. The postures were recorded every 0·33 s during active movement.

Health status evaluation

Silverstein has described subject selection and the general features of the health status evaluation. Shoulder disorders were divided into the following categories: (*a*) degenerative joint disease–acromioclavicular joint; (*b*) degenerative disease–gleno-humeral joint; (*c*) bicipital tenosynovitis; (*d*) rotor cuff tendinitis; (*e*) 'frozen shoulder'; and (*f*) non-specific, if there was no clear pattern on physical examination consistent with one of the above shoulder diagnoses. Disorders of the neck/scapula, elbow/forearm and hand/wrist were also considered.

3. Results

The number of subjects, their ages and duration of employment on their current job were similar for high- and low-risk workers. However, the results of questionnaire analysis and the physical examinations confirmed that cumulative trauma disorders of the neck/scapula, elbow/forearm and hand/wrist were more common among high-risk workers (Table 1).

Table 1.   *Personal characteristics and prevalence rates (as percentages).*

	High-risk jobs	Low-risk jobs
Number of subjects	76	80
Mean age	41	42
Percentage female	88	54**
Positive diagnosis (%) of:		
Neck CTDs	14	3*
Elbow CTDs	5	4
Hand/wrist CTDs	28	11*

$*p < 0.02, **p < 0.001.$

More high-risk workers were female (88% versus 54%) than low-risk workers. The hand forces were somewhat lower in the high-risk workers, while the number of repetitive elements during an 8 hour work shift were higher in the high-risk workers, due mainly to one of the high-risk jobs (Mount point—Table 2). While the high-risk jobs all had elements with (estimated) higher peak forces than for all elements from the low-risk jobs, only one of the high-risk jobs had a long period of high static load on the shoulder (Mount point—Table 3).

The shoulder postures did not differ in terms of the percentage of active cycle time spent in the various levels of flexion or abduction, although only a small percentage of time was spent in flexion or abduction greater than 60° (Table 4). The largest difference was with regard to external rotation. For high-risk jobs the time spent in slight external rotation was 62 and 46% for the right and left shoulder,

Table 2.   *Force and repetitiveness of high- and low-risk jobs.*

	Plant	Adjusted force,[a] right hand (kg)	Adjusted force, left hand (kg)	Fundamental cycles/8 hours
Low risk				
Wax assembly	A	2·60	1·10	2817·00
Cut off	A	14·20	14·50	4310·00
Screw machine	A	9·30	11·10	96·00
Hem leg	B	9·10	10·0	5405·00
Mean		8·8	9·2	3157
High risk				
Punch press	C	6·40	8·0	4095·00
Injection	A	8·90	6·40	802·00
Core press	A	6·30	3·40	1087·00
Mount point	A	0·90	1·10	23126·00
Mean		5·4	3·6	8338·00

[a] Adjusted force = mean force + (variance/mean force).

Table 3.   *Possible confounders.*

	High peak force–shoulder[a]	Long duration of static load[b]
Low-risk jobs		
Wax assembly	No	No
Cut-off	Yes	No
Screw machine	No	No
Hem leg	No	No
High-risk jobs		
Injection	Yes	No
Core press	Yes	No
Punch press	Yes	No
Mount point	Yes (F)	Yes (L)

[a]Less than 70% are predicted to be 'capable' for at least one job element by the Saggital Plane Static Strength Prediction Model for men; (F) is similar for women.
[b]Static load at the shoulder for more than 256% of the job cycle time.

respectively, compared with 39 and 29% for the right and left shoulder for the low-risk jobs (Table 5). Most of the time spent in external rotation was not, however, in extreme rotation.

Table 4.   *Proportion of time in flexion and abduction.*

Posture	Low-risk jobs	High-risk jobs
Right hand		
Neutral		
Abd function $< 20°$, and flexion $< 20°$	33% (0–57%)[a]	40% (8–88%)
Flexion		
20–60°	20% (0–31%)	26% (2–60%)
60–170°	5% (0–15%)	12% (0–32%)
Abduction		
20–60°	41% (12–100%)	21% (0–64%)
61–170°	1% (0–3%)	2% (0–6%)
Left hand		
Neutral		
Abd function $< 20°$, and flexion $< 20°$	39% (12–70%)	48% (0–76%)
Flexion		
20–60°	21% (3–58%)	38% (15–100%)
60–170°	6% (0–14%)	4% (0–15%)
Abduction		
20–60°	19% (5–29%)	9% (0–18%)
61–170°	5% (0–14%)	1% (0–1%)

[a]Figures in parentheses are the ranges, the preceding figure is the mean.

Table 5.   Proportion of time of internal and external rotation.

Posture	Low-risk jobs	High-risk jobs
Left hand		
External rotation		
31–90°	0·2% (0–1%)[a]	4% (0–9%)
0–30°	29% (8–46%)	46% (2–88%)
Internal rotation		
0–75°	71% (55–92%)	51% (3–87%)
Right hand		
External rotation		
31–90°	0·2% (0–4%)	6% (0–18%)
0–30°	39% (20–68%)	62% (40–99%)
Internal rotation		
0–75°	59% (55–92%)	32% (1–60%)

[a] Figures in parentheses are the ranges, the preceding figure is the mean.

4.   Discussion

The differences between the postural characteristics of the high-risk and the low-risk jobs failed to confirm our initial hypothesis that the high-risk jobs would involve significantly greater amounts of time in abduction or flexion greater than 60°. There also was a suggestion that slight external rotation might be a risk factor for rotor cuff tendinitis. While the two sets of jobs appear not to differ on several potential confounding factors, either because of our selection process or chance, they did appear to differ in terms of our estimates of peak forces at the shoulder joint. Thus the explanation for the difference in level of risk between jobs may be due to differing force levels rather than postural characteristics. A recurring observation in our epidemiological studies is that one must consider at the least risk factors of force, repetitiveness and posture in order to understand fully the job-related predictors of the prevalence of upper extremity occupational musculoskeletal complaints and disorders. Future investigations of postural factors should consider external rotation as well as abduction and flexion at the shoulder joint.

Chapter 5
Risk Factors for Work-related Disorders of the Neck and Shoulder—with Special Emphasis on Working Postures and Movements

Åsa Kilbom, Jan Persson and Björn Jonsson

National Board of Occupational Safety and Health, Work Physiology Unit of the Research Department, S-171 84 Solna, Sweden

1. Introduction

According to studies by, for example, Kvarnström (1983) and Westgaard and Aarås (1984), manufacturing work in the electronics industry is associated with a high prevalence of cervicobrachial disorders. These states have been ascribed to the working conditions, with repetitive, manual, short-cycled tasks, raised arms and large demands on accuracy and speed of work (Hagberg 1984 b).

However, the outcome in an individual case varies to a large extent; some employees go without symptoms for many years, whereas others have to be relocated because of serious disorders after a few years only. Since the gross characteristics of the work tasks do not show large variations, some powerful individual factors must exert their influence, either protecting the individual or increasing his/her vulnerability. This study was undertaken in order to identify such individual factors. It was carried out as a cross-sectional study, where an attempt was made to relate disorders from the neck–shoulder–arm region with factors such as individual variations in working technique, physical capacity, medical and work history.

Since any individual risk factor—or protective factor—may exert its effects only over a long period of time, the study was also supplemented with 1 and 2 year follow-ups. The outcome, in terms of deterioration or improvement of disorders, was related to those previously studied individual factors. This paper mainly focuses on the observation method used to study working movements and postures, and the relationships, at the cross-sectional study, between these variables and disorders.

2. Methods

Procedure

At the onset of the study (investigation I) 96 female workers in the electronics manufacturing industry were studied for their medical history and status, emphasizing musculo-skeletal disorders, and an evaluation of their working technique. Their mean age was 31 years, with a standard deviation of 11 years and a range from 17 to 57 years. Moreover, physical working capacity, productivity, sick leave, work history, leisure habits and subjective estimates of stress and work satisfaction were investigated but will be reported separately. No subjects who had been on sick leave or sought medical aid for cervicobrachial disorders during the previous year participated. After 1 year (investigation II) and 2 years (investigation III) a new thorough medical evaluation was done, sick leave and productivity were re-evaluated and the working tasks were checked and any changes were noted. At investigations II and III, 12 and 19 subjects, respectively, were not available, due to childbirth, change of job or sick leave.

Development of a video technique for the analysis of postures and movements of head, shoulder and upper arm (VIRA)

Even though electronics manufacturing work has many common characteristics (such as sitting with forward flexion of neck and raised arms) each task also has its own characteristics in terms of, for example, work-cycle time, number of components assembled, frequency of arm elevations and proportion of the time allocated between preparation, assembling and inspection work. Therefore it was necessary to investigate each worker during her most characteristic tasks. We also had to choose a method with a high degree of accuracy and precision, since no large movements or extreme postures occur. As some movements take place at a very high speed, direct observations had to be ruled out, for instance, OWAS, posture targeting (Corlett *et al.* 1979, Karhu *et al.* 1977). Moreover, since we wanted to analyse durations of postures as well as frequencies, a continuous method had to be chosen. On the other hand, since the aim was an evaluation of neck and shoulder disorders, the observations could be limited to those anatomical areas. Surveys of literature disclosed that no method with the above characteristics was available, and led to the development of VIRA (Persson and Kilbom 1983). The company health service of Ericsson AB (Sweden) participated in the original development and testing of the method (Wernersson 1982).

Procedure of VIRA recording and analysis

The worker is thoroughly informed of the aim of the recording, and of the necessity to work in an ordinary way. In order to facilitate the subsequent analysis, small pieces of coloured tape are positioned in front of the maxillary joint, and 3 cm in front of the protrusion of C7 (corresponds to the centre of gravity and the fulcrum of the head viewed from the side), along the lateral and dorsal sides of the upper arm and on the uppermost part of the acromion. All working tasks during a representative part of the working day are recorded, using videocameras in a rear and a side projection. The side projection is usually taken from the right (most strained side). The lens of the camera is positioned at a height a few centimetres below the shoulder of the worker. In order not to disturb the worker, the camera is placed on a support as far away as possible, and the zoom is used to focus on the head, upper part of the trunk, arm and table. A microcomputer (ABC 80 Luxor) is used for the subsequent analysis.

Each separate posture has a different key on the keyboard, and the clock of the microcomputer is used to measure the duration (each separate occasion and the total of the recording) of each posture. The rear projection is used for studies of upper-arm abduction, and the side projection is used for studies of upper-arm flexion, neck flexion and shoulder elevation, which means that the tapes have to be run four times. The upper arm is studied at rest (hanging vertically or supported) and in 0–30°, 30–60° and 60–90° abduction and flexion, and also in extension and > 90° abduction (Figures 1 and 2). The posture of the head is only recorded as 0–20° flexion or > 20° flexion, and the shoulder only as resting or elevated (Figure 3).

For the routine evaluation the microcomputer is programmed to give:

1. Work-cycle time and number of cycles per hour.

2. Time at rest for arm, shoulder and head. Total number of rest periods as well as average and total duration per work cycle and hour.

3. Frequency of changes in posture between the different angular sectors. Total number of changes as well as number of recordings in each sector during a work cycle and per hour.

4. The duration of each posture in seconds or as a percentage of the work cycle.

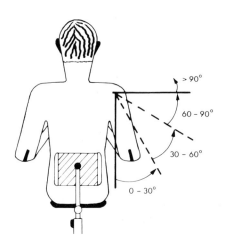

Figure 1.   Rear projection in VIRA recording. The four angular sectors, 0–30°, 30–60°, 60–90° and > 90° abduction, are marked on a glass plate which is positioned in front of the TV monitor.

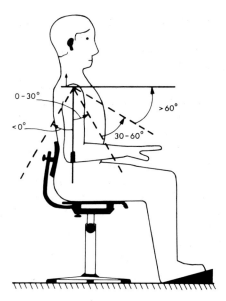

Figure 2. Right side VIRA projection, depicting upper-arm flexion and extension and shoulder elevation.

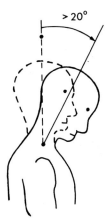

Figure 3. Right side VIRA projection, depicting neck flexion.

The method was designed to study short-cycled repetitive work under visual control, i.e., assuming that the hands are held close to the sagittal plane, and that no heavy objects are handled. If an application for heavier and more varied manual work is desired, the method should be supplemented with an analysis of exerted forces and hand postures. This would, however, make the analysis even more time consuming.

Neck–shoulder–arm disorders

Disorders were separated anatomically between neck, thoracic spine, trapezius, shoulder and arm regions. The character of the disorder (fatigue, ache, pain), its duration and its frequency (seldom, sometimes, frequently or constant) was weighed into a code of severity, where I = non-existent or very slight, II = slight, III = moderate, IV = severe.

Statistical methods

The repeatability of the VIRA method was studied using two-way analysis of variance. For the simple relationship between dependent and independent variables, Pearson's coefficients of correlation have been calculated. For multivariate studies stepwise multiple regression analyses have been performed. The severity of disorders, or deterioration of status between investigation I and II or III, have been used as dependent variables, whereas data on working technique, physical capacity, productivity, etc., from investigation I have been used as independent variables.

3. Results

Repeatability of VIRA

Six members of the laboratory staff were given 30 min of instruction and practice on how to analyse according to VIRA. They then analysed upper-arm flexion from a 2 min recording (side projection) twice, with 2 hours between. An analysis of variance demonstrated no significant differences between investigators or between tests I and II as regards the following variables: work-cycle time, number of times with upper arm at rest, in 0–30°, 30–60°, 60–90° and in extension, and the percentage of work-cycle time with upper arm in 30–60°, 60–90° and in extension. The percentage of work-cycle time spent at rest varied significantly between investigators ($14 \cdot 7 \pm 5 \cdot 5\%$ at test I) but not between tests, and the percentage of time spent in 0–30° flexion varied between investigators ($30 \cdot 7 \pm 2 \cdot 5\%$) and between tests (mean values $30 \cdot 7$ and $32 \cdot 8\%$, respectively, at tests I and II).

Variations in working technique

In order to investigate any differences in working speed ten consecutive work cycles were recorded in all 96 subjects, both in the morning and in the afternoon. The average work-cycle time was 334 s, with a range of 82–822 s. The average difference between morning and afternoon values was 2 s which was not statistically significant ($p > 0 \cdot 05$, $r = 0 \cdot 96$). However, the standard deviation of 156 s was high with respect to the mean.

In order to investigate individual differences in working technique two subgroups of women, A and B, were studied. The 16 women in group A had performed an identical assembling and soldering task on a circuit board for more than 1 year, and the 12 women in group B all performed another assembling task.

Work-cycle time in group A varied between 4·6 and 9·1 min with a mean value of 6·6 min. Corresponding values in group B were 4·4–6·5 min, with a mean value of 5·4. Table 1 demonstrates that the number of changes in posture, as well as the percentage of work-cycle time spent in different angular sectors, varies to a large extent among workers. Thus, some worked in a more relaxed way, seeking support for arms as much as possible, whereas others strained their shoulder and neck muscles during long-lasting static postures with flexed neck and raised arms.

In the entire group of 96 workers the variation was even larger (Tables 2 and 3) as the tasks varied between the subjects. The results indicate that the tasks, as well as interindividual variations in working technique, exert a large influence on the workload.

Neck–shoulder–arm disorders

Even though subjects with reported manifest neck–shoulder–arm disorders during the previous year were excluded, some cases of relatively severe disorders were

Table 1.   Variation in working technique among workers with two identical tasks (groups A and B).

	Group A (Flexion/extension)		Group B (Abduction)	
	Mean	Range	Mean	Range
Total number of changes in posture (between rest, 0–30°, 30–60°, >60°) percentage of work-cycle time in different postures:	261	170–452	123	68–175
Rest (> 2 s)	25	4–42	48	29–81
0–30°	46	31–64	44	17–67
30–60°	11	2–43	5	0–15
>60°	2	0–5	1	0–5

Table 2.   Number of postural changes in 96 women during electronics manufacturing work. Mean values and standard deviations (SD).

	Mean	SD
Flexion/extension of upper arm per work cycle	136	94
Abduction of upper arm per work cycle	85	108
Shoulder elevation per work cycle	31	21
Neck flexion per work cycle	36	35
Flexion/extension of upper arm per hour	1400	807
Abduction of upper arm per hour	832	616
Shoulder elevation per hour	380	339
Neck forward flexion (>20°) per hour	728	365

Table 3.   Percentage of work-cycle time in different postures during electronics manufacturing work.

	Mean	SD
Upper arm at rest > 2 s (from side)	41	18
Upper arm at rest > 2 s (from rear)	46	26
Upper arm flexion 0–30°	33	16
Upper arm abduction 0–30°	36	21
Upper arm flexion 30–60°	16	15
Upper arm abduction 30–60°	18	23
Upper arm flexion 60–90°	4	9
Upper arm abduction 60–90°	4	10
Upper arm extended	11	10
Shoulder elevation	18	13
Neck forward flexion (> 20°)	44	25

detected at the first investigation. The proportion of workers with serious disorders increased gradually over the two observation years (Figure 4), whereas the proportion with moderate to severe disorders increased in investigation II and then decreased slightly (Figure 5). Clinically the disorders could be diagnosed as tendinites and/or myofascial syndromes.

Relations between disorders and working postures and movements

Several significant relationships were obtained between working technique data and neck–shoulder–arm disorders at the cross-sectional study (investigation I) (Table 4). Forward flexion of the neck and elevation of the shoulder, especially, seem to be closely related to symptoms located in the neck and trapezius region. However, the coefficients of correlation are relatively low, indicating that other variables, apart from working technique, also play an important role as risk factors.

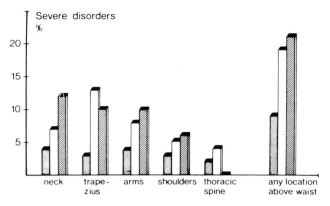

Figure 4.   Rate of severe disorders for the cross-sectional study (spotted columns), after 1 year (unmarked columns) and after 2 years (striped columns).

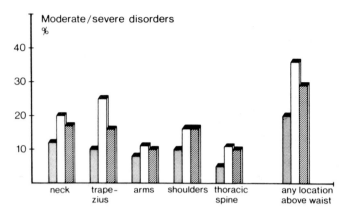

Figure 5. Rate of moderate or severe disorders for the cross-sectional study, after 1 year and after 2 years (columns as Figure 4).

Table 4. Relationship, expressed as coefficients of correlation (r), between disorders in different anatomical regions and working postures and movements. Levels of significance for: $p < 0.05$, $r > 0.20$; $p < 0.01$, $r > 0.26$; $p < 0.001$, $r > 0.33$.

	Neck	Trapezius	Shoulder	Arm	Thoracic spine
Number of postural changes (abduction of upper arm, per hour)	−0·21				
0–30° flexion of upper arm, percentage of work cycle		−0·25			
Shoulder elevation, percentage of work cycle		0·22			
Neck flexion, percentage of work cycle	0·27				
0–30° abduction of upper arm, average time per work cycle	0·24	0·23	0·20		
Neck flexion, average time per work cycle	0·41	0·28			0·25
Upper arm not at rest, average time per work cycle	0·27	0·32			0·25

At the 1 and 2 year follow-up the main analysis was performed using stepwise multiple regression analysis, with deterioration of disorders as a dependent variable. In these analyses any risk factor that had shown a significant relationship with disorders when using coefficients of correlation, was tried as an independent variable, and entered in the analysis until F-values below 3·0 were reached. Tables 5–7 summarize some of the findings. It is obvious that the working-technique variables are important risk factors for deterioration. A further analysis of these data is being prepared for publication.

Table 5.   *Predictors for deterioration to severe disorders (any location) at the 1 year follow-up (in order of importance).*[a]

Percentage of work-cycle time in neck flexion
Average time in neck flexion per work cycle
Number of sick-leave days during 6 months preceding investigation I
Productivity, as a percentage of norm, preceding investigation I
Percentage of work-cycle time with upper arm abducted > 30°
Maximum static strength at abduction (negative)
Regular knitting or crochet in leisure time

---

[a] $R^2_{adj} = 0.46$.

Table 6.   *Predictors for deterioration to severe disorders (any location) at the 2 year follow-up (in order of importance).*[a]

Pain provoked by shoulder movements
Number of shoulder elevations per hour
Number of neck flexions per hour
Stress at work
Participation in pause gymnastics
Headache in connection with work, more than twice per month

---

[a] $R^2_{adj} = 0.35$.

Table 7.   *Predictors for deterioration to severe disorders in the trapezius region at the 2 year follow-up (in order of importance).*[a]

Number of sick-leave days during 6 months preceding investigation I
Number of shoulder elevations per hour
Years of employment in previous strenuous jobs (cleaning, warehouse, catering)

---

[a] $R^2_{adj} = 0.33$.

4.   Discussion

VIRA was developed in order to provide a simple method for the evaluation of working postures and movements in sitting, repetitive work. It proved to be repeatable and easy to learn; even after only 30 min instruction accurate analyses could be performed.

The theoretical basis relies on the simple biomechanical principle, that the arm exerts an increasing torque as it is flexed forwards or abducted. As the method quantifies movements, as well as duration and frequency of postures, it includes dynamic as well as static aspects of the work. The validity of the method depends on the aim of the study where it is used. In the present study both dynamic and static variables seem to be relevant for the given aim, i.e., to evaluate the importance of different risk factors for the development of occupational neck–

shoulder–arm disorders. Thus, neck and shoulder postures and movements, especially, repeatedly turn up as significant indicators of risk.

The disadvantage of the method is that it is relatively time consuming. Even though relevant work cycles may only have a duration of a few minutes, the analysis takes four times as long and sometimes has to be performed in slow motion. An automatic procedure for the analysis would be preferable, but since Sel-spot and similar techniques are so far confined to laboratory surroundings, further technical development of markers has to be awaited (Halbertsma 1983).

Another important aspect of the results is the large variation in working technique between workers, even those who perform exactly the same task. Thus it is not sufficient, in seemingly repetitive jobs, to analyse the technique of only one worker. The relative accuracy of the VIRA measurements disclosed interworker differences that might have been undetected if a cruder method had been used.

One source of uncertainty, when using VIRA variables as risk indicators, is that the working tasks may have undergone subtle changes during the follow-up period. Through interviews at the 1 and 2 year follow-up we controlled whether the tasks were changed or not. Assembling one circuit board may in some cases have been substituted for another slightly different board, but the main task, i.e., assembling circuit boards of a given type, produced at a given department, still remained. Even in the 2 year follow-up several of the working-technique variables appeared as strong risk factors, which supports the view that no large changes in working tasks had taken place.

In conclusion, VIRA proved to be a valuable method for the evaluation of working technique in sitting, repetitive arm work. Its validity was further documented by its ability to identify elements of the working technique which appeared as risk factors for subsequent cervichobrachial disorders. The importance of poor working postures for musculo-skeletal disorders, as documented, for example, by Hünting *et al.* (1980) in accounting work, was thus given further support by this study.

The relationship obtained between disorders and other independent variables will be discussed in a separate publication. In view of the very vague nature of some of these predictors, it is surprising that as much as sometimes 45–50% of the variation in outcome (R^2_{adj}-value) could be explained. Thus it is possible, even with the kind of difficult and semiquantitative dependent and independent variables used in the present study, to investigate the cause–effect relationships in occupational musculo-skeletal disorders.

SECTION 2

METHODS FOR MEASURING
BODY POSTURE

Problems of measurement are of perennial concern to both the research worker and the practitioner. The former seeks precision and the ability to measure a wide range of factors, if possible in parallel and on-line to a computer. The latter would like a simple, non-interfering method which needs minimum expertise and is easily recorded and interpreted. Both seek reliability in their equipment and methods, as well as cheapness, and many of those in each category feel that their answer is just around the corner!

A glance at Armstrong's review, and that of Rohmert and Mainzer in Chapter 18, will demonstrate that the corner is a long way off. Although there have been major advances it is clear that the various purposes for which people require to record posture still produce a wide range of methods, each of which has its protagonists.

Early attempts at posture recording were made to record ballet, and, after some 200 years, have produced effective procedures which are widely taught (see, for example, Hutchinson 1970). These procedures, however, require expert teaching over long periods, even the simpler forms of recording taking some 3 months to learn. Although they present posture to any degree of detail required of the observer, and Labanotation includes a time scale, the recording of forces has not been fully included. One attempt to include effort (Laban and Lawrence 1947) has had its supporters but would appear to have involved subjective assessments of this factor. It is interesting to note that Borg and his co-workers are looking at this possibility again. In recent studies they have noted the feasibility of using observers to rate the amount of force exerted by subjects (Ljunggren 1986). They report the flatness of such ratings, i.e. that observers tend to overestimate the lower force values and underestimate the higher ones. This centralizing tendency for judgements is well known to psychologists and has been a matter for research in the field of work measurement for many years (see, for example, Dudley 1968). In spite of this long history, it is a phenomenon which has not yet yielded to a practical means of control. The most significant attempts at control were the means adopted by work measurement practitioners to use control tests at frequent intervals to 're-calibrate' time-study technicians in their ability to recognize a standard rate of working. On the whole, these were not successful (see, for example, Sury 1968) and, at the time of writing, it seems likely that attempts to control the judgements of forces are likely to be equally unsuccessful. There are a multitude of factors taken into account in judging how much force another individual is exerting, each no doubt being subjectively assessed differently by different observers, and their control seems less likely than the further development of technology to achieve the same purpose.

The choice of one or other of the methods for measuring posture depends on the purposes of the study and on the factors believed to be influencing the postures. Nordin and co-workers, as well as Weber and colleagues, were studying the forward bending of subjects during their normal occupations. The objectives were limited, in that the target was just the measurement of the forward bend in the sagittal plane, or for the Weber study the inclination of the head and a gross measure of the slope of the spinal column. Examples of the practical applications of studies of the postures of selected parts of the body are given by Armstrong, demonstrating that there is no necessity to seek a complete postural analysis in order to attack many real-world problems. A careful analysis of the current situation is the prerequisite to a choice of measurements, but just because measures may appear simple this must not hide the fact that such methods need ergonomics knowledge, care and a considerable amount of patience. This applies, too, to the deceptively simple technique of Burton. He quotes the method as being suitable for examining the effects of different sorts of seats on their users. The measurement may be relatively straightforward but it is the interpretation which provides the difficulty. There may be something to be said for not making measuring systems too easy!

The well known OWAS system, from Finland, provides a broad brush approach to posture recording and analysis which gives opportunities for major inadequacies in work posture to be brought to the attention of investigators. This was its primary purpose and reports from the steel company who were partners in its development showed that it had fulfilled its purpose well. Prior to its introduction there was no method for looking at posture as a factor in work study, it was only investigated when injury or complaint arose. By using the OWAS system as a regular part of every investigation of work method or measurement, each task was posturally evaluated. There are limitations, one of which is its advantage, i.e., its simplicity; but if it is kept in mind that it is not a posture/force measurement system for the detailed study of people at work, but a means for identifying a great number of the inadequate work situations, its value will be appreciated.

Two systems for the detailed recording of posture, presented as poster sessions but not reported in these pages, are worthy of mention. One, by Paradise (1982) and O'Brien and Milner (personal communication) is the use of an instrumented elasticated body stocking (acronym PREMIER). Sixteen major joints or points of flexure are instrumented by sets of strain gauges to record extension, flexion and some rotations. The standard anatomical position is first recorded and then each point of the suit sampled and recorded several times per second. Records can be made on a portable instrument strapped to the wearer's belt or via an umbilical cord directly to a microcomputer. Playback of the recorded postures, which can cover an 8 hour period, can be done very quickly and the desired analyses carried out either on or off line. Work is continuing on a dynamic graphic display of the postures. This device is, as yet, a laboratory research tool.*

The other method bears the acronym CODA (Michelson) and will record the position of eight, or on later versions twelve, pyramidal reflective landmarks in space within a 4 m cube. The accuracy of the instrument is reported as being within 1 mm within this space on all three axes. It will record and transfer to other computers for analysis a virtually continuous record of the movements of these landmarks and the movements can be differentiated, on-line in some instances, to give immediate velocity and acceleration diagrams.

The utility of an earlier version of this instrument, was limited by the presentation of false signals whenever two landmarks were vertically above each other, an effect which was indicated on the records by a special signal. Although relatively reliable there were no provisions for maintenance other than by direct contact with the manufacturers, leading to difficulties in maintaining experimental schedules. The use of the instrument, however, is simple and programming a mini- or microcomputer interfaced with it to process the data is not difficult.†

.ne use of already well-known techniques and methods has not formed part of this discussion since their uses have been widely disseminated. Sections 4 and 5 also consider some measurement techniques, but they are discussed in their own contexts, i.e., measures to assess the effects of posture, and measures related in particular to the study of seating. None of the methods is, of course, exclusive to its use in the context given, since combinations of methods are drawn upon depending on the problem in hand. As was pointed out earlier, it is the problem, and the models adopted to assist with its investigation, which define, or should define, the methods used.

*For further information on PREMIER, contact C. O'Brien, Department of Production Engineering and Production Management, University of Nottingham, U.K.

†For further information on CODA contact D. Michelson of Movement Techniques Ltd., Unit 5, Technology Centre, Epinal Way, Loughborough, Leics., U.K.

Chapter 6

Upper-Extremity Posture: Definition, Measurement and Control

Thomas J. Armstrong

The University of Michigan, 1205 Beal, IOE Building, Ann Arbor, MI 48109, U.S.A.

1. Introduction

Posture is an important consideration in the design of work procedures and equipment; it affects the ability of workers to reach, hold and use equipment and influences how long they can perform their job without adverse health effects, such as fatigue and cumulative trauma disorders. This paper describes some of the reasons upper-extremity posture is important and how it is quantified, measured and controlled.

2. Posture and worker health

Upper-extremity cumulative trauma disorders of the tendons and nerves are major causes of lost time and workers' compensation in many 'hand-intensive' industries (Armstrong *et al.* 1982, Hymovich and Lindholm 1966, Jensen *et al.* 1983, Vihma *et al.* 1982). Although he did not use the term cumulative trauma, Ramazinni (1713) was the first to describe the concept:

> Various and manifold is the harvest of diseases reaped by certain workers from the crafts and trades that they pursue. All the profit that they get is fatal injury to their health, mostly from two causes. The first and most potent is the harmful character of the materials they handle. The second, I ascribe to certain violent and irregular motions and unnatural postures of the body, by reason of which, the natural structure of the vital machine is so impaired that serious diseases gradually develop therefrom.

The concept of a 'harvest of diseases' is an aptly suited term because the modern literature shows that there are numerous nerve and tendon problems associated with repeated or sustained exertions in certain postures. The concept of an 'unnatural posture' arises because these are not the postures that a normal person would voluntarily assume. They are assumed only to perform some task.

The terms 'repetitive trauma' and 'cumulative trauma' are used because these disorders develop over periods of weeks, months and years as a result of repeated action. Some modern investigators object to the term disease because of the lack of foreign bodies; therefore, the term 'disorder' will be used here to avoid controversy.

Most of the papers published on cumulative trauma disorders during the last 100 years deal mainly with diagnosis or treatment. Within these papers are numerous references to occupational factors, such as work posture. A few papers describe studies of pathomechanics; a few epidemiological studies; and a few prevention through the design of work equipment and methods. As a result of these studies, it can be concluded that repeated or sustained exertions in certain postures are associated with certain cumulative trauma disorders, yet little attention has been given to prevention.

Common shoulder disorders include bicipital tendinitis, rotator cuff irritation and thoracic outlet syndrome. These disorders are associated with elevation of the elbows at or above shoulder height (Beyer and Wright 1951, Bjelle *et al.* 1979, Herberts *et al.* 1981, Hoffman 1981, Lord and Rosati 1958, Nicols 1967, Wright 1945). Examples of this posture (see Figure 1 (*a*)) are often seen in overhead assembly-line work and spray painting. Shoulder disorders are also associated with extreme reaching down and behind the torso (Falconer and Weddell 1943, Lord and Rosati 1958, Nicols 1967). Examples of this posture (see Figure 1 (*b*)) are often seen in carrying heavy suitcases and in seated workers reaching for parts.

Extreme flexion of the elbow is associated with cubital tunnel syndrome (Bora and Osterman 1982, Feldman *et al.* 1983, MacNicol 1982, Spanns 1970, Wadsworth and Williams 1973). This posture (see Figure 1 (*c*)) is often associated with assembly and inspection of small parts that must be held close to the eyes. Extreme rotation of the forearm is associated with medial and lateral epicondylitis (Goldie 1964, Hoffman 1981, Kurppa *et al.* 1979, Tichauer 1976). Examples of these postures (see Figure 1 (*d*) and (*e*)) are often seen in workers holding parts and materials in front of their bodies.

Flexion and extreme extension of the wrist is associated with tenosynovitis of the flexor and extensor tendons in the wrist and with carpal tunnel syndrome (Armstrong and Chaffin 1979, Armstrong *et al.* 1984, Brain *et al.* 1947, Phalen 1966, 1972, Robbins 1963, Smith *et al.* 1977, Tanzer 1959, Tichauer 1966). Examples of these postures (see Figure 1 (*f*)) are often found in the use of power tools and materials handling. Ulnar and radial deviation of the wrist are associated with tenosynovitis at the base of the thumb or 'DeQuervains Disease' (Hoffman 1981, Kelley and Jacobson 1964, Lamphier *et al.* 1965, Muckart 1964, Stein *et al.* 1951, Thompson *et al.* 1951, Tichauer 1966, 1976, Younghusband and Black 1963). Examples of these postures (see Figure 1 (*g*)) are often seen in the use of hand tools.

Hand posture is important because it affects how much force the muscle must produce for a given amount of hand force (Chao *et al.* 1976, Swansen *et al.* 1970). Muscle force will be lowest when objects can be held in a power grip posture (see Figure 1 (*h*)); four to five times as much muscle force must be exerted to pinch objects with the finger tips (see Figure 1 (*i*)).

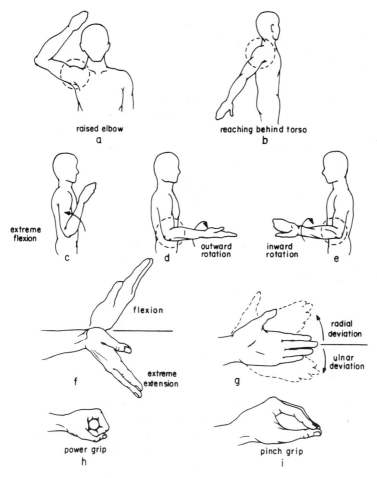

raised elbow
a

reaching behind torso
b

extreme flexion
c

outward rotation
d

inward rotation
e

flexion

extreme extension
f

radial deviation

ulnar deviation
g

power grip
h

pinch grip
i

Figure 1.   Postures commonly associated with cumulative trauma disorders.

Stressful postures are a major cause of disease and lost work in many hand-intensive industries. One of the major goals in the design of work equipment and methods is to avoid the need for workers having to perform repeated exertions in these unnatural or stressful postures. A discussion of design will be facilitated by a discussion of how posture is quantified.

3.   Posture measurement

Posture is defined simply as the position of the body. Common to all schemes for measurement of body position is the use of links and joints to characterize the segments of the body. The geometry of the link–joint system can be described as length of the links and the angular position of the joints. Joint positions can be uniquely described using Euler angles and vector notation (Chao 1980); however, the position of the body or even a limb is difficult to describe rigorously, because there are so many links and joints about which movement can occur and because the centres of joint rotations change with the angular position of the joint. The description can be simplified if the axes of rotation are assumed to be fixed and only those joints with the greatest range of motion are considered. Various schemes for simplifying the link–joint system and describing posture have been proposed.

The Albert–Strasser globographic technique involved locating a sphere about the average centre of rotation (Dempster 1955 a, b). Meridians of longitude and parallels of latitude were fixed on the sphere with respect to the proximal segment of the link. The joint positions about axes perpendicular to the proximal link were measured using protractors and shown on the sphere. The complete range of joint motion was often plotted on a sphere centre about the anatomical joint centre in what were called 'excursion cones', 'joint sinuses' or 'excursion fields'. Additional notations were required to record the position parallel to the axis of the distal link. The axes of rotation were aligned with the standard anatomical position and labelled according to standard anatomical conventions: flexion–extension, abduction–adduction and supination–pronation.

A complete link figure with globographic representation of the joint ranges of extremity links from Dempster (1955 a, b) is shown in Figure 2. Thirteen degrees

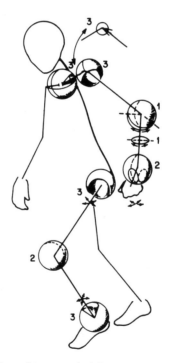

Figure 2.   *Albert–Strasser globographic posture depictions.*
From Dempster (1955 a, b).

of freedom were included between the trunk and the hand. No attempt was made to describe the hand posture.

Because of the simplifying assumptions that are usually made, it is not possible to reproduce joint angle measurements to the nearest degree—especially when the angles are estimated from direct observations or from films. Consequently, the range of joint motion is often divided into a series of ranges or zones (Armstrong *et al.* 1982, Corlett *et al.* 1979, Poirier *et al.* 1984, Priel 1974).

A form for recording upper-extremity postures, which has been used extensively in studies of cumulative trauma disorders by Armstrong *et al.* (1982), is shown in Figure 3. It includes three axes of rotation at the shoulder, two axes at the elbow and two in the wrist. The hand will be discussed separately. Each of the axes is divided into ranges according to the overall range of motion. For example, the 57° of possible wrist motion side-to-side (abduction–adduction) is divided into three zones and labelled U, N or R for ulnar deviation, neutral or radial deviation. The 131° of possible anterior–posterior wrist motion (flexion–extension) is divided into five zones and labelled E, 45E, N, 45F and F for more than 45° extension, 15–45° extension, 15° extension to 15° flexion, 15–45° flexion and more than 45° flexion. In this system the analyst observes the worker and circles the appropriate cell on the table. Additional resolution can be added by circling two adjacent cells.

Armstrong *et al.* (1982) used this postural analysis system to identify possible causes of cumulative trauma disorders in a poultry processing plant. One of the jobs with a high frequency of cumulative trauma disorders was thigh boning. The workers grasped turkey thighs and made two to three cuts along the long axis of the thigh to separate the meat from the bone with a knife in their right hand. A fourth cut was made to separate some of the ligamentous attachments. Two work cycles of wrist posture are plotted in Figure 4. It can be seen that the right wrist is mostly between 15 and 40° of ulnar deviation and that it moves between −15 and 45° flexion to hold and use the knife. The left wrist moves between −15 and 45° extension and −15 and 25° radial deviation to hold the thighs for cutting. This information was then used to identify risk factors of cumulative trauma disorders. As indicated in the earlier discussion, ulnar wrist deviation is associated with DeQuervain's Disease and wrist flexion is associated with carpal tunnel syndrome. An alternative knife design for reducing ulnar deviation and wrist flexion was proposed.

The posture of the hand can also be described on a joint by joint basis. Although this is often desirable for clinical applications, it requires too much time for field work. Simplified systems in which hand position is classified as grasp, three jaw chuck, palmar pinch, tip pinch, lateral pinch and hook grip have been proposed for describing overall hand posture based on observations of people performing manual tasks (Keller *et al.* 1947, Schlesinger 1919, Taylor 1954). It is possible to infer information about the approximate position of each of the joints from these classifications. For example, all of the joints will be highly flexed when the hand is closed around a small cylindrical handle. The finger joints are progressively extended to open the fist as the handle diameter is progressively increased. When pinching an object between the pulp of the thumb and the pulp of

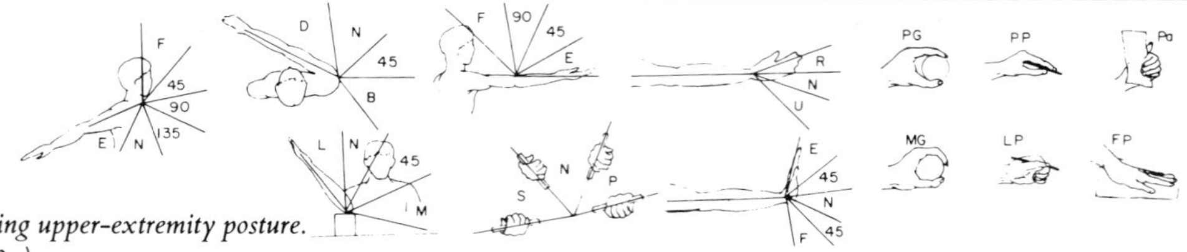

Figure 3.　Form for recording upper-extremity posture.
From Armstrong et al. (1982).

one or more opposing fingers, the first knuckle will be highly flexed; the distal knuckles will be nearly straight. If the object is pinched between the tips of the fingers, all three knuckles will be flexed approximately equally.

Other schemes have been proposed based on functional anatomical consider-ations (Landsmeer 1962, Long *et al*. 1970, Napier 1956). Forceful exertions were classified as power grip or handling; low force manipulations were classified as precision grip or handling. Power grip is defined as grasping with the fist closed and the hand conforming with the shape of the object. Precision grip is defined as grasping with the ends of the fingers. Studies have shown that from four to five times as much force can be exerted in a grasp posture as in a pinch posture (Armstrong *et al*. 1982, Chao *et al*. 1976).

Another system in which grasp and pinch are replaced with palmar grip and digital pinch was proposed by Poirier *et al*. (1984). A third category, pressure, was added to characterize exertions of force with the palm.

A system for describing the position and the load on each finger was proposed by Jacobson and Sperling (1976). A code system is used to indicate the fingers and other parts of the hand which participate in the grip, the relative positions of the fingers, the position of the finger joints, the contact surfaces of the hand and the relationship between the longitudinal axis of the object and the hand. The system provides a detailed description of hand posture but requires a great deal of time and skill to apply.

Another system was proposed by Armstrong *et al*. (1982) for studies of cumu-lative trauma disorders. An encoding form for this system is shown in Figure 3. First, the hand position is indicated as a medial grasp, palmar grasp, pulp pinch, lateral pinch, palm pinch or a press by circling the header in the appropriate cell. Second, the area of contact, with the palm, first, second, third, fourth or fifth digit, is indicated by circling the 0, 1, 2, 3, 4 or 5 in the appropriate cell.

The hand postures recorded in the study of cumulative trauma disorders described above are plotted in Figures 4 and 5. A medial grasp is used to hold the knife handle wih the right hand. The palm and all four fingers are in constant contact with the handle. A lateral pinch is used to hold the poultry thigh with the left hand. The left hand is in use only about 50% of the time. Boning thighs is clearly more stressful for the right hand than the left hand.

A posture analysis can be performed by observing the job; however, it is much easier to use films or videotapes that can be replayed in slow motion or stopped. A significant amount of time and skill are required to perform an analysis. Goni-ometers that can be attached directly to the joints, cinematography and video tracking systems have been proposed (Adrian *et al*. 1965, Chao 1980, Chao *et al*. 1980, Sommer and Miller 1980). Palmar *et al*. (1985) used a triaxial electrogoni-ometer to study wrist flexion/extension and adduction/abduction in activities of daily living. The authors demonstrated that highly accurate measurements of wrist posture could be made under laboratory conditions; however, it would be diffi-cult to use the system in many occupational tasks under field conditions because of the size of the hardware attached to the wrist and arm.

A system for direct measurement of wrist posture under field conditions is

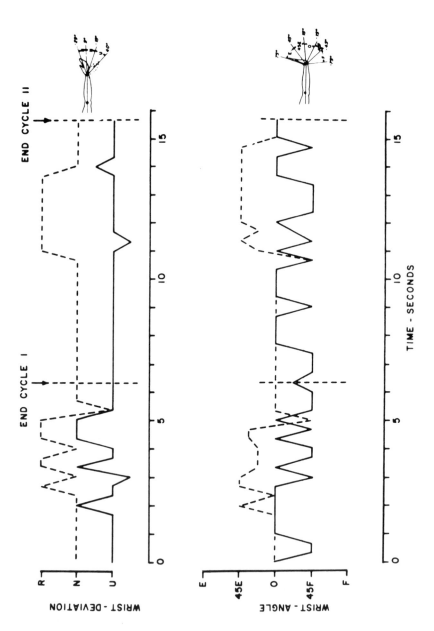

Figure 4.   Right (solid line) and left (broken line) wrist posture recorded in two cycles of thigh boning.

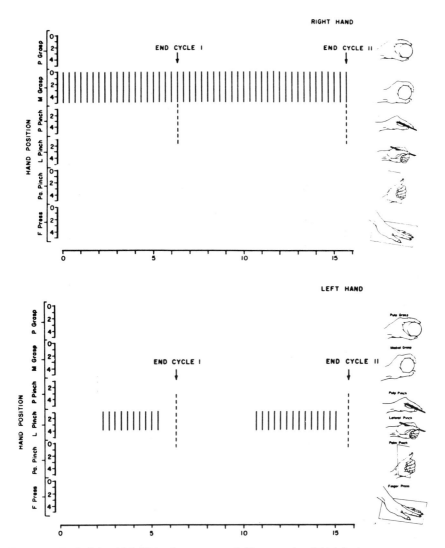

Figure 5. Right (a) and left (b) hand postures recorded in two cycles of thigh boning.

shown in Figure 6. The system uses two linear displacement transducers attached to the back of the forearm and wrist. The system is calibrated empirically to estimate flexion/extension and adduction/abduction to within ±5°.

Direct measurement devices offer great potential because they can be used with computers to collect and reduce the data. Palmar *et al.* (1985) used a computerized data-acquisition system to collect and analyse over 1 million data from ten subjects as they performed 52 personal hygiene, culinary skills, housekeeping, secretarial, carpentry, mechanical and surgical tasks. Such an ambitious effort would not have been possible with manual data-handling methods.

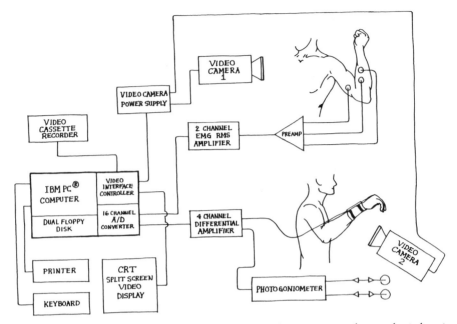

Figure 6.   Computerized data-acquisition system for recording posture via electromechanical goni-ometer and force via surface electromyography.

The system shown in Figure 6 in fact integrates direct posture measurements with surface electromyograms for estimating force and video images for determining work content. Analogue signals from the goniometer and EMG pre-amplifier are entered into the computer via an analogue-to-digital converter. The computer display driver accepts interrupts from a video camera so that the computer display can be synchronized with the video image. The posture and EMG data are also stored in memory and written on a floppy disk for later analysis. Posture and hand-force data for one cycle of a worker loading parts into a broach machine are shown in Figure 7.

Electrogoniometers and computerized data-acquisition systems can be used to make direct recordings of wrist posture under both field and laboratory conditions. Additional work is needed to develop systems for the elbow, shoulder and hand that can be used under field conditions. Use of the computer and integration of EMG data make it possible to develop software for identifying risk factors of cumulative trauma disorders.

4.   *Posture and work equipment design*

Once a stressful posture has been identified, the question becomes how to fix it. Posture can be controlled through the design of the tool and the location of work (Armstrong 1983, Armstrong *et al.* 1982, Tichauer 1966, 1976). For example,

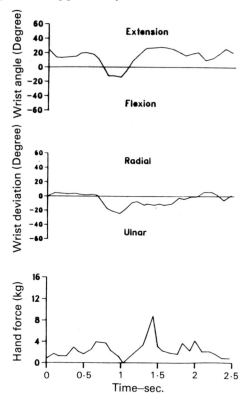

Figure 7. Wrist posture and hand force for loading a broach machine as recorded from an electrogoniometer and surface electromyography.

reaching can be controlled through the height of stock and work surfaces and conveyors. Ulnar deviation and flexion of the wrist can be controlled by using jigs and fixtures to hold parts at proper locations and orientations. When it is not feasible to relocate the part, alternative tools should be investigated. For example, Armstrong *et al.* (1982) showed that ulnar wrist deviation and flexion are required to hold and use a knife for boning thighs (see Figures 4 and 8 (*a*)). It was proposed that a pistol-shaped knife would require less deviation to bone thighs than the traditional straight handle. This proposition was confirmed by making a prototype and testing it in the field (see Figure 8 (*b*)).

The ideal location for a given task depends on the orientation of the work, the shape of the tool and the size of the worker. This can be shown graphically with stick-figure representations of the human body similar to Figure 2. A typical stick figure based on the work of Drillis and Contini (1966), in which the length of each segment is expressed as a fraction of total stature, is shown in Figure 9 (*a*)).

Complete three-dimensional analyses are difficult to perform without elaborate computer algorithms. First-order approximations of posture can be obtained using two-dimensional representations of the body. Two-dimensional figures can be

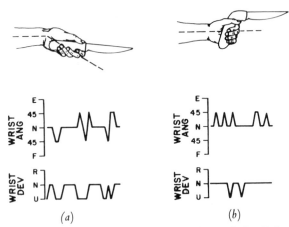

Figure 8. Wrist posture from one cycle of thigh boning with straight-handle knife (a) and one cycle with pistol-handle knife (b).

dressed-up or made to look more lifelike by using silhouettes of each segment. Silhouettes proposed by Dempster (1955 b) for a drawing-board template of an average male are shown in Figure 9 (b). These silhouettes can be scaled to fit other percentiles for the sake of appearance but should not be used to determine clearances.

Stick figures and templates can be used to determine the best work location for a person of given stature performing a given task. The best location is one in which the task can be performed with the elbow at the side of the body, without excessive forearm rotation and without deviating, flexing or fully extending the wrist.

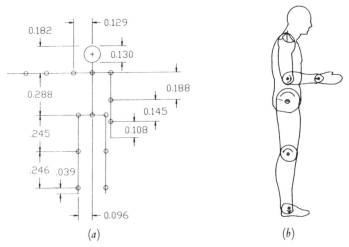

Figure 9. (a) Stick figure proposed by Drillis and Contini (1966). (b) Drawing-board manikin proposed by Dempster (1955 b) for workspace design.

For example, the forearm should be horizontal for work on a vertical surface with a pistol-shaped screwgun. Simulations of this posture for persons with 5th percentile female and 95th percentile male statures and average proportions are shown in Figure 10. The ideal work location varies from 41·5 in for the small female to 50·4 in for the large male.

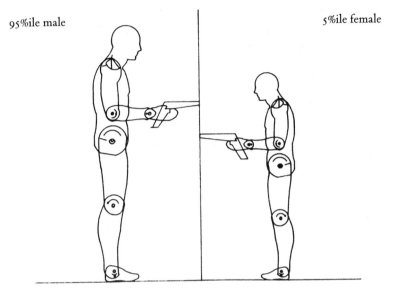

95%ile male

5%ile female

Figure 10.   Optimal heights for work on a vertical surface with a pistol-shaped screwgun. CAD figure proportions according to Drillis and Contini (1966), silhouettes according to Dempster (1955 b).

Two-dimensional figures can also be used to estimate reach limits for various postural constraints. The maximum reach limits for touch can be estimated by drawing an arc in front of the body with a radius equal to the arm + forearm + hand lengths about the shoulder as shown in Figure 11. The maximum reach limits for grasping can be estimated by drawing an arc with a radius equal to the arm + forearm + hand/2 lengths about the shoulder. Reaching overhead can lead to shoulder problems. Acceptable overhead reach limits for repetitive reaching can be estimated by constraining the shoulder to 60° forward flexion as shown in Figure 11. Other constraints can be introduced as necessary.

Both drawing-board manikins and biomechanical models are used to relate posture to work height and orientation for a given stature. Drawing-board manikins have not met with widespread use because they are not available in all desired percentiles and scales; also, they are difficult to manipulate. Biomechanical models have not been widely used because they require specialized skills and computer equipment. It is proposed that many of these difficulties can be overcome by using commercially available computer-aided design systems to manipulate anthropometric stick figures or manikins.

5%ile female

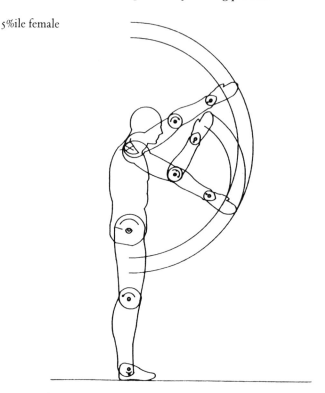

Figure 11.   Reach envelope for tip reach and grasping reach with and without the shoulder con-
strained to 60° flexion.
CAD figure proportions according to Drillis and Contini (1966), silhouette according to Dempster
(1955 b).

Although stick figure and template manipulations can be performed on drawings of workplaces and equipment using standard drafting techniques, they are more easily performed using computer-aided drafting (CAD) systems. CAD systems provide the ability to create engineering drawings that can be stored on disk or tape and can be recalled for later editing or plotting. Most systems do not require extensive computer skills and are considered 'user friendly'. The work described in this paper utilized an IBM PC computer and an AutoCAD computer-aided drafting programme. This system was selected because it provided essential functions at minimum cost and because of the widespread use of microcomputers. Applications developed for the IBM PC and AutoCAD should be easily adapted to more sophisticated systems.

5.   Summary

Posture is an important consideration in the design of work procedures and equipment; it affects the ability of workers to reach, hold and use equipment and influences how long they can perform their job without adverse health effects, such as fatigue and cumulative trauma disorders.

Rigorous descriptions of posture are difficult because joint axes are not stationary and because there are many axes of rotation. Descriptions are often simplified by assumptions of fixed axes of rotation, neglecting joints with small ranges of motion and approximating joint angles. Considerable time and skill is required to perform even the most simple postural analysis. Electrogoniometers and computerized data-acquisition systems show great promise for reducing the work required and increasing accuracy for a postural analysis.

Stressful postures can be controlled through the location and orientation of work and through the size and shape of tools. The use of computer-aided design systems to facilitate application of anthropometric data, and stick figures and manikins for estimating the best work location for a given task, show great promise.

Chapter 7
Dynamic Measurements of Trunk Movements during Work Tasks

M. Nordin, G. Hultman, R. Philipsson, A. Ortelius and G. B. J. Andersson

Department of Orthopaedic Surgery I, Sahlgren Hospital, Göteborg, Sweden

1. Introduction

During the last decade an increased need for basic kinematic and kinetic data for various occupations has led to the development of several methods for recording the body motions and positions of workers in industrial settings. Data obtained by these methods have been helpful in identifying detrimental work situations and in modifying the work environment. Most of the recording procedures involve sampling techniques, repeated observations of the worker at specific intervals, usually 15–30 s, throughout either a work cycle or a defined period of time. The positions of one or more body parts are observed and recorded, and forces and loads acting on these body parts are taken into account (Berns and Milner 1980, Corlett *et al.* 1979, Flux 1977, Holzmann 1982, Karhu *et al.* 1977, Priel 1974).

Few studies have been conducted on trunk motion in workers in industrial settings even though the lumbar spine is the most common site of industrial injury. The primary reason is the lack of simple objective methods for measuring the motion of the spine. The literature reveals only two studies in which spine motion was recorded continuously during industrial work. O'Brien and Paradise (1976) used strain gauges placed on the lower back to measure trunk flexion in workers in various occupations. Snijders and Nordin (1985), in an unpublished study, have used a strain-gauge device mounted on the spine to detect spinal motion and configuration during specific work tasks.

Recently, a simple instrument was developed for the objective and continuous measurement of angles; it allows quantification of the amount of movement and its distribution over time during a work cycle (Hultman *et al.* 1984, Nordin *et al.* 1984). The instrument was adapted for measuring trunk motion in the saggital plane, for forward flexion of the trunk. The purpose of this study was to use this instrument to study flexion movements of the trunk in subjects with various occupations, and to obtain data on how the movements differ among occupations.

74

2. Materials and methods

Movements of the trunk in the saggital plane were measured with a flexion analyser. The instrument consisted of a pendular potentiometer as a transducer, a five-level analogue-to-digital converter, control circuits and nine digital registers (Figure 1). Together, the units form a portable battery-powered system weighing 1 kg that can be worn on the back in a small harness. Although the analyser has the potential for measuring movement of any body segment, it has been adapted for measuring trunk movement in the saggital plane (forward flexion) for the purpose of this study, because trunk motion is most common in this plane and greatly affects the loads on the lumbar spine.

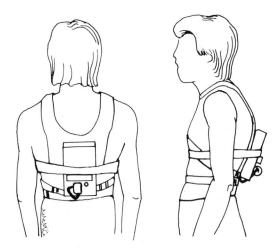

Figure 1. A 1 kg analyser was carried in a harness on the subject's back during the measurement sessions.

The portable unit was placed on the back of each subject with the lower edge of the instrument box at the T12 level. The pendulum potentiometer was adjusted to zero when the subject was standing at ease in an upright position with arms hanging. In this way, the analyser was set to measure forward flexion from 0–90°.

The range of forward flexion is divided into five intervals. The amplitude of those intervals, (a)–(e), each represented a range of 18° (Figure 2). Forward flexion of more than 90° was recorded at interval (e) (73–90°). The analyser recorded the total amount of time that the subject spent in each interval of forward flexion as well as the number of times that flexion changed from one interval to another (borderline passages in one direction of flexion).

Each measurement session lasted for about 60 min. At the end of the session, the time spent in each interval of flexion and the number of borderline passages were read in sequence from a display unit. The mean values (SD) for the time spent

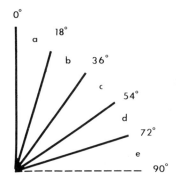

Figure 2.   Forward flexion of the trunk was divided into 18° intervals, (a)–(e), with borderlines between each interval.

in each interval of trunk flexion and the number of borderline passages were calculated for all subjects in each occupation. To provide supplementary data on movements and loads, an observer followed the subjects during the entire testing session, manually recording the number of deep forward flexions and noting the type of activity associated with each deep forward movement, for example picking up an object from the floor or lifting a carton. The observer also recorded the general work routine and the number of heavy lifts during the testing sessions. A heavy lift was defined as one involving an object weighing more than approximately 15–20 kg carried in such an awkward position as 'away from the body' and/or 'with the trunk flexed and rotated'. The observer did not interfere with the subject's work in any way. Six different types of work were studied: dentistry, nurses' aides work, warehouse workers, janitors, riveters and bricklayers.

For each occupation, measurement sessions were scheduled at the time of day which represented a typical workload. All tasks performed were the subjects' actual work tasks. Although each subject was measured only once, testing sessions were held on all five days of the work week to yield an average of a week's workload. Table 1 shows details of the employees studied in each profession. All the participants in the study had worked in their field for more than a year.

Table 1.   Distribution of sex, mean age and mean years in occupation for subjects tested.

	Dentists (n = 10)	Nurses' aides (n = 23)	Warehouse workers (n = 10)	Janitors (n = 6)	Riveters (n = 17)	Bricklayers (n = 4)
Sex (male/female)	7/3	4/19	9/1	1/6	16/1	4/0
Mean age (years)	33	28	34	30	28	×
Mean time in occupation (years)	8	3	10	2	4	3

The dentists' work consisted of regular normal outpatient clinical wɔɪк at the University Hospital. Each dentist was free to sit or stand while treating patients. All dentists had modern equipment which allowed patients to lay horizontally in a supine position during treatment.

The nurses' aides performed bed making and nursing duties in an orthopaedic hospital ward. The patients were mostly geriatric patients in need of help with activities of daily living. The measurement session took place in the ward where the beds could all be mechanically adjusted in height.

The warehouse workers were studied in a warehouse which stocked various kinds of groceries such as fruit, cheese and rice. The job was to load boxes and sacks manually from shelves to hand trucks or power-driven carts and occasionally to drive the carts a short distance (Nordin *et al.* 1984).

The janitors were cleaning in a factory that manufactures cement powder used in industrial products. The janitorial tasks were performed in clerical offices, operation rooms, locker rooms and laboratories (Hultman *et al.* 1984).

Testing of the riveters took place in an industry which manufactures different equipment for heavy manufacturing. Riveting was performed in a standing working position and there was no possibility of adjusting the work height. All tools and supplies for riveting were placed around the employee.

The bricklayers performed hand bricklaying which consisted of taking one brick at a time and adding it onto another. The measurements were made in an outdoors construction area.

At the beginning of each testing session, the function of the device was explained, was attached to the subject and he/she was instructed to work at normal pace. At the end of the session, each subject was asked to subjectively rate his or her perceived workload during the session as normal, greater than normal or less than normal. For the riveters and bricklayers an analogue visual scale was used.

3. Results

The results for all six occupations are presented in Tables 2 and 3, which list the average time (SD) in minutes spent in each of the five 18° intervals of forward trunk flexion and the average number (SD) of borderline passages.

As shown in Table 2 the least amount of time spent in deep forward bending occurred with the dentist, who had an adjustable workplace and where the need for deep forward bending was minimal and usually did not have anything to do with the work. The dentist spent on average 52·2 min (20·5 + 31·7 min) in the first two intervals (0–36°). The first borderline passage (18°) was passed 313 times. Dividing the total recording time by the number of borderline passages showed that the dentist flexed the trunk up to 36° about every 12 s during an hour of work. The data for the dentist further showed that when treating patients, either standing or sitting, dentists spend most of the time moving from interval (b) (19–36°) to enter either interval (a) (0–18°) or occasionally to interval (c) (37–54°). No heavy lifts were recorded by the observer for this group.

Table 2. *Mean time spent in each 18° interval of forward trunk flexion for subjects in six professions.*

Interval of trunk flexion (degrees)	Dentists (n = 10)	Nurses' aides (n = 23)	Warehouse workers (n = 10)	Janitors (n = 6)	Riveters (n = 12)	Bricklayers (n = 4)
(a)	0–18	20·5 (10·6)ᵃ	34·4 (7·4)	27·5 (16·3)	20·8 (8·6)	41·5 (7·1)	20·4 (5·4)
(b)	19–36	31·7 (14·2)	13·6 (3·4)	22·6 (14·1)	11·6 (3·2)	7·8 (3·9)	9·1 (4·1)
(c)	37–54	7·7 (11·0)	7·0 (2·9)	4·8 (2·4)	8·2 (2·4)	4·9 (2·6)	6·4 (2·2)
(d)	55–72	0·8 (1·1)	3·8 (2·5)	2·2 (0·4)	3·9 (1·8)	3·4 (3·1)	7·1 (1·2)
(e)	73	0·4 (0·6)	2·2 (2·1)	4·1 (1·9)	5·6 (4·5)	2·2 (4·1)	17·1 (4·7)

ᵃFigures in parentheses are standard deviations.

Table 3. *Mean number of borderline passages for subjects in six professions.*

Interval of trunk flexion (degrees)	Dentists (n = 10)	Nurses' aides (n = 23)	Warehouse workers (n = 10)	Janitors (n = 6)	Riveters (n = 12)	Bricklayers (n = 4)
(a)	0–18	313 (138)ᵃ	688 (265)	—	368 (78)	215 (190)	392 (118)
(b)	19–36	166 (117)	239 (78)	428 (317)	347 (94)	103 (62)	329 (71)
(c)	37–54	28 (18)	128 (56)	194 (46)	290 (134)	75 (53)	366 (71)
(d)	55–72	8 (8)	70 (54)	153 (45)	213 (138)	44 (62)	406 (71)
(e)	73						

ᵃFigures in parentheses aree standard deviations.

The nurses' aides maintained an upright position for about 34·4 min in interval (a). Deep forward flexion of more than 72° was performed 70 times during 1 hour. Thus, the subjects performed deep forward flexions somewhat more often than once per minute, or about every 53 s. Deep forward bending was not always combined with heavy lifting. About six to ten heavy lifts were performed by the nurses' aides per hour, manually recorded by the observer as performed in a moderate to fully flexed and rotated position when the subject moved patients in and out of bed.

The warehouse workers spent 4·8 min in deep forward flexion with 153 borderline passages beyond 72°. Thus, on average, the warehouse workers performed a deep forward bend every 24 s during 1 hour's work. The deep trunk flexion among these subjects was observed to be always associated with a heavy lift. The borderline passages from 18 to 19° are not displayed in Table 3 as these numbers were not reliable—because of adjacent trucks the pendulum potentiometer moved due to engine vibrations.

The janitors spent a considerable amount of time in a semiforward trunk-flexed position (19–36°), and also about 6 min in deep forward flexion. Due to the type of work they were performing a large number of deep forward flexions beyond 72° were recorded for this group (213 times per hour). This showed that the janitors on average make a deep forward bend about 3·5 times per minute (Figure 3).

Figure 3.   Typical positioning of the trunk when janitor was washing the floor.

In the riveters' work, most of the time was spent in an upright position and little time was spent in a deep forward position. For the bricklayers a considerable amount of time was spent in a deep forward flexed position, more than 17 min per hour.

For all six occupations the standard deviations are great, which suggests that as well as the type of work, the work technique and the movement pattern will influence the registration of borderline passages and the time spent in each interval. For all occupations the numbers of deep forward flexions (> 72°) manually recorded by an observer corresponded very well to the number recorded by the device. For ten subjects followed for 1 hour, the correlation between flexion analyser results and observer's recordings of deep forward flexions was 0·99.

4.   Discussion

In this study the flexion analyser provided a simple objective method for obtaining data on body movements during work. The good correlation for all subjects between the numbers of deep forward flexions recorded manually and the data recorded by the instrument confirmed, to some extent, the accuracy of the system. The instrument is small and light and can thus be worn by the subject without obstruction of the work task. In future studies measurements can be made without the presence of an observer as long as a thorough evaluation of the subject's work routine is first conducted, to assess the work environment, to determine the length of the work cycle and the weight and physical characteristics of the load that is handled. This feature of the system will prevent possible bias in the data that may be caused by alteration in the work routine due to the observer's presence.

Although it would be possible to measure motion in several planes simultaneously by using additional transducers, this extra equipment would make the instrument considerably more complex and not as easy to use. We chose to measure trunk movement in the saggital plane because these movements are the most frequent and therefore produce large loads on the lumbar spine (Schultz *et al.* 1982 a, 1985). In future studies, however, it would be particularly interesting to analyse rotation of the trunk which produces shear loads on the lumbar spine and can be harmful to the intervertebral disc (Farfan 1973).

The flexion analyser used for measuring motion and posture does not provide direct information about loads and forces on the spine. Therefore, the recordings were supplemented by visual observations. Correlations between lifting and forward flexion were made for deep forward flexion only, because this motion was the easiest to observe. Determination of what constituted a deep forward flexion was, however, admittedly subjective. No attempt was made in these studies to correlate lifting with moderate trunk flexion (36–54°).

It was not possible to quantify the loads for the nurses' aides due to the wide variations in patient weight, distance of the load from the centre of rotation in the midlumbar spine and the amount of assistance that the patients provided during the lift. Because the weight of the lifted object was generally known for the warehouse workers, the total loads lifted during an hour's work cycle were estimated to be between 2000 and 4000 kg. No heavy lift was recorded for the dentists, the janitors or the riveters. For the bricklayers, however, a close relationship was found between the amount of bricks laid down and the amount of forward bending. On average, the four bricklayers used an average of approximately 160 bricks per hour, with traditional methods. Comparing this pace to the number of borderline passages means that the bricklayers laid 1 brick per 17–40 s and that the amount of borderline passages in all four intervals are about twice as many as the number of bricks laid.

Because of the relatively high standard deviations present for all occupations, the mean values are to be taken as representative figures for the occupations. The high standard deviations indicate that trunk flexion differs greatly among individuals during work.

Although the tasks performed and the work environments were similar between occupations, the work pace varied considerably. Motion patterns were slower for some subjects and faster for others. A slower motion pattern does not necessarily mean that the work is less efficient or that it results in lower productivity, since personal habits and motion behaviour have have a strong influence on individual patterns of movement. Variations in body posture due to variations in work environments and differences in the number of trunk flexions among subjects in each occupation, especially in the first two intervals (0–36°). The high number of borderline passages beyond 18° could suggest that slight trunk flexion may occur with each step of walking for some individuals, or may even occur during talking or gesturing. In recent studies, work technique and workplace design have been shown to significantly modify the objective values of the amount of forward trunk flexion (Hultman *et al.* 1984).

Our system for objectively measuring trunk motions should prove very useful for analysing how workplace designs, work techniques and work organization affect loads on the lumbar spine. Modifications based on these data may result in significant reductions in the number of deep trunk flexions, in the number of highly repetitious trunk motions and in the period of time during which static position is maintained. Such modifications may aid in optimizing the work environment from the view of workers' health relating to spinal problems.

Chapter 8

Development and Evaluation of a New Instrument for the Measurement of Work Postures; in Particular the Inclination of the Head and the Spinal Column

J. Weber and A. van der Star

Twente University of Technology, P.O. Box 217, 7500 AE Enschede, The Netherlands

and C. J. Snijders

Erasmus University, P.O. Box 1738, 3000 DR Rotterdam, The Netherlands

1. Introduction

Recent studies have revealed that approximately half of the working population is employed in offices. Due to increasing automation this number will be even higher in the near future (Grandjean 1977, Peters 1976). Distribution of their sitting and standing working postures is reported as 75 and 18%, respectively (Zuidema 1976), and the time spent on diverse activities is as following (Fucique 1967):

reading	15%	absent	21%
writing	32%	other	10%
conversation	22%		

Many working conditions force workers to maintain a certain posture. In most cases the furnishings of the offices have negative influences inducing unfavourable postures, which often lead to a number of somatic complaints (Burandt and Grandjean 1963). These complaints are proven to be generated by the prolonged maintenance of the same posture and are related to a static loading of the muscular system (van Wely 1970). The complaints occur predominantly in the head, neck and thoraco-lumbar regions.

From different sources a number of measures can be found that help to avoid unfavourable postures. One should be aiming for:

1. A working situation in which the trunk and head are to be held upright as much as possible.

2. A good allocation of tasks, which creates sufficient bodily exercise.

3. Relief of the shoulder girdle by good support of the arms by armrests.

4. Avoidance of a too great relative forward flexion of the head with respect to the trunk.

5. A lumbar support of the back by a well-dimensioned and positioned backrest.

(Dul 1979, Grandjean 1977, Gutmann 1968, 1977, Steen 1964.)

The purpose of this study is to analyse and quantify the effect of such measures. This report deals with the influence of an inclined tabletop and the illumination level on sitting posture during reading activities. For this purpose a new type of inclinometer was developed and tested in a pilot study (Weber 1984).

2. Methods and materials

Ergonomics literature describes a number of methods for the assessment and recording of postures. In this case a method using two body-mounted inclinometers was chosen to record the position of the head and the trunk. A similar method has already been used to record the position of the trunk in the sagittal plane (Nordin 1982, Snijders and Philippens 1985). In the present study an instrument was needed to make possible continuous recordings of flexions in the frontal and sagittal planes. This led to the development of a new type of inclinometer: the biplane inclinometer (Nordin 1982, Weber 1984). This instrument consists of two tumbling weights on which miniature loadcells are mounted. Stops on the lip of the cell prevent the rotation of the weights. The resultant force on the lip is proportional to the angle that the tumbling weight takes with respect to the vector of gravity (see Figures 1 and 2).

The inclinometer simultaneously measures the angles with the vector of gravity in the sagittal and frontal planes. The recorded angles differ less than 3°

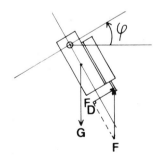

$$F_D = C \cdot \sin \varphi$$

Figure 1. *Working principle of the inclinometer.* F_D = *the resultant force on the lip;* G = *force of gravity.*

Figure 2.   Close-up of the inclinometer. Dimensions: 42 mm × 26 mm × 24 mm. Weight: 0·4 N.

from the exact values on a total scale of 60°. Because of the small dimensions and low weight of the inclinometer, it can easily be mounted on the subjects. Preliminary tests showed that they were not hindered by the presence of the instrument.

The experiment was conducted with 12 male subjects. They are characterized by the following data:

Mean age 22·75 years, SD 3·62 years
Mean body length 181·82 cm, SD 2·14 cm

Their visual acuity was better than 1·25 (checked with a set of Landolt rings). A test for presbyopia was also performed.

The experiment was held under the following conditions. The tabletop height was set according to data using bodylength as the parameter (Diffrient *et al.* 1974) and the chair was adjusted separately for each subject. The experiment took place in a normal office which was specially prepared, making low illumination levels possible. The task consisted of reading intensively a text typed in IBM Prestige Elite 72 lettertype. To stimulate the intensive reading by the subjects, two multiple-choice tests were performed.

Previous research on the relationship between posture and tabletop inclination indicated that most effect could be expected in a range of tabletop inclinations of $\alpha = 0–30°$ (Dul 1979, Höfling 1972, Hooykamp 1976). Three specific inclinations were chosen: $\alpha = 0$, 15 and 30°, and the illumination (E) was set at three levels: low, 40 lx; normal, 600 lx; and high, 1500 lx. In order to avoid sequential effects, the sequence of the setting of the $\alpha–E$ combinations was randomized for each subject.

As a reference the neutral position of the head and trunk was used, defined as when the subject was sitting upright, his eyes focused on a distant point at eye level. In this position, the forces exerted by the postural muscular system, compensating for gravity, are minimal (Dul *et al.* 1982, Grandjean 1977, Rizzi and Corelli 1975, Schoberth 1962). The neutral position also appears to be very reproducible (Dul 1979).

Two biplane inclinometers were used; one mounted on the subject's head (H), the other on the back at the level of the 7th cervical vertebra (C7) (see Figure 3). While the subject was maintaining the neutral position, the inclinometers were set to 0. Every flexion leaving the neutral position in the frontal and sagittal planes could then be measured and recorded.

For deflections from the neutral position the following parameters were used:

Ψ_H and Ψ_{C7}: ante- and retroflexions in the saggital planes. $\Psi > 0$ indicates a forward flexion.

Φ_H and Φ_{C7}: lateral flexions in the frontal planes. $\Phi > 0$ indicates a flexion to the right.

Axial rotations of the spinal column contribute partly to the lateral flexions. This is inherent to the curved form of the cervical spine.

For each $\alpha - E$ combination, the duration of the recording was about 9 min. To allow for adjustment to the new inclinations (about 6 min—Höfling 1972), and adaptation to the new levels of illumination, only the mean values of Ψ and Φ as

Figure 3.   Inclinometers mounted on a subject.

recorded in the last 3 min of recording were taken into account. The signals from the inclinometers were recorded on a four-track x–t recorder.

3.   Quantities to be measured, experimental variables

Apart from proper adjustment of the chair to the subject's body dimensions, there are indications that the use of inclined tabletops has a favourable effect on posture by inducing a more erect position of the head and the trunk (Gutmann 1973, Höfling 1972) as well as an improvement of the sense of comfort (Eastman and Kamon 1976). In this experiment the inclination of the table top, α, was chosen as one of the variables. A desk specially designed for this purpose was used. The height and inclination of the tabletop could be varied by two DC motors between, respectively, 630–850 mm and 0–65° (Weber 1984).

The position of the head and trunk is adapted in such a way that the reading conditions of the ocular musculature are optimized (Gutmann 1973, Höfling 1972, Snijders *et al.* 1977). The reading distance is therefore important, and a number of factors affect this. They can be discriminated as intrinsic and extrinsic factors. The most evident intrinsic factors are: (*a*) visual acuity; (*b*) range of accommodation; (*c*) adaptational ability; (*d*) age; (*e*) body dimensions; (*f*) convergence of the optical axes; and (*g*) the wearing of spectacles and their focal power. Extrinsic factors include: (*a*) magnitude of objects/letters (critical detail); (*b*) illumination; (*c*) reflections; (*d*) contrasts; (*e*) vibrations; and (*f*) working situation (environment, dimensions, tasks).

The recommendations of the Dutch Institute for Illumination, concerning interior illumination, take the following factors into account: classification of the visual task (by critical detail), duration of the task, contrasts, error-risks and age.

Using a symbol height of 1·5 mm and a reading distance of 40 cm, the angle of vision will be 13′ of arc. For reading tasks the critical detail is approximately one-fifth of the symbol height (in minutes of arc), in this case, 2·6 (Leebeek 1983). According to the recommendations the minimal illumination should be 250 lx (under optimal conditions). DIN 5035 recommends 500 lx for offices and normal reading tasks.

Taking into account the relationships between, on the one hand, illumination and the reading task and, on the other hand, the reading task and the working posture, illumination (E) was chosen as a second variable in this experiment. The other influential factors can be disregarded by a good selection of the test subjects and by maintaining the extrinsic and environmental factors within acceptable bounds.

In summary, then, the experiment investigated the following hypotheses:

1. With increasing inclination, α, of the tabletop, the subjects will adopt a more erect posture.

2. With increasing illumination, E, the subjects will come to a more erect posture.

4. Results

During the whole experiment every $\alpha - E$ combination was offered twice to the 12 subjects. These values are represented, with corresponding standard deviations, in graphical form in Figures 4–7). The mean values of the angles have been processed by the r.m.s. method; the equation of the line of regression and coefficient of correlation, r, are marked in the corresponding graphs. The values of Φ_H and Φ_{C7} (lateral flexions) were nearly zero, and are omitted. To test the hypotheses, a t-test was performed on the Ψ-values.

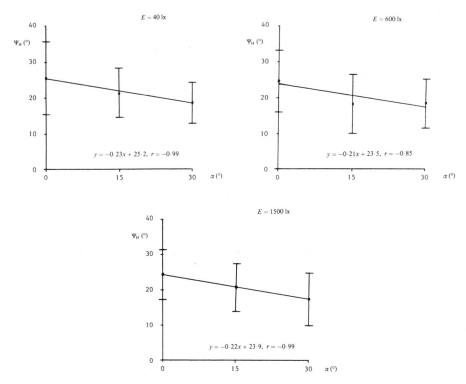

Figure 4. Forward flexion of the head (Ψ_H) as a function of tabletop inclination (α) at $E = 40$, 600 and 1500 lx.

5. Discussion

As mentioned above, the lateral flexions varied only slightly around the neutral position (Φ_H and $\Phi_{C7} \simeq 0°$). If we look at the position of the head in the sagittal planes (Ψ_H):

1. Ψ_H as a function of α at $E = 40$, 600 and 1500 lx. In table 1 the inclination α and illumination E are displayed together with the calculated p-value of the t-tests

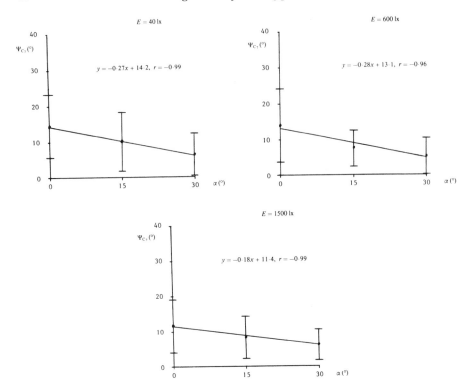

Figure 5. Forward flexion of the back (Ψ_{C7}) as a function of tabletop inclination (α) at $E = 40$, 600 and 1500 lx.

($p < 0.05$), the calculated mean values ($\overline{\Psi}_H$) and the decrease of $\overline{\Psi}_H$ ($\triangle \overline{\Psi}_H$). The influence of the inclination of the tabletop on Ψ_H is evident. By increasing α from 0 to 15° and from 0 to 30°, the decrease of Ψ_H at every illumination level is significant. In the case of an increase of α from 15 to 30°, the effect is less significant, and at $E = 600$ lx is not significant.

2. Ψ_H as a function of E at $\alpha = 0$, 15 and 30°. The effect of illumination on Ψ_H is, except in one case, negligible; the t-tests reveal no significant relations.

Looking at the position of the trunk in the sagittal planes (Ψ_{C7}):

1. Ψ_{C7} as a function of α at $E = 40$, 600 and 1500 lx (see Table 2). In all cases, a significant effect of the inclination on Ψ_{C7} can be observed. Here also, the discrimination of the effect of increasing α from 15 to 30° is less clear, though in all cases the t-test gives significant relationships.

2. Ψ_{C7} as a function of E at $\alpha = 0$, 15 and 30°. The effect of illumination on Ψ_{C7} is marginal; in all but two cases no significant decrease of Ψ_{C7} can be observed.

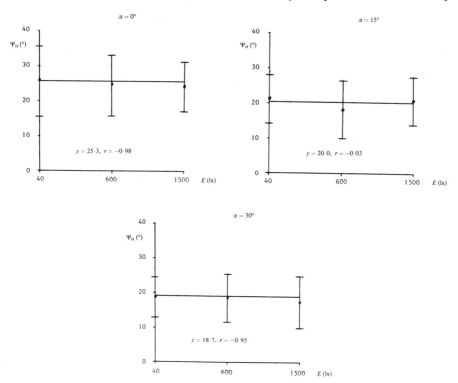

Figure 6. Forward flexion of the head (Ψ_H) as a function of illumination (E) at $\alpha = 0, 15$ and $30°$.

6. Conclusions

The inclination of the tabletop has a clear effect on the positions of the head and the trunk in the sagittal planes. The most explicit decrease of the forward flexions Ψ_H and Ψ_{C7} is achieved by an increase of α from 0 to 15°. However, the effect of illumination on reading posture appeared to be negligible. This indicates that in this experiment, *for the subjects selected* (young men with good eyesight) *and the lettertype used*, illumination was not of such importance that it shows effects on posture. This leads to the conclusion that the first hypothesis was valid and the second hypothesis, for this population and under these conditions, was not.

The results concerning the effect of inclined tabletops, obtained by this method of measurement with inclinometers, agree with what was reported in the literature (mostly based on photographic methods). The instruments designed functioned properly and appear to be well suited for the recording of working postures. At Twente University of Technology the method will find further application in a broad-scale research programme for the assessment of working postures.

Table 1. Forward and lateral flexions of the head with varying illumination and inclination.

α (°)	$\overline{\Psi}_H$ (°)	p	Ψ_H (°)
Ψ_H as a function of α at $E = 40$ lx			
0–15°	25·4–21·1	0·011	4·3
15–30°	21·1–18·5	0·018	2·6
0–30°	25·4–18·5	0·001	6·9
Ψ_H as a function of α at $E = 500$ lx			
0–15°	24·6–18·3	0·000	6·3
15–30°	n.s.[a]	0·946	—
0–30°	24·6–18·3	0·000	6·3
Ψ_H as a function of α at $E = 1500$ lx			
0–15°	24·1–20·6	0·003	3·5
15–30°	20·6–17·2	0·022	3·4
0–30°	24·1–17·2	0·000	6·9
Ψ_H as a function of E at $\alpha = 0°$			
40–600 lx	n.s.	0·600	—
600–1500 lx	n.s.	0·595	—
40–1500 lx	n.s.	0·421	—
Ψ_H as a function of E at $\alpha = 15°$			
40–600 lx	21·1–18·3	0·030	2·8
600–1500 lx	n.s.	0·068	—
40–1500 lx	n.s.	0·634	—
Ψ_H as a function of E at $\alpha = 30°$			
40–600 lx	n.s.	0·935	—
600–1500 lx	n.s.	0·347	—
40–1500 lx	n.s.	0·335	—

[a]Non-significant.

Table 2. Forward and lateral flexions of the back with varying illumination and inclination.

α (°)	$\overline{\Psi}_{C7}$ (°)	p	Ψ_{C7} (°)
Ψ_{C7} as a function of α at $E = 40$ lx			
0–15°	14·3–10·0	0·002	4·3
15–30°	10·0–6·3	0·001	3·7
0–30°	14·3–6·3	0·000	8·0
Ψ_{C7} as a function of α at $E = 600$ lx			
0–15°	13·8–7·5	0·000	6·3
15–30°	7·5–5·2	0·947	2·3
0–30°	13·8–5·2	0·000	8·6
Ψ_{C7} as a function of α at $E = 1500$ lx			
0–15°	11·5–8·2	0·004	3·3
15–30°	8·2–6·1	0·038	2·1
0–30°	11·5–6·1	0·000	5·4
Ψ_{C7} as a function of E at $\alpha = 0°$			
40–600 lx	n.s.	0·723	—
600–1500 lx	n.s.	0·125	—
40–1500 lx	14·3–11·5	0·052	2·8
Ψ_{C7} as a function of E at $\alpha = 15°$			
40–600 lx	10·0–7·5	0·022	2·5
600–1500 lx	n.s.	0·437	—
40–1500 lx	n.s.	0·081	—
Ψ_{C7} as a function of E at $\alpha = 30°$			
40–600 lx	n.s.	0·449	—
600–1500 lx	n.s.	0·390	—
40–1500 lx	n.s.	0·885	—

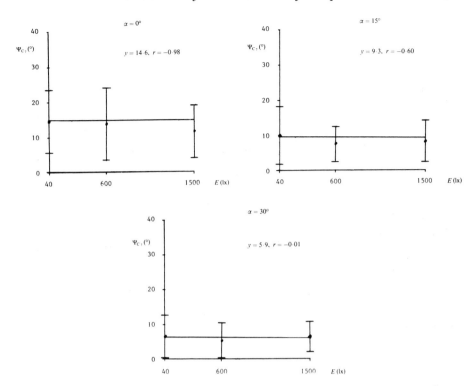

Figure 7. Forward flexion of the back (Ψ_{C7}) as a function of illumination (E) at $\alpha = 0, 15$ and $30°$.

Chapter 9
Measurement of Regional Lumbar Sagittal Mobility and Posture by Means of a Flexible Curve

A. K. Burton

Department of Life Sciences, Huddersfield Polytechnic, Huddersfield HD1 3DH, U.K.

1. Introduction

Posture and/or mobility are constant features of human existence which become especially significant in the ergonomic design of workplaces, where postures may be held for considerable lengths of time, or repetitive movements through a finite range may occur. The relationship between occupation and musculo-skeletal disorders is well documented in recent review papers (Andersson 1982, Anderson 1980, Snook 1982), whilst the particular subject of the effects of working postures has been discussed by Corlett and Manenica (1980).

The evidence available suggests that musculo-skeletal symptoms are likely to develop in workers whose work environment produces high levels of dynamic and/or maintained postural load on their musculo-skeletal system. Furthermore, it has been shown that alterations to the physical working conditions can reduce loads on, say, the spine and will produce a significant reduction in (spinal) symptoms, e.g., Corlett and Bishop (1978), Aarås (1982), Hultman *et al.* (1984).

The ergonomist's task in designing safe physical working conditions will, among other things, require tools that enable measurement of the posture and movement of the spine. Physiological recordings, whilst undoubtedly of value, are not generally applicable to studies in the field. What is needed, according to Corlett and Manenica (1980), is a method of recording which is reliable and repeatable and could be used by technicians in the field. They describe a 'body diagram' graphical method of postural recording which has been used successfully. However, this technique provides only limited information concerning spinal posture.

The object of the study reported here was to develop a simple technique capable of recording the various postures adopted in work tasks, together with the actual movement occurring in regions of the lumbar spine whilst achieving those postures.

2.   The technique

The device chosen for investigation was a draughtsman's flexible curve capable of bending in one plane only, and maintaining an adopted shape which can be transferred to paper. A number of such devices were obtained through normal commercial outlets and, following preliminary tests for accuracy of shape reproduction from inanimate objects, one was chosen for the investigation (Jakarflex, U.K.). The optimal length for lumbar spine recording is considered to be 350 mm, though other sizes may be more suitable for other spinal regions.

A permanent mark is made 20 mm from one end of the flexicurve and two brass sliders are fitted which may be moved along the length of the device to correspond with bony landmarks (Figure 1).

To make a record of the midline contour of the lumbar spine and the relative positions of upper and lower sections, the flexicurve is applied to the dorsal aspect of the spinous processes (SP) and gently pressed to conform to the spine. The S2 mark on the flexicurve is placed over the SP of the second sacral segment, and the sliders positioned over the SPs of the fourth lumbar (L4) and twelfth thoracic (T12) vertebrae (Figure 2). The resultant curve is traced onto paper with the S2, L4 and T12 points marked. Tangents to the curve at these three points are drawn and the angles formed by the L4 line with the S2 and T12 lines are measured.

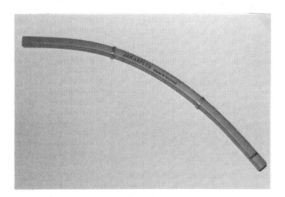
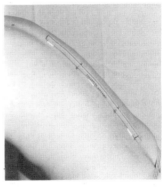

Figure 1.   *Jakarflex flexicurve with 'sliders'*   Figure 2.   *Flexicurve applied to spine.*
and permanent reference mark.

These angles are thus a measure of the movement of the upper (UL) and lower (LL) lumbar sections from a hypothetical straight line, i.e., they are an actual measure of the degree of lumbar lordosis or kyphosis existing in the posture under study. Recordings taken at both extremes of a range of movement will permit: (*a*) the overall range to be quantified (the sum of UL and LL angles from both traces); (*b*) the calculation of the relative contribution to that range of UL and LL sections; and (*c*) the determination of extent of lordosis or kyphosis (Figure 3).

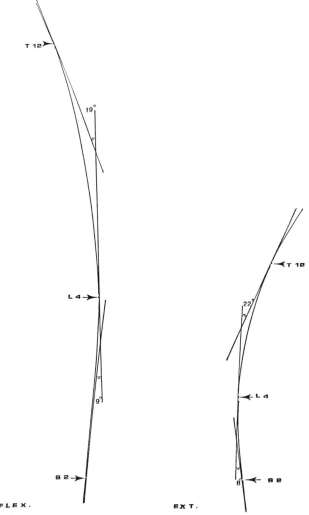

Figure 3.   Flexicurve traces made in flexion and extension, showing tangents with upper and lower angles.

3.   Validation

In order to determine the reliability of the flexicurve technique for recording the actual movement taking place in the spine a comparison with X-rays was made. For ethical reasons this was confined to one volunteer subject.

The subject adopted a forward bending posture (flexion) and a flexicurve (which is radio-opaque) was taped to his lumbar spine with lead markers at T12 and S2 (this study was done prior to the decision to include L4). An exposure was made and the flexicurve removed to enable a trace to be made on paper. The same procedure was followed for backward bending (extension), and the whole process repeated for slightly different postures. The X-rays revealed good conformity of the flexicurve to the spinous processes whilst the lead markers allowed tangents to be drawn at T12 and S2 (Figure 4). Tangents were drawn on the corresponding traces at the same points. The results (Table 1) showed good agreement between the measurements.

The radiographs were also subjected to a superimposition method for measuring total sagittal mobility (Begg and Falconer 1949). The overall range of movement from T12 to S2 was within 1° of that obtained from the traces (Table 1).

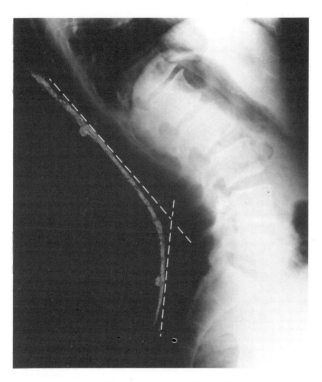

Figure 4. Radiograph of spine and flexicurve. Angle subtended by tangents to spine at T12 and S2 was compared with the resulting trace from the flexicurve.

Table 1. Angles measured from two sets of X-rays compared with flexicurve method (degrees).

	First X-ray	First trace	Second X-ray	Second trace
Tangents				
Flexion	31	32	34	35
Extension	46	47	44	45
Superimposition				
Flexion and extension	79	80	80	80

In order to investigate the ability of the flexicurve to record differences in mobility/posture between UL (T12/L4) and LL (L4/S2), a plastic articulated skeleton was employed. The lumbar spine was manipulated into two different postures of flexion and extension with approximately the same overall angle (T12/S2) but with different upper/lower regional bias. Traces were made for each posture and are shown in Table 2 to be capable of recording a difference in UL/LL bias.

Table 2. *Flexicurve measurements from plastic articulated skeleton manipulated into different postures (degrees).*

	Total ROM[a]	FUL	FLL	EUL	ELL
Position 1	47	24 (51%)[b]	0 (0%)	20 (43%)	3 (6%)
Position 2	46	13 (28%)	12 (26%)	9 (20%)	12 (26%)

[a]ROM, overall range of motion; FUL, flexion upper lumbar; FLL, flexion lower lumbar; EUL, extension upper lumbar; ELL, extension lower lumbar.
[b]Figures in parentheses are proportions of ROM.

In clinical practice a frequently used method of recording lumbar sagittal mobility employs an inclinometer, originally described by Loebl (1967) which has been shown to be a repeatable and valid measure of lumbar mobility and posture. The flexicurve technique was tested against this method (Figure 5). Analysis of data from 20 subjects displayed good correlation between the two methods ($r > 0.80$).

Intra-observer repeatability of the flexicurve technique was assessed by one examiner measuring 15 subjects twice in the same positions within 30 min. The statistical tests used were Pearson's correlation coefficient and the Technical Error of Measurement (Hamill *et al.* 1973). The results are shown in Table 3 and display good agreement. Interobserver error studies will be reported elsewhere. The intra-subject repeatability of the method was assessed from flexicurve traces of 16 subjects made at the same time on different days, in standardized postures of flexion and extension. The results (Table 4) reveal good test/retest agreement and indicate that the method is capable of repeatedly recording the spinal mobility required to adopt given postures.

Table 3. *Intra-observer repeatability of flexicurve meaurements (degrees).*

	Test				Retest			
	FUL[a]	FLL	EUL	ELL	FUL	FLL	EUL	ELL
Mean	13·6	7·2	16·8	14·3	13·4	6·8	17·2	14·0
Correlation coefficient (r)					0·95	0·97	0·96	0·97
Technical error of measurement					1·35	0·85	1·40	0·89

[a]Key for variables as Table 2.

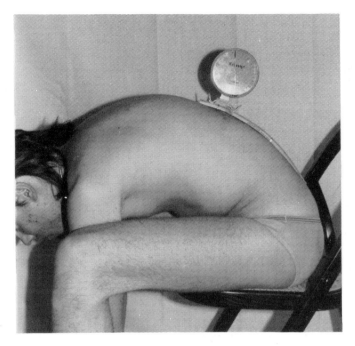

Figure 5.   Pendulum goniometer (inclinometer) measurement of angle between T12 and S2 was compared with flexicurve trace.

Table 4.   Intrasubject repeatability of flexicurve meaurements (degrees).

	Test				Retest			
	FUL[a]	FLL	EUL	ELL	FUL	FLL	EUL	ELL
Mean	16·4	5·0	20·1	6·8	17·0	4·4	21·0	6·5
Correlation coefficient (r)					0·98	0·95	0·96	0·81
Technical error of measurement					1·07	0·95	1·51	1·32

[a]Key for variables as Table 2.

4.   Applications

In order to illustrate possible applications of the flexicurve technique to ergonomic assessment of work postures, three situations were examined and the resultant lumbar characterizations are shown in Figure 6.

   1. The movement required in the lumbar spine to reach forwards (110 cm) across a cutting table was measured in an existing industrial situation (Figure 6 (*a*)) and following raising of the table by 10 cm (Figure 6 (*b*)).

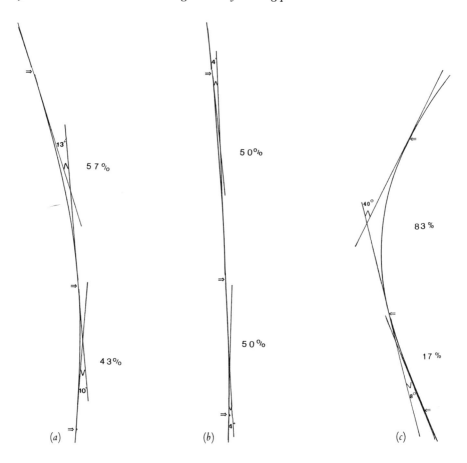

Figure 6. Flexicurve traces of lumbar posture from subject: (a) leaning over cutting table and (b) after raising table by 10 cm; (c) lifting drum of wool from above head height and (d) when standing 18 cm higher on a platform; (e) typing on 'standard' typist's chair and (f) on a stool with forward-sloping seat.

2. The lordotic posture adopted in the lumbar spine whilst lifting a drum of wool from above head height onto a spinning mule in an existing industrial situation (Figure 6 (c)) and following positioning of the operative on a platform (18 cm) (Figure 6 (d)).

3. Comparison of lumbar postures at the end of a period of typing on a 'standard' typist's chair (Figure 6 (e)) and on a stool with a forward-sloping seat (Figure 6 (f)).

These examples are offered simply as illustrations; no statement of ergonomic benefit is necessarily implied.

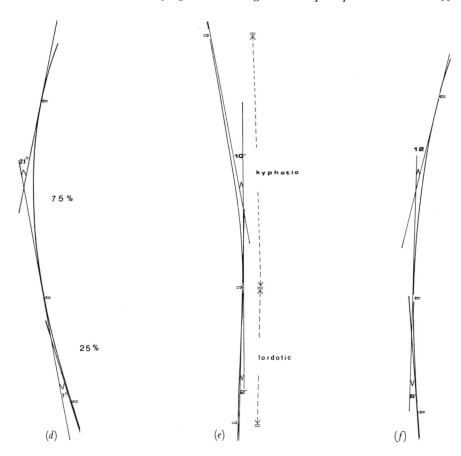

(d)　　　　(e)　　　　(f)

5. Conclusions

Measurement of lumbar sagittal posture and mobility by means of flexicurve traces is shown to be valid and repeatable. It is comparable with existing methods but has the advantage of characterization in terms of upper and lower regions, which may enable more precise estimations of the forces acting on the spine in various work situations and between different anthropometric groups.

The method described is non-invasive, inexpensive and relatively quick to perform (approximately 3 min) in both field and laboratory situations.

Although the technique has been developed for the lumbar region it is likely that it could be readily adapted for other areas of the spine.

Acknowledgements

Jakarflex: Jakar International Ltd., 2–6 Friern Park, London N12 9BX, U.K.
Inclinometer: Medesign Ltd., Clock Tower Works, Railway Street, Southport, U.K.

Chapter 10
Method to Measure Working Posture Loads at Working Sites (OWAS)

Pertti Heinsalmi

Posts and Telecommunications, Helsinki, Finland

1. Introduction

The OVAKO working postures analysing system (OWAS) was developed in the Finnish steel industry (OVAKO) during 1974–1978. The main reason was to improve working methods on the basis of identifying and eliminating harmful working postures. In the background there were increasing amounts of sick leave and premature retirement due to musculo-skeletal disorders.

The method was created by collecting information on all possible working postures and then standardizing them with regard to trunk (four classes), arms (three classes) and legs (seven classes). Three classes for the effort required, or load, were also added. The method's feasibility was tested and it was found that observers trained for 5 days were able to utilize the standardized codes correctly.

In order to measure the potential damage due to each of these 252 (4 × 3 × 7 × 3) different posture combinations, a large study was made. Several different professional groups estimated individually the strain which each posture causes or may cause. From these results all the postures were classified into four action categories according to their strain.

The system was tested in OVAKO factories and then it was donated to the Finnish National Fund for Research and Development and the Rationalization Committee for the major Finnish trade unions. These started a project to test and improve the system for general use. The results presented here are based on that project.

2. Method description

Field observations are usually made at 30 s intervals by a work-study technician who tries to obtain an overall picture of each working posture at the moment of observation. A major objective is to obtain a total picture of working postures and the total work burden. The back, legs, arms and the load and/or effort involved in the work are observed. The postures are defined as follows.

Back:
1. Back straight.
2. Back bent.
3. Back twisted.
4. Back bent and twisted.

Arms:
1. Both arms below shoulder level.
2. One arm at or above shoulder level.
3. Both arms at or above shoulder level.

Legs:
1. Sitting.
2. Standing with both legs straight.
3. Standing with the weight on one straight leg.
4. Standing with both legs bent at the knee.
5. Standing with the weight on one bent leg.
6. Kneeling.
7. Walking.

Load/effort:
1. No effort or effort less than 10 kg.
2. Load or effort of 10–20 kg.
3. Load or effort of 20 kg or more.

For certain tasks extra postures were needed (sitting on the ground, creeping, lying, climbing, and hanging). These were added and used if needed. Because these postures were very rare, no categorization according to their potential harm was made. When using the OWAS system for standing and dynamic work, most difficulties were overcome and the system could be utilized well. In sedentary work, problems were different. Long-lasting postures and static load are important. Information about the posture of the neck was also found necessary. However, because the observation of sedentary posture itself was easier these factors could be included without any decrease in accuracy. The neck had five different alternatives: free (1), bent forwards (2), bent sideways (3), bent backwards (4) and twisted (5). The reported strain was tested and categorized separately for every neck posture. For arms an extra posture was coded, when workers held both their arms statically below shoulder level (4).

Action categories

The postures were divided into the following four 'action' categories:

1. Postures considered normal with no particular harmful effect on the musculo-skeletal system (blue).
2. Postures with some harmful effect on the musculo-skeletal system (the stress is so light that no immediate action is necessary but should be dealt with in future planning) (green).

3. Postures with a distinctly deleterious effect on the musculo-skeletal system (the working methods involved should be changed as soon as possible) (violet).

4. Postures with an extremely deleterious effect on the musculo-skeletal system (immediate measures should be taken to abolish these postures) (white).

Training

A 5 days' course, including observation theory, exercises and tests, was arranged for everyone using the OWAS method. Only persons with an acceptance test result of over 75% were allowed to work in industry.

Use at the working site

After necessary information had been provided for both employees and employers, including the objectives, an OWAS study was made by trained observers. A working group was established at shop-floor level. It had normally three or four members; supervisor and employee (both from the working site itself), the observer and the occupational nurse. This group studied the results and made written concrete proposals of improvements for that worksite. All these proposals were dealt with by a committee which had representatives of employers, employees and occupational safety and health personnel, etc. Improvements were made on the basis of the committee decisions and the working place was subsequently retested by OWAS.

3.   Accuracy and reliability of the method

During the project the accuracy of the method was tested in six factories in different branches of industry (steel, textiles, meat, mining, wood and light metal industries). For the calculations of accuracy, a total of 4567 observations were made, on about 100 people. Accuracy, defined by crude agreement (CA) between observers, was good at over 90% overall, although less for some postures. When the worker was moving a slight tendency was found to observe the posture as being more difficult than it actually was. The only real difficulty in observation was in differentiating between straight and twisted back postures.

Coverage

In order to test the coverage of the OWAS method, a further 59 891 observations were made on 231 people in the six factories. Each posture combination could be found in real work in industry. In certain branches some extra postures (mentioned above) could be found. As expected most of the postures were classified under action category 1 (normal work—about 82%). About 16% of the results were in action category 2, about 2% in category 3 and the most harmful category, 4, had only 0·2% of the postures.

Reliability was tested by one person over several types of work, and also by several persons for the same work. The results obtained from the method were similar when made by different observers, and also by the same observer at different times.

Relationships to health

During the total project the states of health of the musculo-skeletal system of 231 persons were examined twice. Their sick leaves were recorded over altogether about 350 man years. The load of the working postures was calculated on the basis of the estimation of the strain of each posture. The frequency of postures was multiplied by the strain figure. When compared with sick leaves we found that, for women, there was a significant relationship between harmful postures and musculo-skeletal-related sick leaves, especially back diseases ($\chi^2 = 9 \cdot 63$, $p < 0 \cdot 01$). There was no significant relationship for males. Furthermore, changes in the state of the musculo-skeletal system had a negative relationship with individual strain figures. Also, a relationship was found between poor arm postures and sick leaves related to the neck and shoulder region.

Thus, we found that there exists a general relationship between poor postures (identified by OWAS) and sick leaves and also between poor postures and changes in the state of the muscular system of a worker. (The influence of a *single* posture could not be measured because the study period was relatively short and the number of people was too small.) The reason for the fact that the relationships were more prominent in women could be explained by the greater age of the women examined.

4.   Applications of the method

The OWAS method has now been applied in about 20 cases. In general, both employers and employees have been satisfied. The effect on productivity has been generally positive, with no negative effects found. In some instances the related increases in productivity have been large.

The OWAS method was used in the Finnish mining industry. A drilling unit had a proportion of postures in action categories 2–4, taken together, of 5·4%. When the job was examined it was found that the drilling machine was just in front of the worker and in order to see properly he had to twist himself. By moving the steering equipment about 20 cm away a new set of postures was allowed. The total proportion of postures in action categories 2–4 is now 1·0%.

Another observation at the drilling unit was the postures of the worker's head, which was turned backwards in 16% of all the observations. The reason was the illumination. The helmet lamp was the only source of task lighting and it had to be directed to the viewed object by the posture of the head. When more general illumination was added the proportion of backward-facing postures decreased to 0·3%. The expense of both the above changes was negligible.

Another example of the use of the OWAS method is from welding. In this case the method was used in the plant planning phase. The objects being welded were large units belonging to heavy trucks and lorries. This was previously done by the worker climbing up to the machine to do the work from there. This took time, many accidents occurred, it was inconvenient and only young healthy men could do the job. The old postures were measured by OWAS. In the new plant planning phase a model was built and future working postures were predicted by the OWAS method. From these ideas a prototype wooden manikin was built, and OWAS analyses were made with that.

When the planning phase was finished and the new welding unit was built, the work could be carried out in sitting and standing postures and the large units could be moved into position as required. The postures were better, and the task time saved was about 90%. All the participants, the employer and the employees, were satisfied that the work could be done in a far more productive and convenient way.

SECTION 3

MODELS OF POSTURE

This set of seven papers deals with muscle behaviour under various working conditions and some models which can assist in the evaluation of body loads, as well as fatigue and recovery levels.

Two papers, in particular, introduce information concerning muscle loading which should have major impact on practical workspace design. Hagberg points out that minimizing muscular loads is not necessarily the best way of achieving low levels of discomfort or disease (although this criterion is widely used—see the introduction to the next section). He indicates that change of activity, even where the new activity may require higher muscle loads, may be more beneficial. The 'obvious' direction of minimizing load is taken up also by Sjøgaard. She shows that where muscles are lightly loaded the endurance time can depend heavily on motivation. Even more importantly, fatigue in lowly loaded contractions appears to be different in nature to high force contractions, in that it is not a matter of lack of energy in the muscle. She suggests that it arises from changes in muscle potassium concentration which reduce the coupling between the excitation potentials and the contractile behaviour of motor units. Thus, regardless of the level of voluntary contraction which is recommended, there is *no* level of effort which can be continuously maintained without interspersed rest pauses.

Allied to this concept of a lowered level of transmission of neuromuscular stimulation is the interpretation by Marek and Noworol of the differences in posture between two groups of VDT operators doing stimulating and less-stimulating tasks. Arousal levels as a result of the task are postulated to affect muscle tone, both positively and negatively, and were shown to affect posture in the first 2 hours of the experimental task. After 2 hours the subjects' postures tended towards the same form, due to compensatory reactions to alleviate the under- or overstimulation.

The nature of coordinated muscle behaviour and its effect on fatigue is discussed by Dul. Here a range of models is tested to see which gives a best fit to experimental data, taking into account in the models a wide range of possible forms of coordination. The possibilities of changes in the ways in which different muscle groups coordinate during a given posture or activity are discussed and experimental evidence for such changes illustrates that simple biomechanic calculations or the use of one postural loading model can be deceptive. Nevertheless, he illustrates the utility of his more complex models of muscular force production in the opportunities they present to demonstrate the effect of more- and less-efficient coordinations of muscular activity and the implications these have for training in, for example, materials handling tasks. These tasks have for long lent themselves to direct biomechanical assessment and Jäger and Luttman illustrate a typical use for such models. It would appear that the biomechanic model used is a static one, i.e., it does not include inertia loads. The criteria for maximum load, also, are drawn from the literature and relate to the torque at the lumbo-sacral disc. The calculated values illustrate how changes in the task can reduce loading to below the criterion levels given and demonstrate the utility of the approach.

The effects of posture on coordinated muscle activity and force output was studied by Nag, using, *inter alia*, some unusual postures which included sitting without a backrest and reclining. He investigated both arm and leg work and has shown elsewhere (Nag *et al.* 1982) the uses of this information in the design of hand- or leg-powered invalid carriages.

Whilst the energy output and coordination of muscles are of major importance for the design of work and the study of posture, we must return again to the study of the nature and manifestation of muscular fatigue if a full picture is to be gained, for ultimately it is fatigue which limits performance. Milner *et al.* present a model for the prediction of recovery from both maximal and submaximal isometric exercise which does not use the percentage of the MVC of other models, but considers the form of the recovery curves which arise when submaximal loads have been imposed. The basic data were derived from experiments and the constants for the theoretical model derived therefrom. The model was then used to estimate recovery times for an entirely different posture and compared with experimental results. The numbers of subjects used were small, but the predictions were shown to be statistically significantly the same as for the experimental data.

When engaging in field studies where severe loading is experienced the investigator will usually face a dilemma: should an ameliorating strategy be pursued or should a recommendation for a quite new approach be put forward? There is a temptation to see the technology as fixed, often because the investigator concerned is not familiar with it and cannot assess how flexibly it can be used. The result will be, effectively, to fit the person to the job, it may be a better and a safer fit, but that is what it is.

Ergonomics really requires the other approach, to take the person as fixed and modify the technology. Typical is the problem of repetitive work. Lighter forces, better postures, rest pauses, etc., can be introduced, but these are attacking the results of such work, which are the muscular damages arising. It is often believed that the repetition itself cannot be attacked because the work can be done no other way.

This, of course, is a failure of the imagination which is reinforced by the assumed rigidity of the job and organization structures. In a text on posture it is entirely right that job design should be brought into the discussion, for there is no way that a good ergonomic job can be done by repeating the errors of Taylorism, that a person at work is just a mechanism to provide forces and displacements. One of the papers in this set (Sjøgaard) comments on the role of motivation in performance, almost as an aside. The contribution of constrained work activities to muscular complaints is well recognized and documented, but only reducing the constraints will change the effects.

Hence, the uses and interpretations of material in this and other sections of the book will be most successful where they are seen as contributions to the total job design process. To view them as sources of data for maximum threshold limit values, as means for pursuing factional interests of 'management' or 'workers', or as interesting scientific findings which form the basis for further research would be to misuse and undervalue what they can tell us. Their validity, ultimately, lies in how effectively they contribute to the betterment of working life.

Chapter 11

Optimizing Occupational Muscular Stress of the Neck and Shoulder

Mats Hagberg

National Board of Occupational Safety and Health, Medical Division, Box 6104, S-900 06 Umeå, Sweden

1. Working posture and disorders

Only 8 years ago Hadler (1977) referred to the exposure–effect relationship for work-related musculo-skeletal diseases as 'anecdotal'. Recent research has shown that muscular stress on the shoulder–neck is an important causative factor of shoulder–neck disorders (Table 1).

The pathophysiological mechanisms involved in the work-related shoulder–neck disorders are largely obscure. Degenerative tendinitis of the rotator cuff tendons could be caused by a working posture involving elevated arms and, thus, impairment of circulation due to static tension and humeral compression against the coracoacromial arch. Working tasks with repetitive arm movements may evoke shoulder tendinitis or tendovaginitis probably due to friction (Herberts *et al.* 1984).

Painful disorders of shoulder–neck muscles are characterized by the presence of one or more discrete areas which are tender and from which pain may radiate when presure is applied. An appropriate term for these disorders is myofascial syndrome (Hagberg and Kvarnström 1984). Different hypotheses regarding the pathogenesis of occupational shoulder myofascial syndrome include mechanical failure (e.g., ruptures), local ischaemia and metabolic disturbances (Hagberg 1984 a).

In order to avoid muscular disorders due to poor postures, the most important ergonomic action is a properly designed workstation. However, the reality today is that most workstations are designed without reference to ergonomic consider-ations. Furthermore, redesign is often impossible, because of economic factors or the production system. Nevertheless, the physical work conditions can be improved by optimizing muscular load, using: (*a*) supports; (*b*) changing work organization; and (*c*) by selection of the worker. These optimizing procedures will be discussed below, with special reference to the muscular load on the shoulder and neck.

Table 1. Disorders of the shoulder and neck, related to working postures. A sample of recent literature.

Working posture	Disorder	Proposed aetiology	Occupational task	Study design	Reference
Hands at or above shoulder level	Supraspinatus tendinitis	Humeral compression of rotator cuff tendons Static tension of tendons with impaired circulation	Shipyard welding plate workers	Cross-sectional	Herberts et al. 1984
Hands at or above shoulder level	Degenerative tendinitis of rotator cuff tendons	Humeral compression of rotator cuff tendons Static tension of tendons with impaired circulation	Assembling trucks	Case referent	Bjelle et al. 1979, Hagberg 1984 b
Abduction or flexion on the shoulder joint more than 60°	Degenerative tendinitis of rotator cuff tendons Chronic regional myalgia (myofascial syndrome)	High muscular load on shoulder muscles during work Localized muscle fatigue	Assembling trucks	Case referent	Bjelle et al. 1981
Hands kept between elbow and shoulder height	Myofascial syndrome	Continuous low static load on shoulder muscles Localized muscle fatigue	Electronic assembling	Case reports	Hagberg and Kvarnström 1984
Flexion of the head and neck	Neck pains	Not stated	Accounting-machine operating	Cross-sectional	Hünting et al. 1983
Constant sitting with static strain in the shoulder–hand region	Tension neck	Not stated	Data-entry operators	Cross-sectional	Kukkonen et al. 1983
Half abduction and flexion in the shoulder	Cervicobrachial disorders	Continuous muscle tension	Cash-register operators	Cross-sectional	Sällström and Schmidt 1984
The arms raised to some extent	Shoulder muscular tenderness	Static fatigue of the trapezius muscle	Film-rolling workers	Cross-sectional	Onishi et al. 1976

2. Optimizing the muscular load on the shoulder and neck, using arm supports

Modern keyboards are usually equipped with wrist supports. A support at the wrist when the arm is slightly abducted and flexed as in the working posture for VDT or typewriting would, from a biomechanical view, reduce the torque in the gleno-humeral joint and thus the scapular elevators, e.g., the descending part of the trapezius. However, astonishingly, preliminary results have shown an increase in the load of the descending part of the trapezius muscle when wrist supports are used (Bendix and Jessen 1984, Olssen and Åhlen 1979). In both studies EMG activity from the descending part of the trapezius muscle was measured during typing. Although the load on the trapezius muscle was higher with a wrist support than without, the subjects preferred the working posture with the support. It is possible that structures other than the descending part of the trapezius muscle are optimized when wrist supports are used, e.g., forearm muscles. When wrist supports are supplied to patients with a shoulder–neck problem, not only is an extensive introduction of the supports necessary but also a close follow-up of the workstation should be made by the ergonomist.

In the electronics industry there is an association between the working posture in sitting assembly work and myofascial syndrome in the shoulder–neck. It has been suggested that the static contraction of the descending part of the trapezius muscle is the cause of the disorders (Hagberg and Kvarnström 1984, Jonsson *et al.* 1981). Elbow supports or forearm supports have been suggested as a possible way of optimizing the working posture in electronic assembling (Westgaard and Aarås 1984). A Swedish company planned to improve working conditions in sitting assembly work by placing height-adjustable forearm supports on the chairs in their assembly workstations. The load-reducing effects on the shoulder and neck thus produced were to be documented by electromyography (Norin and Seijmer-Andersson 1982). To the investigators' surprise they found that the muscular load on the descending part of the trapezius muscle was higher when working with a forearm support, for all the workers examined but one. These results were obtained despite the fact that the subjects were well trained, the supports carefully adjusted and the investigators' prior view that 'forearm support is a good ergonomic aid'. Many of the workers experienced the forearm supports as helpful even though an increased load on the descending part of the trapezius muscle was recorded during work. The electronic assembly work studied consisted mainly of small movements of the hands and forearms, however, the forearm support may have been an obstacle when the forearm was to be moved, thus forcing an elevation of the scapula.

Ekholm *et al.* (1983) showed, in a mock-up workstation for assembling printed circuit boards, that forearm support reduced the muscular load on the shoulder and neck. The supports used were movable horizontally and this may provide an explanation for the different results obtained in these two investigations. When considering elbow or forearm support in sedentary work, careful note has to be taken of arm movements in order to determine whether and what type of forearm support to supply, if any. Ekholm *et al.* (1983) suggested that a *suspension force* is preferable to elbow support in postures where the thoraco-lumbar spine is kept vertical or inclined backwards slightly. Furthermore, an anterior head support (applied to the forehead) produced no reduction of shoulder–neck load in forward flexed postures. Suspension of the forearms in a sling is becoming widespread in Sweden for working postures where the arms are held with a forward flexion in the shoulder.

3. *Optimizing neck–shoulder muscular load by pauses and job rotation*

Since the work of Rohmert (1960 a) the importance of interrupting static contractions in order to enhance local work capacity has been recognized.

The muscular load level that can be sustained for 1 hour in an isometric contraction by the elbow flexors is as low as 8·8% of the MVC (maximal voluntary contraction (Hagberg 1981 b). Surprisingly, a similar but dynamic exercise is as fatiguing, whith a 1 hour limit of 7·6% MVC (Figure 1). This is true if the dynamic contractions are slow and isotonic; the dynamic contractions referred to in Figure 1 were continuous, eccentric and concentric flexions of the elbow flexors at an angular velocity of 30°/s. Thus dynamic contractions can also be regarded as

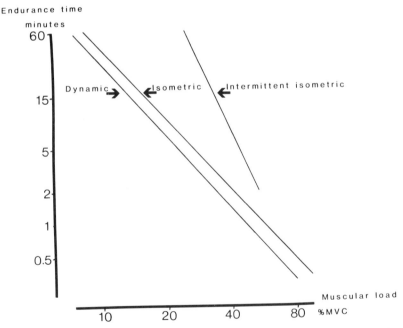

Figure 1.   Endurance time (minutes) versus muscular load (% MVC) in sustained isometric, dynamic and intermittent isometric exercise. Note the logarithmic scales.
Data from Hagberg (1981 a).

'static' in work if they are isotonic. For a working posture with continuous repetitive forward flexions in the shoulder, EMG fatigue development in the descending part of the trapezius muscle was correlated to the load as glenohumeral torque and similar to isometric exercise of the elbow flexors (Hagberg 1981 b). This implies that repetitive arm elevations may induce a 'static' load on the shoulder–neck muscles. However, if a muscle contraction is interrupted by pauses even as short as 2 s, endurance time and work capacity will increase (Figure 1). One commonly held early theory of the development of muscle fatigue concerned the accumulation of metabolites and a drop in the pH level inhibiting glycolytic enzymes. This has been questioned by Miller and Edwards (1984), who report muscular fatigue development despite normal pH level in patients with McArdle's syndrome. Other possible available mechanisms include, for example, intracellular potassium depletion (Sjøgaard 1983). It is, however, beyond dispute that the recovery of muscle function with relaxation is dependent upon the perfusion of the muscle. Recovery of muscular endurance may be delayed for several days when a subject is exposed to strenuous eccentric contractions (Hagberg 1984 b, Komi and Viitasalo 1977).

Unfortunately, there are only a few studies in which the shoulder muscular effects of pauses and micropauses (2–60 s) have been studied. Rohmert and Luczak (1973) reported that optimum performance was obtained with short rest periods, but short rests had no effect on the subjective feeling of fatigue in VDT work. Data entry operators who tried micropauses of 10–15 s every tenth minute over a working day perceived less fatigue in the shoulder–neck than on an ordinary workday (Figure 2).

Job rotation is often proposed as a compensatory measure for poor working

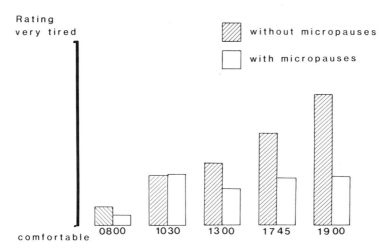

Figure 2. Perceived fatigue in the neck, during a workday with and without micropauses, among data-entry operators.
Data from Ehnström (1981).

postures in short cyclic work tasks. The effects of job rotation on reducing occupational shoulder muscular stress are unknown. One possibility is that a work task
demanding a low static load on shoulder–neck muscles may lead to selective
fatigue of low threshold motor units. If the job is then changed, a change in the
load will cause the recruitment of other motor units and may decrease the fatigue
rate. If this hypothesis is correct, even a change of work involving an increase of
the load would reduce the muscular strain.

4.   Optimizing neck–shoulder muscular load by selection of workers

The major aim of workstation design is that no selection of workers should be
necessary. However, there are many industries where selection of workers does
occur, due to strenuous working postures, for example, on automobile assembly
lines involving work tasks where the hands have to be kept above shoulder height.
Selection of workers through isometric strength tests was found in one report to be
a means of controlling medical incidents on strenuous jobs in building tyres
(Keyserling *et al.* 1980). Occupation may serve as a modifying factor in the
symptoms of rheumatic diseases (Anderson 1974). Working patients with chronic
diseases such as ankylosing spondylitis are known to have a general myopathy and
increased fatiguing of shoulder muscles (Hagberg *et al.* 1986, Hopkins *et al.* 1983).
Thus, these patients are probably more susceptible to static loads as factors in
developing muscle pain syndromes. Acute viral infections, for instance the
common cold, affect muscular function (Friman 1976). Friman (1978) also found
that it took more than 4 months to recover muscular endurance after acute viral
infections. The incidence of cervicobrachial syndromes at an industrial health care
centre in an industrial plant was reported to be related to acute infections (Cardell
and Melin, 1981). It is possible that acute infections may predispose a subject to
occupational shoulder muscular disorders in two ways. First, during acute
infections, when the muscle is infiltrated with viruses it may have a reduced
tolerance to stress. Secondly, after acute infections, reactive tendinitis or myalgia
may occur due to prior occupational degenerative changes.

Chapter 12

Muscular Coordination in Working Postures

J. Dul

TNO—Institute of Preventive Health Care, P.O. Box 124,
NL-2300 AC Leiden, The Netherlands

1. Introduction

Body posture and body movement involve a coordinated action of skeletal muscles across the joints. The major joints in the human body are crossed by several muscles with similar functions. For example, at least three muscles are able to contribute to knee flexion. Strictly speaking, one muscle only would be sufficient to realize such an action. This, what we can call 'anatomical redundancy' in the musculo-skeletal system, provides the body with considerable versatility in selecting the muscles and their activation levels to produce a desired output. Two types of muscular coordination have been observed: (*a*) *recruitment*, in which only a few redundant muscles are active; and (*b*) *synergism*, in which several muscles are active simultaneously. The first type of muscular coordination has been observed at the onset of a muscle action (Kuo and Clamann 1981). However, most postures and movements involve muscular synergism, and it appears that there is considerable intra- and inter-individuality (e.g., Townsend *et al.* 1978, Walmsley *et al.* 1978). This versatility in selecting the muscles and their activation levels can be explained biomechanically. The mechanical analysis of an activity involving several muscles normally results in an indeterminate problem. The number of unknown forces exceeds the number of equilibrium equations. A unique solution for the muscle forces cannot be obtained. There is an infinity of mechanically possible coordination modes. In this paper we define a coordination mode as the set of individual muscle forces of the muscles performing the activity.

The question arises as to whether certain coordination modes are more 'efficient' for maintaining working postures than others, and if so, whether the differences are appreciable. Many working postures involve static muscular contractions. A constant output must be produced for a long period of time. Ideally, the central nervous system selects an efficient mode of muscular coordination, such that the constrained posture can be maintained for as long as possible. In this paper we address the efficiency question by analysing static-isometric knee flexion at 20% MVC (maximum voluntary contraction). The knee was selected as an example of a redundant system because the muscle data required were readily available from earlier studies (Dul 1983, Dul *et al.* 1984 a, b). Ten coordination modes (both recruitment and synergism modes), which are computed on the basis of biomechanical models from the literature, are compared. We discuss the characteristics of these modes, and each mode is evaluated by computing the period of time that the postural flexion task could be maintained under the given coordination mode. This time is called here 'postural endurance time'. The computation of the postural endurance time is based upon the relationship between endurance time and force of individual muscles, as measured experimentally in animals. According to the above definition, predicted coordination modes with long postural endurance times are more efficient for maintaining working postures than coordination modes with short endurance times.

Other linear MINISUM criteria have been used less frequently. Mode 2 is based upon minimization of total muscle stress ($\Sigma F_i/A_i$), where A_i is the physiological cross-sectional area (Crowninshield 1978, Crowninshield *et al.* 1978). Mode 3 is based upon minimization of total muscle ratio ($\Sigma F_i/F_{imax}$), where F_{imax}) is the maximum possible force in an individual muscle (Pedotti *et al.* 1978). Modes 2 and 3 are recruitment modes as well. The muscles are selected according to the mathematical product of moment arm and physiological cross-sectional area (b_iA_i), and of moment arm and maximum force (b_iF_{imax}), respectively.

In recent years, non-linear MINISUM optimization models have been employed. These models predict synergism (modes 4–6, Table 1). The redundant muscles are active simultaneously. Pedotti *et al.* (1978) defined an objective function consisting of the sum of squared forces (ΣF_i^2). In the resulting synergism mode relatively more force is allocated to muscles with large moment arms. The same authors also defined the sum of squared ratios of muscle force to maximum force ($\Sigma (F_i/F_{imax})^2$). Now there is relatively more force in muscles with a large moment arm and with a large maximum force. Crowninshield and Brand (1981 a) selected an exponent with a value of 3 and used the sum of cubed muscle stresses ($\Sigma (F_i/A_i)^3$) as the objective function. In this mode the load sharing depends on the moment arm and the cross-sectional area. Non-linear MINISUM optimization problems can be solved analytically using the Lagrangian multiplier technique, as has been shown by Dul *et al.* (1984 b). The resulting expressions for the load sharing between the muscles are given in the last column of Table 1.

Recently, Dul *et al.* (1984 a) have formulated a 'minimum fatigue' optimization model which employs a MINIMAX objective function. This approach is based upon the assumption that each muscle exhibits an endurance time–force characteristic which decreases continuously with increasing load (specific forms of these characteristics are given below), and that the system seeks to maximize the minimum individual endurance time (T_i) of the muscles involved. An analytical solution for the load sharing was derived and is shown in Table 1 (mode 7). In this synergism mode relatively more force is allocated to muscles with large maximum possible force (F_{imax}) and to muscles that are fatigue resistant (muscles with a large fraction of slow-twitch fibres; p_i large). The load sharing does not depend on the moment arms, although the absolute force levels do depend on this variable.

The last three modes of Table 1 (modes 8–10) are synergism modes as well.

Here the load sharing is not the result of an optimization approach, but is based upon 'intuitively reasonable' distributions that were selected *a priori*.

2. Methods

Selection of coordination modes

Two strategies were employed to sort among the many possible coordination modes and to select only ten modes for further analysis. Firstly, a biomechanical optimization approach was employed. This approach selects a coordination mode which is optimal in a mathematical sense. An objective function which supplements the equilibrium equations is formulated. This function represents the physiological principles of the muscular coordination, in the sense of physiological 'cost' to be minimized. Minimization of this cost function (or objective function), subject to the constraints of the equilibrium equations, limits of forces, etc., results in a unique and optimal coordination mode. Hence, the coordination mode that is selected in this manner depends on the objective function that was employed. In this paper, the first seven modes are based upon seven different objective functions from the literature. In the second strategy to select coordinations modes, a load-sharing principle is formulated *a priori*. For example, it can be assumed that the redundant muscles share the load according to their maximum possible forces. Three such modes are considered below.

The first mode that is based upon the optimization approach is the "principle of minimal total muscular force" (MacConnaill 1967), which postulates that no more total muscular force than is both necessary and sufficient to maintain a posture or perform a movement would be used. Accordingly, this would minimize an objective function consisting of the sum of muscle forces (ΣF_i), where F_i is an individual muscle force. This linear MINISUM objective function has been widely used for a variety of muscle groups (Barbenel 1972, Hardt 1978, Kralj 1969, McLaughlin and Miller 1980, Patriarco *et al.* 1981, Pedotti *et al.* 1978, Penrod *et al.* 1974, Seireg and Arvikar 1973, Yeo 1976). The linear optimization problems can be readily solved with the Simplex method (Dantzig 1968). The resulting coordination mode (mode 1 in Table 1) is of the recruitment type (Dul *et al.* 1984 b). Only a few muscles are active. When the load increases there is an orderly recruitment of muscles. The muscles are recruited according to their moment arms (b_i). First, the muscle with the longest moment arm is recruited and when this muscle reaches its force limit, the muscle with the next longest moment arm is selected, etc.

Knee flexion

Each of the ten muscular coordination modes has been applied to predict possible muscular actions across the human knee joint during static-isometric flexion at 20% MVC. It is assumed that after the selection of a certain coordination mode, the co-ordination does not change during performance of the task. The system is depicted

Table 1. Muscular coordination modes.

		Coordination mode	
Mode	Criterion	Type	Description
1	ΣF_i	Recruitment	Muscle selection according to b_i
2	$\Sigma F_i/A_i$	Recruitment	Muscle selection according to b_iA_i
3	$\Sigma F_i/F_{imax}$	Recruitment	Muscle selection according to b_iF_{imax}
4	ΣF_i^2	Synergism	Load sharing according to $F_i/F_j = b_i/b_j$
5	$\Sigma (F_i/F_{imax})^2$	Synergism	Load sharing according to $F_i/F_j = (b_i/b_j)(F_{imax}/F_{jmax})^2$
6	$\Sigma (F_i/A_i)^3$	Synergism	Load sharing according to $F_i/F_j = (b_i/b_j)^{0.5}(A_i/A_j)^{1.5}$
	min $\{T_i\}$	Synergism	Load sharing according to F_i/F_j $= \{F_{imax}/(F_{jmax})^{p_i/p_j}\}F_j^{(p_j-p_i)/p_i}$
8	$F_i/F_j = b_i/b_j$	Synergism	Load sharing according to $F_i/F_j = b_i/b_j$
9	$F_i/F_j = F_{imax}/F_{jmax}$	Synergism	Load sharing according to $F_i/F_j = F_{imax}/F_{jmax}$
10	$F_i/F_j = A_i/A_j$	Synergism	Load sharing according to $F_i/F_j = A_i/A_j$

schematically in Figure 1 (*a*). An 'average' man is sitting upright with his thigh in the horizontal position. The hip and the ankle are fixed externally. The angle between the thigh and the lower leg (knee angle) is constant (140°). This is approximately the angle of maximum isometric force (Murray *et al.* 1980). The task that is performed is the sustained pressing of the lower leg against a resistance The level of effort is 20% MVC. This corresponds to a muscular knee moment (*M*) of approximately 17 N m. After a specific period of time this task cannot be performed anymore, due to fatigue in the knee flexion muscles. The major muscle groups that can contribute to knee flexion are listed in Table 2. The 'short hamstring' is the short head of the biceps femoris, which is a one-joint muscle. The two-joint muscles that cross the knee and the hip are the long muscles of the hamstrings and are discussed below as the 'long hamstring'. The two heads of the gastrocnemius are grouped into the 'gastrocnemius'. The groupings are shown in Figure 1 (*b*). Note that grouping of muscles is *not* essential; however, it is done here

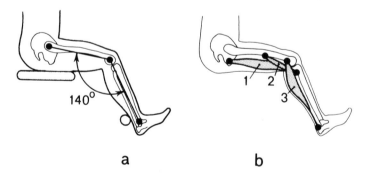

a   b

Figure 1. (a) Static-isometric knee flexion. (b) Muscle groups for knee flexion.
1, long hamstring; 2, short hamstring; 3, gastrocnemius.

for simplicity of the analysis, and clarity. The resulting muscle force of each group is assumed to act along the line between the average points of origin and insertion. These points were estimated from data given by Seireg and Arvikar (1973). The corresponding moment arms (b_i) of the muscle groups for a knee angle of 140° are given in Table 2.

Table 2.   Knee-muscle data.

	Long hamstring ($i = 1$)	Short hamstring ($i = 2$)	Gastrocnemius ($i = 3$)
Moment arm b_i (cm)	4·4	3·2	2·8
Physiological cross-sectional area A_i (cm²)	40	10	25
Maximum force F_{imax} (N)	1200	300	750
Fraction slow-twitch fibres s_i (—)	0·67	0·67	0·48

Other muscle data are given as well. Values for the physiological cross-sectional area of each muscle were obtained from data of Pedotti *et al.* (1978). The maximum possible muscle force (F_{imax}) was calculated by multiplying the cross-sectional area by the maximum possible stress. A stress value of 30 N/cm² was selected. Estimation of the values for the fraction of slow-twitch fibres (s_i) is based upon data of Johnson *et al.* (1973); s_i is large for muscles that are fatigue resistant (see below).

For the mechanical analysis of knee flexion it is assumed that the extensors of the knee are not active during knee flexion. The moment equation with respect to the knee joint centre (where the resultant joint forces interact) for 20% MVC is

$$4·4F_1 + 3·2F_2 + 2·8F_3 = 1668 \text{ N cm} \tag{1}$$

F_1, F_2 and F_3 are the muscle forces in the long hamstring, the short hamstring and the gastrocnemius, respectively. A coordination mode for knee flexion is the set of values for F_1, F_2 and F_3. Equation (1) can be solved by using one of the expressions for muscular coordination (Table 1). For example, for coordination mode 4 the load sharing between the long and the short hamstring is $F_1/F_2 = 1·38$, and the load sharing between the long hamstring and the gastrocnemius is $F_1/F_3 = 1·57$. Substituting these equations in Equation (1) yields $F_1 = 196$ N, $F_2 = 142$ N and $F_3 = 125$ N. Hence, this set of muscle forces is the quantitive description of muscular coordination mode 4 at 20% MVC. A description for the nine other modes can be obtained similarly.

Postural endurance time

For each coordination mode, the postural endurance time is computed as follows. First, for each individual muscle the *individual* endurance is computed. From animal experiments we know that the endurance time of an individual muscle (T_i) is an

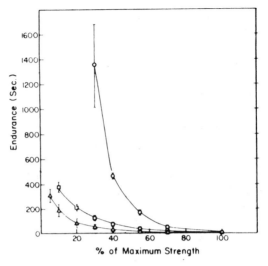

Figure 2.   The relationship between muscle force and endurance time of three cat muscles.
$\bigcirc$, *soleus (98% of slow-twitch fibres);* $\square$, *medial gastrocnemius (37%);* $\triangle$, *plantaris (26%).*
Adapted from Petrofsky and Lind (1979).

exponential function of muscle force (F_i), with the maximum muscle force (F_{imax}) and the fraction of slow-twitch fibres (s_i) as muscle parameters. Figure 2 shows experimental data from Petrofsky and Lind (1979). The endurance time was measured in three cat muscles with different fibre-type composition at several levels of muscle force. Muscle force is given relative to maximum muscle force.

The three muscles are the soleus, the medial gastrocnemius and the plantaris, having fractions of slow-twitch fibres of 0·98, 0·37 and 0·26, respectively (Ariano *et al.* 1973, Close 1972). The relationships of Figure 2 can be quantified using a least-squares curve-fitting technique (Dul *et al.* 1984 a). Equation (2) represents the data points

$$T_i = T_{imax}(F_i/F_{imax})^{p_i} \tag{2}$$

T_{imax} and p_i are muscle parameters. T_{imax} is the endurance time at maximum contraction $(F_i = F_{imax})$. It turns out that T_{imax} is approximately the same for each muscle (13 s). p_i indicates how much the endurance time decreases with muscle forces; p_i is always negative and depends on the fraction of slow-twitch fibres (s_i) in the muscle. The following relationship was found:

$$p_i = -0·25 - 3·6s_i \tag{3}$$

It is assumed that Equations (2) and (3) are valid for all animal and human skeletal muscles. This assumption is reasonable since the muscle force is scaled by dividing it by the maximum force, and since the anatomy and physiology of animal and human tissues are similar. Values for s_i and F_{imax} for the human knee flexors are listed in Table 2. From the individual endurance times we can compute the postural endurance time. The postural endurance time is the maximum duration that a group of individual muscles can maintain the required output. The muscles that contribute to this output may have different *individual* endurance times (T_i). The *postural* endurance time (T_p) of a set of muscles is equal to the shortest individual endurance time. In other words, the muscle with the shortest endurance time is exhausted first and determines the postural endurance time. Because of this muscle, the desired output cannot be produced anymore. The postural endurance time of a group of *n* muscles can be written as:

$$T_p = \text{minimum}\{^1 T_i\} \qquad (i = 1, 2, \ldots, n) \qquad (4)$$

For a given coordination mode, the postural endurance time can now be computed. First, the three muscle forces are computed using Equation (1) and one of the expressions given in the last column of Table 1 (see above). Then the three individual endurance times (T_i) are determined using Equations (2) and (3), and, finally, the postural endurance time (T_p) is computed using Equation (4).

3.   Results

For each coordination mode the muscle forces (F_i), individual endurance times (T_i) and postural endurance time (T_p) for knee flexion at 20% MVC are given in Table 3. The muscle forces (second column) are given as a percentage of the maximum muscle force. The endurance times are expressed in minutes.

The results show that the computed postural endurance times for the different modes (last column of Table 3) range from 1·6 to 12·4 min. The recruitment modes (modes 1–3) are moderately efficient with postural endurance times of

Table 3.   *Comparison of ten muscular coordination modes for knee flexion at 20% MVC.*

Mode	Muscular force (% of max. force)			Individual endurance time (min)			Exhausted muscle	Postural endurance time (min)
	LH	SH	GA	LH	SH	GA		
1	32	0	0	4·9	∞	∞	LH	4·9
2	32	0	0	4·9	∞	∞	LH	4·9
3	32	0	0	4·9	∞	∞	LH	4·9
4	16	48	17	28·9	1·6	7·8	SH	1·6
5	46	8	18	1·7	178·0	6·4	LH	1·7
6	24	11	15	10·7	84·0	9·7	GA	9·7
7	22	22	13	12·4	12·4	12·4	LH, SH, GA	12·4
8	16	48	17	28·9	1·6	7·8	SH	1·6
9	20	20	20	16·7	16·7	5·4	GA	5·4
10	20	20	20	16·7	16·7	5·4	GA	5·4

LH = long hamstring; SH = short hamstring; GA = Gastrocnemius.

almost 5 min. Certain synergism modes (modes 4, 5 and 8) are less efficient (not more than 2 min), whereas other synergism modes (modes 6, 9 and 10) are more efficient (up to 10 min). The most efficient mode is synergism mode 7 which is based upon the minimum-fatigue optimization model (more than 12 min).

The variability in postural endurance time is due to the fact that the force levels in the individual muscles, and therefore the individual endurance times, are quite different for most modes. The three recruitment modes (modes 1–3), which are based upon linear MINISUM optimization models, are identical. In these modes only the long hamstring is the muscle that is selected to perform the activity while the other two muscles are inactive. This is because the long hamstring has a larger flexion moment arm than the other two flexors (b_i in mode 1, see Tables 1 and 2), it has a larger product of moment arm and cross-sectional area (b_iA_i in mode 2) and it has a larger product of moment arm and maximum force (b_iF_{imax} in mode 3). In all three modes the long hamstring is the only muscle that is active (32% of maximum force). The corresponding individual endurance time is 4·9 min (Equation (2)). The short hamstring and the gastrocnemius are not active, such that their individual endurance times are infinite. Hence, the postural endurance times of the three recruitment modes are 4·9 min (Equation (4)). After this period of time the flexion task cannot be performed anymore due to fatigue in the long hamstring. The long hamstring is the exhausted muscle (see the last but one column in Table 3).

In the three synergism modes that are based upon a non-linear MINISUM optimization model (modes 4–6), the three knee flexors are active simultaneously. In mode 4 the load sharing depends only on the moment arm, but in modes 5 and 6 also on the maximum force and the cross-sectional area, respectively. The levels of effort in the individual muscles differ considerably. For example, in mode 4 the effort in the short hamstring is 48%, in mode 5 it is only 8% and in mode 6 it is 11%. The corresponding individual endurance times are 1·6, 178 and 84 min, respectively. In mode 4 the short hamstring will be exhausted, in mode 5 the long hamstring and in mode 6 the gastrocnemius.

Synergism mode 7 is based upon a non-linear MINIMAX optimization model. In this minimum-fatigue approach the system seeks to maximize the postural endurance time. Therefore it is not surprising that mode 7 is the most efficient mode. In the minimum-fatigue mode, the individual endurance times are the same and are equal to the postural endurance time (Dul 1983). Hence, all muscles fail to produce the required output simultaneously. This is only true in the case where one joint (e.g., the knee) is loaded. In a more complex load situation involving two joints (e.g., knee and ankle) only a few muscles have equal endurance times (Dul and Townsend 1985, Townsend and Dul 1985).

Modes 8–10 are the synergism modes that are based upon an *a priori* model. Mode 8 is identical to mode 4. The *a priori* coordination principle of mode 8 is the same as the coordination of mode 4 that was predicted mathematically (Table 1). Mode 9 and mode 10 are identical as well. In mode 9 the distribution of muscle force is according to the maximum muscle force; in mode 10 it is according to the cross-sectional area. These two variables are linearly related (Crowninshield and Brand 1981 a).

4. Discussion

The order of magnitude of the computed postural endurance times corresponds to endurance times of muscle groups that have been measured experimentally in humans. From experiments on arm, leg and trunk muscle groups Rohmert (1960 a) and Monod and Scherrer (1965) derived expressions for the average postural endurance time of a group of muscles. This 'law of Monod and Rohmert' (Monod 1972) does not take into consideration possible differences in coordination modes. According to this law the average postural endurance time at 20% MVC is about 8 min. The order of magnitude of this value corresponds to the postural endurance times for the different coordination modes given in Table 3, which were computed from the characteristics of individual muscles within the muscle group (Equations (2) and (4)). At present the measurement of coordination modes (i.e., of the set of individual muscle forces) in humans is difficult. EMG measurements may be particularly useful for widentifying the presence or absence of muscle activity, and, hence, to differentiate between recruitment modes and synergism modes or between different recruitment modes. However, it remains to be seen whether the EMG is useful for quantifying load-sharing patterns in synergism modes. Several authors have expressed scepticism about the possibility of using the EMG signal as a quantitive measure of muscle force, particularly during activities with non-isometric muscular contractions (Bouisset 1973, Crowninshield and Brand 1981 b, Perry and Bekey 1981, and elsewhere). Nevertheless, Bouisset states that "it seems reasonable to envisage that in future it may be possible to use the value of the instantaneous surface EMG signal as an indirect complex indication of the force developed by the muscle". Several encouraging attempts have been made to quantify this relationship during well-defined activities (e.g., Hof 1984).

Recently, techniques have been developed to measure muscle forces in cats *in vivo* (Gregor *et al.* 1981, Walmsley *et al.* 1978). This allows the direct measurement of synergistic coordination modes. Figure 3 shows the data points for the (instantaneous) synergism between the medial gastrocnemius and soleus muscles in a cat for postures and movements (Walmsley *et al.* 1978).

The curves in Figure 3 are computed synergism patterns according to the computed coordination modes given in Table 1. Values for b_i, $F_{i\max}$, A_i and s_i, which are required for these computations, were obtained from the literature. Comparison of the measured synergism with the computed synergism shows that during posture and slower movements, the curve which corresponds to mode 7 fits the data best. This suggests that for these muscular activities, in cats, a coordination mode similar to the minimum-fatigue mode is adopted. It is interesting to note that, according to the previous results, the minimum-fatigue mode is more efficient than other coordination modes. Because of the good fit between the experimental data and the force predictions of the minimum-fatigue model, this model may be useful for force predictions during human posture and movement. For example, we have employed the minimum-fatigue approach to predict individual ankle muscle forces in humans during standing and walking and, in particular, the mechanical effects of muscle tendon transfer surgery (Dul *et al.* 1985 b).

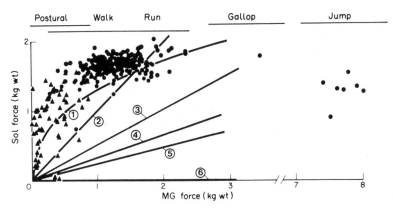

Figure 3.   Measured (instantaneous) synergism between the soleus and the medial gastrocnemius (MG) muscles during cat posture (▲) and locomotion (●). The curves are the computed synergism patterns for different coordination modes. 1, mode 7 (minimum fatigue); 2, mode 4 (ΣF_i)^2 and mode 8 (b_i/b_j); 3, mode 9 (F_i/F_{imax}) and mode 10 (A_i/A_j); 4, mode 5 ($\Sigma (F_i/F_{imax})^2$); 5, mode 6 ($\Sigma (F_i/A_i)^3$); 6, mode 1 (ΣF_i), mode 2 ($\Sigma (F_i/A_i)$) and mode 3 ($\Sigma (F_i/F_{imax})$). Measurements from Walmsley et al. (1978).

The results regarding the efficiency of muscular coordination modes were based upon the assumption that during performance of the knee-flexion task, no changes take place in the coordination mode. However, when these changes took place, computed postural endurance times may be different. For example, suppose a recruitment mode with two stages is adopted. In the first stage only the long hamstring is selected to perform the task, as in modes 1, 2 or 3. According to Table 3 the corresponding level of effort in this muscle is 32%, and the postural endurance time is 4·9 min. The second stage starts after 4·9 min. Suppose that in this stage the muscular coordination has changed into a recruitment mode without activity of the exhausted long hamstring, and with activity of the gastrocnemius. Then, the effort produced by the gastrocnemius must be 79%. The corresponding postural endurance time of the second stage is 0·3 min. Now, the total postural endurance time is 5·2 min (increase of 6%). A further increase could be realized when in the second stage the exhausted muscle has sufficient time to relax, such that the first stage could start again. Hence, it may be possible to increase the postural endurance time by changing the coordination mode during performance of the task.

It is interesting to note that Sjøgaard et al. (1985 b) could demonstrate alternating recruitment of various parts of the knee extensors during long-term, low-level static contractions, using intramuscular pressure and EMG recordings. Another interesting observation has been made by Smith et al. (1982) who found that during a repetitive industrial materials handling task by experienced industrial workers, small variations in lifting pattern occurred. They suggested that these variations allowed the body to adapt to muscular load sharing and to reduce localized muscle fatigue. It may be hypothesized that changes in muscular coordination may be initiated by changes in posture.

In the above discussion, a coordination mode is considered to be efficient when the postural endurance time is long. For example, the minimum fatigue mode (mode 7) is more efficient than the sum of squared forces mode (mode 4) because the endurance time is eight times longer. However, in certain applications (e.g., prevention of low-back pain) it may be more useful to employ the load on the joints as a criterion for efficiency. An evaluation of the resultant knee-joint force showed that in that case the recruitment modes are slightly more efficient than the synergism modes. The joint forces are 315 and 334–371 N, respectively, while the joint force for synergism mode 5 is considerably larger, namely 613 N. Hence, the joint forces for the most efficient modes (modes 1–3) are about half that in the least efficient mode (mode 5). A composite efficiency criterion (low joint forces and long endurance time) could be formulated as well.

When simple techniques to measure human coordination modes become available, several interesting applications lie ahead. Since it appears that certain coordination modes are considerably less fatiguing than others, modes adopted during work might be compared by computing the postural endurance time, or any other efficiency measure. When muscles are used inefficiently, employees might be trained to adopt an efficient coordination mode. One problem might be the learned automatisms in muscular coordination. To overcome these automatisms intensive training programmes, which, for example, include instructional feedback (Dul *et al*. 1985 a), may be required.

5. Conclusions

Analysis of ten different muscular coordination modes shows that the endurance time of a posture depends on muscular coordination. With efficient coordination (the minimum fatigue coordination mode) a posture can be maintained eight times longer than with inefficient coordination.

This large theoretical difference in efficiency between coordination modes justifies research on possible applications. One such application is the (re)training of workers to use their muscles efficiently.

Chapter 13
A Model to Predict Recovery from Maximal and Submaximal Isometric Exercise

N. P. Milner, E. N. Corlett and C. O'Brien

Department of Production Engineering and Production Management,
Nottingham University, University Park, Nottingham NG7 2RD, U.K.

1. Introduction

Attempts at predicting the amount of recovery attainable after different durations of isometric exercise have been produced by several researchers (Monod and Scherrer 1965, Rohmert 1960 a). These models require the estimation of the approximate forces applied, expressed as a percentage of an individual's maximum voluntary contraction (MVC). The current trend in industrial tasks, however, is that the main source of loading is that imposed by the body itself (Corlett and Manenica 1980). In such cases where no external loads exist it may be difficult to estimate the load in terms of a percentage of the MVC. For this reason an attempt has been made to produce a model that requires only information about the length of the holding time and rest period to estimate the level of recovery. This model is derived from a theoretical base and validated by experiment, as described below.

2. Theoretical boundaries to a recovery model

Figure 1 represents a simple theoretical model of recovery after maximal and submaximal isometric exercise. The y-axis represents the level of recovery and the -axis represents the duration of the holding time, expressed as a percentage of an individual's maximum holding time (MHT) in a given posture. The level of recovery in this model is defined as the amount of remaining endurance to perform a given exercise, expressed as a percentage of the MHT. If subjects are completely rested then the length of time that a posture will be maintained is equal to the MHT in that posture, and if the posture is held for this time, then we may judge them to have been totally recovered at the start of the period. Directly after holding a posture maximally (i.e., for 100% MHT) subjects will be in a state of zero recovery and if they are asked to hold the posture again it will be impossible to do so.

    Let us now consider what happens under submaximal conditions. Imagine a subject who holds a posture for 75% MHT. At the moment the exercise ceases the

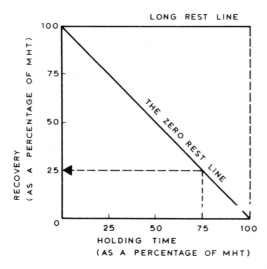

Figure 1.   Theoretical recovery levels after maximal and sub-maximal exercise and zero rest or a very long rest.

level of recovery will be 25% because this amount has not been used during exercise. Similarly, after 50% MHT the recovery level will be 50%, and after 25% MHT it will be 75%. In all these conditions, as the level of recovery is stated immediately the exercise ceases it is not influenced by subsequent rest. These points, shown in Figure 1, can be joined by a diagonal line and called the zero-rest line.

Now, what if a subject were given a very long period of rest, say 1 week? Regardless of how long the posture was held, one would assume total recovery. From a theoretical point of view then the top line of the square in Figure 1 may be seen as a very long rest pause. The space between the top line and the diagonal line represents the area of concern of this paper. For any percentage of MHT full recovery is possible providing the rest period is long enough. However, it is unlikely that it will take as long to recover as the exercise periods become smaller and smaller fractions of an individual's MHT. At each percentage of the MHT, the more rest that is provided the greater will be the level of recovery, until full recovery is reached. Thus, the amount of recovery attainable is an interplay between exercise and rest.

Previous work

Barbonis (1979) carried out an investigation of the effects of different periods of rest on the level of recovery after maximum static exercise. His experiments showed that the longer the rest period the greater the amount of recovery. However, the rate at which an individual recovered during rest was rapid to begin with and then slowed, such that full recovery was not observed even after a rest pause 12 times the exercise period. This finding is supported by Kilbom (1984,

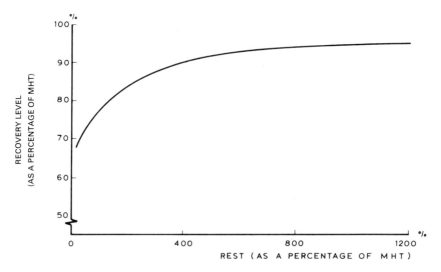

*Figure 2.   Changes in the level of recovery with increased rest after maximal exercise.
From Barbonis (1979).*

personal communication) who estimated that total recovery after maximal exercise may take days. Barbonis's relationship between recovery and rest after maximal exercise can be seen in Figure 2.

From Figure 1 we already know that after maximal exercise and zero rest the level of recovery is zero, so the curve must pass through the origin. We also know the level of recovery for zero rest after submaximal exercise as stated earlier. Therefore, the relationship between rest and submaximal exercise must be located above the curve in Figure 2. In order to observe the nature of the relationship between submaximal and maximal exercise, rest and recovery the following experiment was undertaken.

3.   Experimental description

Subjects and materials

Twenty-four male student volunteers were used in the study. Their personal details are summarized in Table 1. The subjects selected all stated that they were in good health and had no history of musculo-skeletal problems.

Experimental posture

The experiment required subjects to hold a forward stooped posture (see Figure 3) for different durations of time. The posture was such that each subject was stooping to reach for a point, in front of the body, equal to the distance from the acromion to the knuckles and at half shoulder height.

Table 1.   *Summary of subject details* ($N = 24$).

	Age (years)	Weight (kg)	Height (cm)	Shoulder height (cm)	Shoulder reach (cm)
Max.	22·00	83·00	190·00	150·00	75·00
Min.	19·00	61·50	172·00	140·00	66·00
Mean	20·58	71·27	180·70	148·80	70·80
SD	0·99	6·23	4·72	4·73	2·71

The hand position was fixed by two hand controls suspended on a length of string between two stands. This provided a practical method of both precisely dictating the hand position whilst preventing subjects from using the hand controls to support any of their body weight.

Experimental task

The hand controls operated a computer game which was played on a PET computer positioned directly in front of the subjects. The choice of experimental task was the result of a number of smaller studies which showed that maximal endurance could be significantly influenced by the choice of task (Milner *et al.* 1985). Briefly, the task required the subject to land an object on the back of a randomly moving figure, at the bottom of the screen, using two hand controls. The game had no discrete stages and could only be stopped by the subjects depressing both hand controls simultaneously for several seconds. The computer

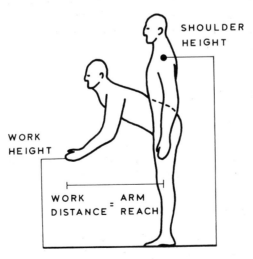

Figure 3.   *Experimental posture. The posture was normalized between subjects by setting it at half shoulder height and at arms length.*

recorded their performance on the task but no feedback of results was provided to the subjects.

Heart rate and subjective levels of discomfort were also recorded during exercise and rest but the method of recording these and the associated results are not reported in this paper.

Method

Each subject was required to perform five separate trials. Each trial consisted of a period of rest (for the subject to relax) followed by the first period of exercise, the duration of which was set by the experimenter. After exercise the subject was given a set period of rest and then was required to exercise again for as long as he could.

On the first visit to the experiment each subject was instructed to hold the posture for as long as possible in the first exercise period. This voluntary maximal endurance was taken to be the MHT. The subjects were told that the criterion for the MHT was to exercise until the discomfort they experienced was judged as intolerable. Subjects were instructed to remember this level of pain and to stop whenever they experienced it in any of the other experimental trials. In the remaining four trials subjects were required to hold the experimental posture for 25, 50 and 75% of this MHT, and also to rehold the posture maximally. These four conditions were presented in a 4 × 4 Latin square design (after Winer 1970), to randomize any order effects. Therefore, the first trial that each subject performed was always the MHT, in order that the submaximal holding times (SMHTs) could be calculated and also to provide a comparison with the second MHT, to determine if any changes occurred during the experimentation.

During rest the subjects were seated in a comfortable chair adjusted to their own preference. Magazines were provided to allow them to read quietly when seated so that they did not become bored. Two lengths of rest were investigated at each level of holding time in the experiment. These corresponded to 100 and 50% of the observed exercise period. However, it was felt that ten trials per subject (i.e., the four levels of holding time, each with two levels of rest, plus the MHT condition repeated) was too many from a practical point of view. For this reason the experiment was designed as an incomplete repeated measures design with each subject receiving different combinations of holding time and rest balanced through the 24 subjects by using an incomplete Graeco-Latin square design shown in Table 2.

4. Results

Due to the general incomplete design, analysis of variance was performed on holding time and recovery using the P4V BMDP programme (BMDP 1984), which treated the empty cells as missing data and weighted their influence accordingly.

Table 2. *Scheme of assigning rest pauses and holding time conditions to subjects. The stars (*) indicate the conditions performed. Each subject performed each level of holding time with one level of rest period.*

MHT (%)	Rest (% MHT)					
	100	75.0	50.0	37.5	25.0	12.5
100	*		*			
75		*		*		
50			*		*	
25					*	*

The results of this analysis showed that recovery was significantly affected by both holding time ($p < 0.0001$) and rest ($p = 0.0212$). Stepwise regression analysis was also performed on the results to observe the relative amounts of variance explained by each variable in turn. When the mean effects in the data and those attributable to rest were removed a significant amount ($p = 0.0004$) of the error term could be further explained by the duration of the holding time. This was not the case when the variables were regressed in the reverse order; the probability of rest accounting for the variance in the residual error term, after the effects of holding time had been removed, was not significant ($p = 0.1991$). Interactions between rest and holding time could not be carried out on the data due to the number of cells with missing data.

The first MHT and the second MHT, which was performed in a randomized order in the experiment, were compared for differences. A t-test showed no significant changes in the length of the MHT ($t_{obs} = -0.329$, significant at 0.7435) indicating that if any training effects occur (Vanderhoof *et al.* 1961) then they do not occur within the five trials undertaken by each subject.

5. Modelling

The recovery data were plotted against rest for each condition of rest and holding time. This showed that in some instances the variation in the level of recovery under a given condition was quite marked. For this reason more emphasis was placed on describing the underlying trends in the data than on minimizing the unexplained variance.

Framework of an empirical model

The mean levels of recovery observed in the experiment described above can now be superimposed on the general form of Figure 2 (see Figure 4). Each curve represents a different level of the MHT and is made up of two observed points and one theoretical point. The theoretical points are the conditions of zero rest given by Figure 1.

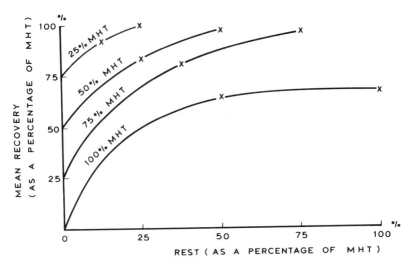

Figure 4.   The relationship between recovery, duration of holding time and rest.

The formula for the model may be expressed thus:

$$\text{recovery} = (\text{MHT} - \text{HT}_{\text{obs}}) + f(\text{rest}) \qquad (1)$$

Where rest is zero, the level of recovery is equal to the amount of remaining recovery described in Figure 1. As rest is provided so the level of recovery rises.

However, the gradients of each of the four curves in Figure 4 are not the same. Despite the shorter levels of holding starting at higher levels of recovery, at zero rest the rates of recovery are slower than for the longer holding times. This can be illustrated by Figure 5 where the step function $(\text{MHT}-\text{HT}_{\text{obs}})$ has been removed. It shows that rest and the step function are insufficient to account for the rate of recovery. The curves therefore require a further function which appears to be related to the duration of the holding time. Several different forms of the rest term were tried in order to find the best way to represent the four holding times by a single term.

Model description

The best fit model was of the form:

$$\text{recovery} = (\text{MHT}-\text{HT}) + \text{HT} \exp(-a\text{HT}/\text{rest}) \qquad (2)$$

The reason for using this form was that the shape of the curves in Figures 4 and 5 suggested a negative exponential component; that is, as the value of rest increased the amount of recovery increased but at a diminishing rate. The exponential component also prevented the model calculating recovery values in excess of the MHT and also balanced the units on either side of the equation, permitting recovery to be calculated in real time units or as percentages of an individual's MHT.

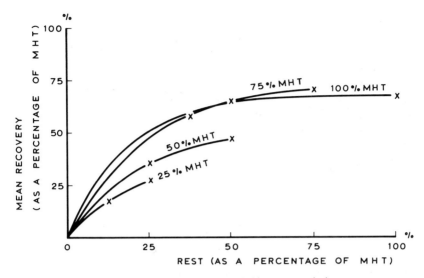

Figure 5. The relationship between recovery, rest and holding time with the instantaneous recovery values removed. The figure shows the rate of recovery after different durations of holding time.

The value of *a* was calculated by simplifying the relationship, taking natural logarithms (Equation (3)) and then using regression analysis.

$$\ln\left[(MHT - HT)/HT\right] = aHT/\text{rest} \tag{3}$$

The regression results are given in Table 3.

The regression constant derived from the analysis gives the final model as

$$\text{recovery} = (MHT - HT) + HT \exp\left(-0.164HT/\text{rest}\right) \tag{4}$$

The equation was further checked by performing a *t*-test on the scores predicted by the model and those observed in the experiment. No significant differences were observed ($t = 0.008$, $p = 0.9937$). The results showed very close agreement in the means (predicted mean = 8.636, observed mean = 8.639) but the model predicted a tighter distribution of scores than those observed (standard deviations: predicted = 2.46, observed = 3.21).

Table 3. *Regression analysis based on Equation (3), where* Y = *ln[MHT − HT)/HT] and* X = HT/rest.

Equation	*t*-value (X)	R^2
Y = 0.164X	−6.76	0.38

6. Repeatability and applicability

The main experiment was repeated with six subjects who had taken part in the study, selected on a random basis. The experiment was designed to examine three levels of holding time (33, 66 and 100% MHT) and three levels of rest (25, 50 and 100% MHT). Each subject performed each level of holding time with a different amount of rest in a controlled order given by a Graeco-Latin square design (Winer 1970).

The observed levels of recovery were compared with those predicted by the model using a *t*-test, which showed that the observed and predicted recoveries were not significantly different ($t = 0.814, p = 0.42$).

In order to further test the model an experiment was carried out with six subjects who had no prior experience of static endurance experiments. They were required to hold an erect posture and perform the task described in the main experiment at head height (see Figure 6). The holding times and rest pauses were the same as in the previous study. A *t*-test was again performed on the pairs of predicted and observed recoveries. It was again found that no significant differences existed ($t = 0.969, p = 0.3375$) between the model and experimental data.

7. Discussion

The model given in Equation (4) appears to be a good predictor of recovery after maximal and submaximal exercise and rest. However, it should be noted that it is based on data which contain an unavoidable experimental error. The error is due to the need to estimate the MHT on a given day in order that a percentage of the MHT may be calculated for experimental purposes. From our earlier studies we found that the MHT does vary from day to day, but that these changes are not con-

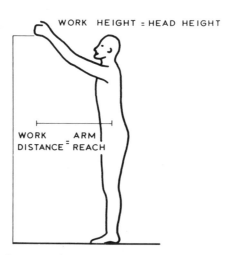

Figure 6. Experimental posture. The posture was normalized between subjects by setting it at head height and at arms length.

sistently up or down and therefore no significant changes will be found. However, it does mean that on a given day what is estimated to be 25% of an individual's MHT may in fact be closer to 20 or 30% of their MHT and this will be the same in the case of all submaximal holding times. Although when the mean performance of the group as a whole is considered the effect of these diurnal fluctuations is minimized, the effect on the statistical analysis is a large increase in the variance. Regression analysis was performed on the observed and predicted values in each of the three studies described above and the R^2 values, which indicate the amount of variance explained by a regression, were 38·3, 51·0 and 56·8%, respectively. It is suggested that a large part of the variance left unexplained is due to the error which arises when estimating the percentages of the MHT.

Despite these weaknesses in the model it does show that full recovery is possible providing the holding time is a relatively small fraction of the maximum endurance capacity of the individual for that posture. This was shown in the analysis of variance where holding time was found to have a greater effect than rest.

Where an individual is working to the limits of endurance capacity it has been found that full recovery is not possible within a rest period 12 times the MHT (Barbonis 1979) and may continue for several days (Kilbom *et al.* 1983). The recent research suggests that this is most likely to be caused by actual structural damage to the contractile properties of the muscle (Friden *et al.* 1983, Kuipers *et al.* 1983, Newham *et al.* 1983)

However, this only accounts for the latter part of the recovery curves. The early steep aspects of the curves are most probably due to biochemical changes that occur within the muscle during exercise which have not been fully restored during the rest pauses allocated. Most of the processes which occur during exercise are quickly reversed during post-exercise hyperaemia.

The rate of recovery was estimated to be more rapid after longer holding times even though the actual level of recovery attained was not as great. It is suggested that the higher rates of recovery are due to the increased rate of metabolite removal and restoration of energy stores facilitated by the elevated blood pressure (Humphreys and Lind 1963, Kilbom *et al.* 1983, Morioka 1964, Sharkey 1966) and heart rate (Kilbom *et al.* 1983, Lind and McNicol 1967, Rohmert 1960 a, Tuttle and Horvarth 1957) which rise with the duration of holding time.

8. Closing remarks

The model presented gives a close estimate of the interaction between static exercise and rest and illustrates the serious body loadings which arise from tasks which require the maintainance of single postures for long periods of time. For precise estimates of recovery, the model is limited due to the problem of estimating the duration of MHT on a given day. When used in practical situations, this may not be a serious limitation as extreme and repeated loading can lead to musculo-skeletal damage, and conservative estimates would be required. Additionally, the model was found to be repeatable and applicable to another posture.

Chapter 14
Intramuscular Changes during Long-term Contraction

Gisela Sjøgaard

Danish National Institute of Occupational Health, Baunegaardsvej 73,
DK-2900 Hellerup, Denmark

1. Introduction

Working postures are maintained by contractions of the skeletal muscles. During heavy work high-intensity contractions have to be performed, which cause muscular exhaustion within a short time. Naturally, frequent resting periods are demanded. Heavy work still exists, but due to the developments in modern technology an increasing number of heavy jobs are being replaced by physically light work. In many respects this is an advantage, but a drawback is that, consequently, many monotonous jobs have developed. Low-load muscular work can be maintained for long periods of time (hours) without rest. The muscles are not exhausted but may well be fatigued. Light working tasks are often performed in constrained postures, where certain muscles or groups of muscles maintain the same tension for many hours day after day. These contractions involve a large static component and there is evidence that this could be a significant factor in the development of, for example, occupational cervicobrachial disorders (Maeda 1977).

2. Maximal endurance time

The relationship between force and maximal endurance time was mapped out in a very extensive study by Rohmert (1960 a). Based on this relationship and the heart-rate response, guidelines were given for acceptable contraction and rest periods throughout an 8 hour working day. If a muscle contracts at a high percentage of its maximal voluntary strength (% MVC) it is exhausted within seconds or minutes. The loads studied by Rohmert ranged from 20 to 100% MVC, and it was indicated that 15% MVC might be sustained for an 'unlimited' period of time. Later studies which included low-level contractions ($< 15\%$ MVC) showed, however, that the magic limit of 15% MVC could not be maintained even for 1 hour (Björksten and Jonsson 1977, Hagberg 1981 b). The contraction level had to be as low as 8% MVC (range 5–13% MVC) in order to be maintained for 1 hour. Power function regression analyses of endurance time versus contraction level have been performed for continuous isometric contractions with maximal endurance times of 1 hour or less (Hagberg 1981 b).

In Figure 1 this regression is extrapolated to 8 hours, and, based on the ranges for values up to 1 hour, rough estimates of the ranges up to 8 hours are drawn also. This is a drastic extrapolation, but illustrates well that Rohmert's guidelines cannot simply be extended to include low percentages of maximal voluntary strength also. The maximal endurance time for contractions above 20% MVC can be determined within an error of ±10%, and with only minor differences between different muscle groups or subjects (Rohmert 1960 a). However, this is not the case for low-level isometric contractions. Muscle structure and muscle fibre composition may play a role (Hulten *et al.* 1975), and perhaps even more important may be the motivation factor. A well-motivated subject is generally able to maintain a low-level contraction for "just a few more minutes". For example, 5% MVC has been found in some muscles—but not all—to cause exhaustion within 1 hour. From the regression line (Figure 1) it can be calculated that—as a mean—5% MVC may be maintained for 3 hours; most likely some subjects can be pushed to maintain 5% MVC for even more than 8 hours for some of their muscles. Thus, for low-level contractions the maximal endurance time does not seem to be a valid criterion in the assessment of general guidelines for acceptable contraction limits.

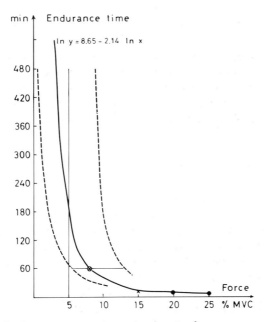

Figure 1.   Maximal endurance time versus isometric contraction force.
●, *from Rohmert (1960 a);* ×, *From Björksten and Jonsson (1977);* ○, *from Hagberg (1981 a).*
Based on data in the latter study the full drawn curve is extrapolated and the broken lines indicate ranges.

3.   Fatigue index

A method which has been widely used in the study of muscle fatigue is electro-myography (EMG). Analysis of the EMG amplitude and frequency has revealed that during fatiguing contractions the amplitude increases and the frequency simultaneously decreases (Lindström *et al.* 1977). The EMG fatigue index shows similar changes during high and low contraction levels; only the rates of change of amplitude and frequency are dependent on the contraction force (Hagberg 1981 b). During contraction levels below 15% MVC the fatigue index may demonstrate fatigue in some muscles but not in others (Becher *et al.* 1983, Hagberg 1981 a); and during very low contraction levels it has been indicated, by EMG analysis, that muscles are even able to recover from previous fatiguing contractions (Körner *et al.* 1984).

Such data suggest that there may exist non-fatiguing contraction levels. It must be noted, however, that the EMG signal may be affected by several physiological variables, for instance temperature (Petrofsky and Lind 1980), muscle acidity (pH) (Mortimer *et al.* 1970) and intra- and extracellular electrolyte concentrations (Palla and Ash 1981), and also by the recruitment pattern of fast- and slow-twitch muscle fibres. Based on the EMG signal we do not know which physiological changes have occurred, but even worse is that changes in some physiological variables may influence the EMG signal in one direction and others into the opposite. Thereby the EMG signal may, in theory, remain unchanged despite significant physiological changes within the muscle. Thus, it is not safe to rely only upon the EMG signal in the study of constrained working postures. It is increasingly important to extend our knowledge regarding possible intramuscular changes during long-term, low-level contractions. In order to decide upon which variables we should focus our attention, we must consider the possible mechanisms of fatigue.

4.   Mechanisms of fatigue

Muscle fatigue is defined as the failure to maintain a required or expected force or power output (Edwards 1983). Fatigue may occur within the nervous system, especially in the motorneurones and at the neuromuscular junction. Such causes of fatigue will not be considered here, but only the local fatigue within the muscle tissue. Even with this limitation different mechanisms of fatigue may exist, depending on the particular form of muscular activity. The concept of high-frequency fatigue and low-frequency fatigue with a long lasting (>24 hours) element of fatigue (Edwards *et al.* 1977) indicates that different mechanisms may be responsible during high- and low-level contraction forces. The development of muscle fatigue during high contraction forces is related to an increased intramuscular pressure, which causes a restriction of the muscle blood flow. At contraction levels of around 15–20% MVC the blood flow starts to become relatively impaired and is completely arrested at around 60% MVC. Impaired blood flow means limited substrate supply and accumulation of metabolites, which in turn impair the rate of energy turnover. During low-level contractions energy turnover is low and lack of energy is unlikely to be a primary cause of fatigue. In line with this, low-frequency fatigue does not seem to be related to intracellular acidosis or lack of energy but is rather to be explained by an impaired excitation–contraction coupling (Edwards 1983). The excitability of muscle fibres depends on their membrane potential, which again depends upon intra- and extracellular electrolyte concentrations (mainly potassium). Muscle fatigue could then be localized on its cell membrane, and in addition to the changes in muscle metabolism it would be of interest to study the electrolyte changes within the muscle.

5. Experimental model

The model developed to study the interplay between the mechanical performance of the muscle and intramuscular changes, such as tissue pressure, blood flow, muscle metabolism, electrolyte fluxes and EMG amplitude and frequency, is shown in Figure 2. One-legged knee extension can be performed as either isometric or dynamic exercise and simulates an 'isolated muscle preparation' in man. An experimental programme was undertaken, with the participation (after their informal consent) of healthy male subjects aged 20–30 years. In each of the experimental procedures reported here, six to eight subjects were studied. The mechanical force exerted by the muscle was measured using a strain-gauge device via a strap placed around the ankle, which, for dynamic exercise, was connected to the crank of a bicycle ergometer. The femoral artery and vein were catheterized for blood sampling and blood-flow measurements were made using the thermodilution technique. A cuff beneath the knee was inflated to 200 mmHg before each measurement in order to arrest blood flow from the lower leg. Teflon catheters were introduced into the muscle for intramuscular pressure measurements, and surface electrodes were fixed to the skin above each intramuscular pressure recording, for

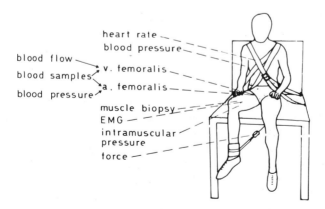

Figure 2.   Schematic illustration of the experimental situation.
Adapted from Saltin et al. (1981).

EMG recording. Muscle biopsies were obtained from the vastus lateralis part of the quadriceps muscle. Blood and muscle samples were analysed for substrates, metabolites (Lowry and Passonneau 1972), electrolytes and water content (Sjøgaard 1983, Sjøgaard and Saltin 1982). Additionally, cardiovascular changes, such as heart rate and blood pressure, were recorded.

6. Results

Using this model (Saltin *et al.* 1981), results similar to those previously reported in the literature were obtained during high isometric contraction levels. These were, increased intramuscular pressure, restriction of blood flow and impaired muscle metabolism (accumulation of lactate). In addition, it was found that the potassium concentration in the venous effluent blood from the contracting muscle increased during 15, 25 and 50% MVC, and decreased as soon as the muscle relaxed. The course of these changes over time closely followed those of heart rate and blood pressure, and a positive correlation was found between venous potassium concentration and heart rate (Figure 3). Reflex control of the circulation by afferent nerves from skeletal muscle has been demonstrated (Longhurst and Mitchell 1979), and potassium has been recognized as a potent stimulator of some of the small nerve fibres coming from the interstitial spaces of the skeletal muscles (Hnik *et al.* 1969). The heart-rate response may thus follow any significant changes occurring in the contracting muscle, and therefore Rohmert's use of the heart-rate response as an indicator of local muscle fatigue may well be a relevant choice.

Long-term dynamic exercise

The limitation of the model is that the venous concentration of substances such as potassium only reflects the magnitude of the interstitial concentration if the muscle

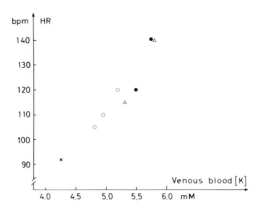

Figure 3.   Heart rate versus venous blood potassium concentration at rest (×), during 15% MVC(○), 25% MVC(△) and 50% MVC(●).
Adapted from Saltin et al. *(1981).*

is well perfused, as is the case during dynamic exercise. During moderate dynamic exercise, the venous potassium concentration increased from 4·5 to around 5 mM and remained above its resting value, as well as above the arterial concentration throughout a 2 hour period of exercise (Sjøgaard and Saltin 1983). As a consequence, potassium was lost from the muscle cell. When the muscle was exhausted after 2 hours, it had lost approximately 10% of its potassium content and gained 10% water, as determined in the muscle biopsy analysis. The intracellular potassium concentration had therefore decreased from around 170 to below 140 mM. The membrane potential, which is mainly dependent on the intra- and extracellular potassium concentration (Figure 4), decreased by roughly 10 mV from its resting value of around 90 mV. If the muscle was exhausted by high-intensity dynamic exercise lasting only a few minutes, similar changes were observed in the potassium concentration gradient across the cell membrane (Sjøgaard *et al.* 1985 a).

An attempt to stimulate the Na–K pump by infusion of terbutalin during moderate dynamic exercise had no effect on the potassium release from the contracting muscle (Rolett *et al.* 1985). However, other tissues, such as resting skeletal muscle, had increased their potassium uptake, since the arterial potassium concentration dropped from 4·5 to 3·5 mM. The potassium release from the exercising muscle may then be a well-regulated mechanism and not a consequence of a limited Na–K pumping capacity.

Long-term isometric exercise

During low-level isometric contractions the energy turnover within the muscle is low and only a small fraction of the muscle fibres have to contract simultaneously. One might speculate whether during long-lasting contractions there exists some orderly rotating recruitment pattern, so that each fibre only contracts for a few

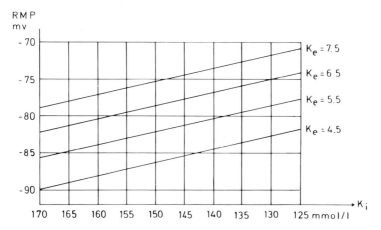

Figure 4. Resting membrane potential (RMP) versus intracellular potassium concentration, K_i, for various intracellular potassium concentrations, K_e. Calculated from $RMP = 61·5 \log (K_c + 0·01\ Na_e)/(K_i + 0·01\ Na_i)$.

seconds or minutes and then relaxes in order to recover. Such a pattern would in reality allow the muscle as a whole to contract for an unlimited period without fatiguing. During 5% MVC (range 4–7% MVC) maintained for 1 hour, we could in fact demonstrate alternating recruitment of the various parts of the knee extensors based on intramuscular pressure and EMG recordings (Sjøgaard *et al.* 1985 b). The mean intramuscular pressure stayed at a low level (10–45 mmHg) throughout the exercise, but, in most recordings, large fluctuations occurred from resting values up to 90 mmHg. EMG recordings confirmed that the changes in intramuscular pressure may in fact be related to this alternating recruitment of muscle fibres. Another indication that 5% MVC may be non-fatiguing was the high muscle blood flow throughout the contraction period. Within a few minutes of contraction, blood flow increased to a constant level of 1·58 (1·25–2·22) l/min or 0·67 (0·51–0·84) l/(min kg). No hyperaemia was observed when the muscle relaxed; on the contrary, blood flow almost immediately decreased. This blood flow supplied the muscle with sufficient substrate for the increased metabolism, since muscle glycogen content and lactate release were at their resting levels at the end of the contraction. Also, the muscle temperature remained at a constant level after an initial increase. At a depth of 3 cm the temperature increased from 34·2 to 36·6°C within 5 min of contraction and remained at this level throughout the 1 hour contraction period.

In contrast to these findings, the rating of perceived exertion (RPE) increased significantly during the 1 hour contraction. The subjects were asked specifically to rate the fatigue they experienced in their knee-extensor muscle and this increased from 2 to 4 (10-graded Borg scale) during the 1 hour contraction. Three MVC test-contractions performed before and after the 1 hour 5% MVC, confirmed that the muscle was fatigued, although not exhausted, since the MVC had decreased by 10%. In many EMG recordings the changes in amplitude and frequency also indicated local muscle fatigue according to the fatigue index. A disturbance in the potassium homoeostasis could then be a factor responsible for the impaired mechanical response, as has been reported in previous studies. The venous potassium concentration increased from 4·4 to 4·7 mM within a few minutes of contraction. Simultaneously, the heart rate increased by about 10 beats/min, which agrees with the results depicted in Figure 4. The venous potassium concentration remained above its resting value as well as above the arterial concentration throughout the contraction period. The total potassium loss was around 5%, and as the muscle water content had simultaneously increased by 10%, the intracellular potassium concentration had decreased significantly (Sjøgaard and Saltin 1985).

7. Conclusion

It is concluded that human skeletal muscles are not adapted for continuous long-lasting isometric activity. Indeed, no matter how low the energy turnover is within the muscle, resting periods are needed for the muscle to recover. Guidelines for low-level contractions should therefore not deal only with how much percentage of MVC is acceptable. To recommend a reduction in workload from, for instance, 6 to 3% MVC will not help much physiologically; also it is not practicably usable. Instead we need guidelines for 'maximal acceptable limits of monotony'. This means that we must focus our attention on what periods of time a muscle or group of muscles can tolerate maintaining or repeating low tensions. This is especially true if the same low load is imposed on the muscles day after day.

Chapter 15
Biomechanical Model Calculations of Spinal Stress for Different Working Postures in Various Workload Situations

Matthias Jäger and Alwin Luttmann

Institut für Arbeitsphysiologie an der Universität Dortmund, Ardeystrasse 67, D-4600 Dortmund 1, F.R. Germany

1. Introduction

Many work activities often require the manual handling of goods. High stress commonly appears during such manipulation activities especially at the spine, and in the long run orthopaedic diseases are established in many persons in such jobs, as epidemiological studies have shown (e.g., Luttmann *et al.* 1985). So it is of interest to occupational health, work physiology and ergonomics specialists to determine spinal stresses during manual handling tasks.

Postural stress depends not only on the mass of the manipulated load but on the body mass and the working posture, too. In order to determine 'non-permitted' working postures, the torques and forces acting at the lower spine were calculated with the help of biomechanical models and were compared with limits given in the literature. In this paper, biomechanical model calculations are introduced for work involving unloading heavy goods vehicles, transporting dustbins and brick laying. From these calculations, concerning vertical or horizontal lifts and horizontal or diagonal pulls, postures are derived which are likely to exclude a high health risk for working persons.

2. Methods—anthropometric conditions and biomechanical models

The torques and forces acting at the lumbo-sacral joint (L5/S1) were calculated using biomechanical analysis. This joint was chosen specially because the disc is frequently injured in people who move heavy weights at work (Junghanns 1979). During manual handling tasks, external forces generally act at the hands although their height and direction may differ depending on the task. If a person holds a load it has only a vertical component, but if the load is being lifted an additional force due to acceleration must be considered. During horizontal movements, the weight of the load acts vertically and the acceleration component horizontally, so that the combined external force has an oblique direction.

Postural stress, the forces on the spine, varies depending on the posture. Thus a forward bending posture results in larger forces than an upright trunk. In consequence, a 'bad' posture and simultaneous high hand forces can bring about a postural stress which exceeds the mechanical strength of a vertebral body or inter-vertebral disc, which may then result in irreversible diseases.

The load on the spine due to these external forces is increased by the torques at the lumbo-sacral joint, of the various segments of the body superior to it. An additional load arises due to the tension in the back muscles which is needed to support those parts above the lumbo-sacral joint. The resistive torque provided by these muscles enables the forward bending posture to be maintained, but produces a further compressive load on the spinal discs. (The shear forces, as produced by the components orthogonal to the spinal column, are not discussed in this paper.) Clearly, activities which are performed asymmetrical to the median sagittal plane must be considered in a three-dimensional way (for example, bringing an object from beside the body to the front of the body). In order to calculate the torques and forces at the lumbo-sacral joint, it is necessary to know the body height and weight, the link lengths, the segment weights and the location of the centre of gravity of each segment. The relative anthropometric and gravimetric data were taken from Dempster (1955 b), but the absolute values were adapted to European conditions (body mass 74 kg, body height 1·76 m). For more details concerning the assumed biomechanical models see Jäger *et al.* (1983) and Jäger and Luttmann (1985).

3.   Results—application of the biomechanical models

The biomechanical models were applied to estimate the postural stress for various postures. In the first example, common lifting activities which are usually performed two-handed and are symmetrically related to the median sagittal plane are analysed. This is followed by three examples discussing specific professional activities: an oblique lift in the sagittal plane which occurs during the transport of dust containers; horizontal movements as found in brick-laying tasks, requiring a three-dimensional analysis; and, finally, in the fourth example, horizontal pulling is analysed with special emphasis on the effects of acceleration.

Vertical lift in the sagittal plane (general aspects)

The results of two-dimensional computations for two-handed lifts in the median sagittal plane are shown in Figure 1. Body and load were assumed to move with constant velocity, so that no acceleration forces appear and the reaction force at the hands is vertical and equal to the weight of the external load. Therefore, these values of the lumbo-sacral torque and force are the minimum values with which something can be lifted. The trunk inclination varied between 0 and 90°. The mass of the lifted load was assumed to be a maximum of 50 kg, with the minimum weight 0 kg. This latter relates to the torques and forces at the lumbo-sacral disc which are influenced solely by each posture.

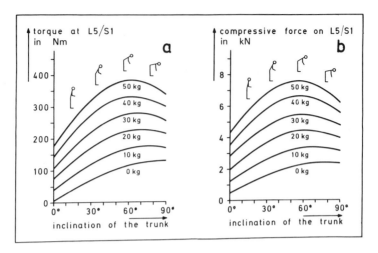

Figure 1.   Stress values at the lumbo-sacral joint of a person who holds, lifts or lowers a load at a constant velocity, two-handedly and symmetrically to the median sagittal plane by inclining the trunk (for more details of the arm movement, see the text). (a) Lumbo-sacral torque versus trunk inclination. (b) Lumbo-sacral compressive force versus trunk inclination.

Figure 1 shows that both torque and force are larger than zero even when the external load is zero and the person is standing upright. This is because the centre of gravity of the upper part of the body is forward of the lumbo-sacral joint. With increasing trunk inclination, the values of the torque and force follow a sine curve due to the forward shift of the centre of gravity. With the addition of a load, the values grow in proportion to the weight, the curves moving to higher values.

Also, with increasing external load, the maximum torque and force at L5/S1 occur at smaller values of trunk inclination. In the movement studied here the arms were bent at the elbow, the upper arms hanging vertically and the forearms held orthogonally in relation to the trunk. Thus, although the lever arm of the centre of gravity of the upper part of the body reaches its maximum at a trunk inclination of nearly 90°, the lever arm of the hand forces do so at a substantially smaller angle (about 60° of the trunk inclination). Hence, the shift of the maxima occurs.

Summing up it may be concluded that both the lumbo-sacral torque and the compressive force on L5/S1 grow with an increasing trunk inclination. If a load in the hands must be lifted in addition to erecting the trunk, torque and force increase even more. In cases of heavy external loads, postural stress may be a multiple value of the stress produced by body weight only.

Oblique lift in the sagittal plane (during the transport of dust containers)

In this study, the transport of empty 1·1 m³ dust containers, of a mass of 170 kg, is investigated in relation to the spinal stress for dustmen. It is in this task that oblique lifts occur.

Electromyographic ' of usual work cycles showed that the strongest muscular strain an vhen, after emptying the dust into the cart, the container is reet on to the pavement across a kerb-stone (Jäger *et a* ³ container is transported by one or two persons, sch Figure 2. A so-called 'main working person' pulls This lift can be performed in various working postur n the one hand, the dustman may approach the con with his knees bent as much as possible, in this case nk. On the other hand, the working person n ned the upper part of his body. The lifting a n working person only, but a 'co-worker' t the container by skilfully using his body re of the co- worker is as shown on the le foot-board with his body close to the con postures, however, will be between the ex

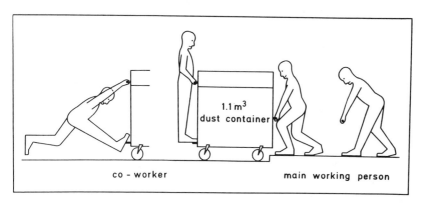

Figure 2. Typical workload situation during refuse removal jobs: two persons are manipulating a dust container across a kerb-stone. The so-called 'main working person' is lifting the container with a posture between the two extremes, as shown on the right-hand side. The so-called 'co-worker' may help his colleague by skilfully using his body weight and working in a posture between the two extremes, as shown on the left-hand side.

The lumbo-sacral torque (T_{L_5}) and the compressive force on the lumbo-sacral disc (F_{L_5}) of the main working person are shown in Table 1. These values suppose a constant velocity movement; so the given data represent minimal values of torque and force acting at the lower spine.

If the main worker lifts the container with straightened knees, the torque values are 110–125 N m larger than when working in the other extreme posture. The lumbo-sacral torque is more than 300 N m, when the lifting person is working

Table 1.   Stress values at the lumbo-sacral joint of the so-called 'main working person', who is pulling the container across a kerb-stone and alternatively working in one of his extreme postures. He may work 'without' any help, or with help by the so-called 'co-worker' through different postures. (Only the minimal values are tabled; they are calculated for moving with a constant velocity without any acceleration of container or men.)

		Main working person		
		T_{L5} (N m)	T_{L5} (N m)	F_{L5} (kN)	F_{L5} (kN)
Co-worker	Without	364	239	8·1	5·8
	(figure)	315	195	7·0	4·8
	(figure)	200	90	4·4	2·4

alone or with only a little help from the co-worker. With one of the two in an advantageous posture, the torque can be lowered to about 200 N m. The main worker with bent knees and the co-worker leaning backwards will produce the smallest postural stress; in this case the torque at L5/S1 is only a quarter (90 N m) of the worst case (364 N m). Comparable results were calculated for the compressive force on L5/S1. The difference between the two extreme postures of the main working person is about 2 kN for each of the three possible activities of the co-worker. The influence of his various postures is as follows: without any help the compressive force is about 1 kN larger than in the case of just a little help. But when he is leaning far back, F_{L5} is lowered by about 2·5 kN. With both workers in their most advantageous posture, the lumbo-sacral force is less than a third of the highest value (2·4 kN compared with 8·1 kN).

Another activity of the co-worker is also included in the computations above. When the co-worker pushes the container in the direction of his colleague, he does not produce any tilting torque on the dustbin and therefore does not help the main working person in lifting the container across the kerb-stone. Consequently, the lumbar stress would be the same as in the 'without' case in Table 1.

Horizontal move in a transverse plane (during brick laying)

Light objects are usually moved by one hand only. In most cases this activity is done non-symmetrically in relation to the median sagittal plane, since only one hand holds and translates the load, whereas the other hand is empty.

During brick laying, horizontal movements are typical. For example, the bricks to be set are positioned on one side of the working person. He brings the load single-handedly from the side into the area in front of his body brick by brick. This horizontal move in a transverse plane is examined for two different postures, stooped and upright.

The calculated values of the compressive force on the lumbo-sacral disc against arm rotation for single-handed loads of up to 10 kg are depicted in Figure 3. The angle of arm rotation is 0° with the arm pointing to the side, and it reaches 90° when the upper arm is pointing forwards. Simultaneously, the bending of the elbow increases. However, the arm is not entirely lifted, but only 30° relative to the vertical.

The curves in Figure 3 demonstrate that a stooped posture produces a stronger lumbar stress (in fact more than doubled) than an upright posture. The maximal stress for a man in an upright posture is 2·0 kN, the minimum for a stooping man is 2·6 kN. Corresponding curves for the same weight, but with different postures, lead to differences of about 2–3 kN.

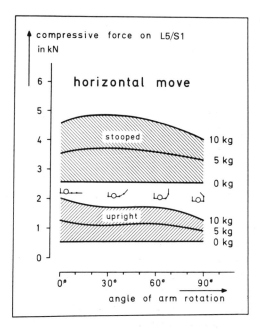

Figure 3. Compressive force on the lumbo-sacral disc of a person who holds a load single-handedly and turns it from the side into the front of the body at a constant velocity; the posture was supposed to be either upright or entirely inclined.

The influence of various weights of the load is about 0·8–1·6 kN at 'upright', 1·5–2·3 kN otherwise. In stooped postures, the lever arm of the external load consists of two parts, namely the distance between the lumbo-sacral joint and the shoulder, and the projection of the arm length. With an upright trunk, however, only the projected arm length represents the lever arm. Hence, torque and compressive force are influenced to a lesser extent in an upright position.

In addition, the curves show that the compressive force depends on arm location. An arm pointing sidewards produces larger values than an arm positioned in front of the body. The maximal force on L5/S1 is produced in a median arm position. That is due to a similar curve for the values of the lever arm of the external load during arm rotation.

Horizontal pull in the sagittal plane (during unloading of sacked goods)

Sacked goods are often manipulated by hand. In the building industry, for example, cement is usually sacked in 25 or 50 kg paper bags. In order to avoid damage to the bags, they are usually picked up from the lorry as shown in Figure 4; they are pulled to the body, symmetrically to the median sagittal plane. The reaction force at the hands acts horizontally. Its value depends on the mass of the sack, on its acceleration and on the resistance caused by friction. (The frictional coefficient was estimated from laboratory experiments to be about 0·7.)

Both the hand force and the working posture affect the load on the spine. Generally, the sacks to be handled are situated at a height of 0·7–1·4 m and at a distance of 0·3–1·2 m, and the posture adopted depends on the position of the sack in this space. As the location of a sack can be associated with a certain posture, the postural stress may be estimated in relation to the weight, acceleration and position of the sacks.

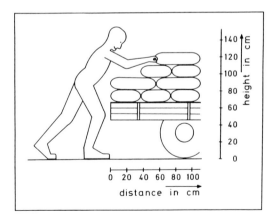

Figure 4.   Typical workload situation during the unloading of trucks: a person is pulling a sack, positioned at a certain distance and a certain height, to the body. The working posture depends on the location of the sack to be pulled.

Having determined the posture for each sack position and the associated lumbo-sacral torque values, the points in the matrix of distance and height which result in the same stress value are connected, to give 'isotorque' curves. In Figure 5 this was done for a lumbo-sacral torque of 135 N m (for an explanation of why that particular value was chosen, see the later discussion). Each isotorque curve represents a boundary between the two areas of possible sack locations corresponding to a torque of either more or less than 135 N m.

Figure 5 (a) shows the isotorque curves for two common sack weights, 25 or 50 kg. In these calculations an acceleration of the sacks of 5 m/s² was assumed. The curves are approximately linear. The one for the lighter sacks lies higher within the matrix than the curve for the heavy sacks, since, although the hand force is smaller owing to the smaller weight, a larger lever arm for the hand force still results in a constant lumbo-sacral torque of 135 N m.

Figure 5 (b) shows the isotorque curves for various accelerations of the pulling activity; the maximal acceleration is supposed to be 10 m/s², the lowest one zero. In the latter case a sack is not accelerated but is pulled with a constant velocity. With increasing acceleration, the isotorque curvers have a 'lower' position within the

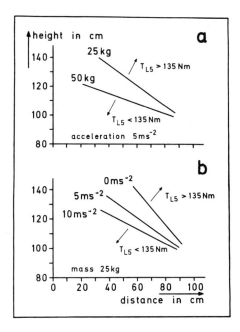

Figure 5. *Lumbo-sacral torque during sack pulling (as illustrated in Figure 4): the location of a sack was associated with a certain posture; the hand force was calculated from the weight of the sack to be pulled, its acceleration and the friction. From both posture and hand force, the torque at L5/S1 was estimated for each point of possible sack location, given by distance and height. The points resulting in a value of 135 N m were connected to give the so-called 'isotorque curves 135 N m'. (a) Isotorque curves of 135 N m for sacks of different weights, accelerated at 5 m/s². (b) Isotorque curves of 135 N m of a sack weighing 25 kg for different accelerations.*

matrix of possible sack locations, as illustrated. Because the hand force is proportional to the acceleration of the manipulated load, the hand force increases in the same way as the acceleration. Therefore, the lever arms have to be smaller to produce the same torque at L5/S1.

All isotorque curves run obliquely within the matrix of sack positions. Since all points of one curve correspond to the same weight and acceleration of the sack, the oblique line can be traced back to different postures of the pulling person. Thus, sacks lying at shoulder height may produce the same postural stress as sacks at abdominal height lying further away.

4. Discussion

The spinal stress was quantified, with the help of biomechanical computations, for selected common workload situations. The calculated data are compared with criteria for stress given in the literature. As a result, advantageous working postures can be differentiated from disadvantageous ones, and those postures representing a high health risk can be determined and should be avoided.

The severity of a manual activity can be classified into four groups after Tichauer (1978): light, medium-heavy, heavy and very heavy. The heaviest case is when the lumbo-sacral torque exceeds 1200 in-lb ($=$ 135 N m). Then "the work is very heavy, cannot always be performed on a continuing basis for the entire working day, and requires great care in recruitment and training". Evaluation of compressive forces on a lumbo-sacral disc is much more difficult, because unequivocal data for spatial movements are not available. For example, NIOSH (1981) gives limits only for lifting activities in the sagittal plane. But values of mechanical security of vertebral bodies of intervertebral discs of the lumbar spine may give some support to the calculated stress values, since, in the lumbar region, compressive forces must be transmitted for the most part through the spinal column.

During experimental compression of spinal elements (Decoulx and Rieunau 1958, Perey 1957, Sonoda 1962, Wyss and Ulrich 1954, and others), the spongy part within the vertebral body yields first, then microfractures occur in the middle of the end-plates and, finally, nucleus material leaks into these interstices (Farfan 1973). The values of forces causing irreversible failure of lumbar vertebral bodies or intervertebral discs range between 4 and 10 kN, depending on age and sex (Wyss and Ulrich 1954).

A small lesion of spinal elements is not usually painful in the beginning and cannot be diagnosed in many cases, but it may later have a negative influence on the proper functioning of the spine, when strongly stressed. To avoid irreversible damage to the spine, activities resulting in compressive force values larger than 4 kN should be avoided, or they should be changed in such a way that this limit will not be exceeded.

The estimated stress at the lumbo-sacral joint during holding, lifting or lowering a load at a steady speed was given in Figure 1. Applying Tichauer's criteria for handling activities to these data, it can be noted that the lumbo-sacral torque of 130 N m for a man in a stooped posture and without any external load, nearly reaches the upper limit 'allowed' for a whole day's work (135 N m). As can be seen the weight of the upper part of the body is able to produce a relatively strong postural stress. Therefore, activities entailing periods of long-lasting stooped postures should be avoided or changed. Additional external loads lead to torque values above Tichauer's limit. Considering these limits, loads of a weight of 50 kg should never be lifted with the trunk leaning forwards. Additionally, such heavy loads also exceed the limit derived for compressive forces, namely 4 kN. Finally, loads heavier than 15 kg should not be lifted in a back-lift procedure, but with a leg lift (Davis and Troup 1964). During leg-lift activities the trunk remains in an upright posture, and the spinal stress is lower owing to the shorter lever arms of the external load and upper-body segments.

From the above discussion, several practical consequences can be derived for the work activities described in this paper. Transport of the large and heavy dust containers across kerb-stones should not be performed by one person only. Both the lumbo-sacral torque and the lumbo-sacral compressive force exceed the stress limits to a great extent, no matter in what posture the 'main working person' lifts the container. When the 'co-worker' helps his colleague by generating a small tilting torque at the container, the torque and force at L5/S1 are still larger than 135 N m and 4 kN, respectively. Only when both dustmen are in an advantageous posture are the values below the limits of 90 N m and 2·4 kN, respectively. In consequence, the main working person should bend his knees, bring them close to the container and should lift the load with an upright trunk. The co-worker should pull the container at a special grip by hand, push it away with his foot and should skilfully use his body weight. Excessive spinal stress can be avoided using these optimal working postures.

During horizontal movements, posture has a great influence on spinal stress also. The curves (in Figure 3) of the compressive force at the lumbo-sacral joint illustrate that only in a stooped posture will the limit of 4 kN be exceeded, although the external load is not even unusually heavy. However, there is a leaning forward by the heavy upper part of the body, and therefore the disadvantageous lever arms are dominant. With the same arm rotation, but with an upright trunk, the force values are cut by half. Consequently, working postures with a relatively small trunk inclination should be preferred.

Advantageous postures during brick laying can be guaranteed by skilful arrangements of the bricks. For example, the bricks may be located at the height of the hip so that the person may grasp the bricks with an upright trunk—mainly by moving only his arm.

The calculated results for horizontal moving, given in Figure 4, do not reach the high values of the previous workload situations. This is because the external force is not usually as large with single-handed jobs (up to 0·1 kN, equivalent to 10 kg) as with two-handed activities (up to 0·5 kN when lifting, up to 0·8 kN for transporting the container across a kerb-stone). When the working person handles an external load of 10 kg horizontally and in a stooped position, the compressive force at L5/S1 ranges between 4·0 and 4·8 kN. The values, given in Figure 1 for lifting 10 kg, reach 3·4 kN maximally, however. The difference of about 1 kN can be traced back to the following two reasons. First, non-symmetrical work produces torques in the frontal plane in addition to the ones in the sagittal plane, which also appear during two-handed lifting activities in the sagittal plane. Secondly, the upper arm in the above example of a 'horizontal move' was supposed to be slightly spread aside, whereas in the other case it hangs downwards. Therefore, the lever arm of the external load is increased and so results in higher external torques and, finally, in a stronger compression of the lumbo-sacral disc.

The lumbo-sacral torque when pulling sacked goods horizontally (Figure 4) was evaluated with the help of the criterion from Tichauer (1978). The isotorque curves of 135 N m (Figure 5), each representing a spatial boundary, define the area of locations from where sacks might be pulled. In the case of sacks of 50 kg they should not be taken from a height above 1·2 m, and sacks of 25 kg not from above 1·4 m, as can be seen from Figure 5 (a). Additionally, they should not be positioned lower than about 1 m, since then sacks will not be pulled but must be lifted, which would result in a stronger postural stress (see Figure 1 also). Sacks lying further away require working postures with an inclined trunk and so result in torque values larger than 135 N m. Therefore, the working person should not pull the sack from such positions, and a co-worker may help in putting the sacks closer to the man who grasps them. He may then work with a better posture and, therefore, the spinal stress will be lowered. Furthermore, the pulling person should not manipulate the sacks jerkily, since increased acceleration leads to larger stress values and, in addition, reduces the area of 'allowed sack positions' (Figure 5 (b)). The same is true for heavy sacks. In such a case, two men should manipulate each sack together, leading to smaller lumbar stresses.

In conclusion, during handling activities not only the weight of the loads determines the spinal stress but, to a substantial extent, the working posture does so too. Biomechanical model calculations have shown that some tasks should not be performed in disadvantageous postures. To avoid a high health risk for working people, heavy loads must at least be moved so that advantageous postures can be adopted. These are derived, from the model, for each task.

Chapter 16
Influence of Posture on Muscle Contraction Behaviour in Arm and Leg Ergometry

Pranab Kumar Nag

National Institute of Occupational Health (Indian Council of Medical Research), Ahmedabad 380 016, India

1. Introduction

Studies on how muscles work are, in the final analysis, concerned with the optimization and management of manual work. This is in view of the fact that the standardization of manual work in industry and other places of human employment is targeted to achieve improvements in working methods and conditions. The force applications in most occupational tasks are the result of a combination of various limb positions, and are largely effected through successions of isotonic and isometric contractions of muscles of the limbs and trunk. The resulting strain of a given type of muscular work is indicative of the cumulative influence of either of these components of work.

When an activated muscle shortens, the contraction is called isotonic (dynamic), in which at least a quarter of the chemically available energy is utilized to perform work. Dynamic work allows muscle groups of the limbs involved to contract and relax alternately. Energy expenditure increases during the first minutes of work, and then levels off as the energy demand reaches a level adequate to meet the requirements of the task. An oxygen deficit results which is repaid after the cessation of work. The oxygen debt is repaid in at least two phases—the fast or alactoid and slow or lactoid components of recovery.

On the other hand, when the muscle contracts without any change in its length, the contraction is called isometric (static). During static contraction the force exerted remains constant although the muscular motor units remain active throughout the contraction. Due to the accumulation of acidic metabolites, static work induces rapid muscular fatigue.

Extensive studies of the effect of work patterns and postures on cardiovascular responses (Bevegard *et al.* 1966, Lind *et al.* 1975, Nag 1982, Vokac *et al.* 1975) reveal that there is an inherent difficulty in identifying the extent of musculature involved in isometric and/or isotonic work in performing a given muscular movement. Obviously, the magnitude of the physiological responses due to either components of the work are difficult to determine. Based on laboratory investigation (Kilbom and Brundin 1976, NIOSH 1982, Sanchez and Monod 1979) on selective isotonic and isometric kinds of work, it has been observed that the physiological strains due to combined work are not simply additive. Since the cardiorespiratory responses are interdependent and responsive to both static and dynamic work, the predominant component of work is greatest in the resulting physiological responses. From studies of the electromyography (EMG) of trunk muscles, when measured amounts of dynamic workloads are imposed on the legs while the arms are in sustained static contraction, this author (NIOSH, 1982) observed that the static work facilitates activity of the rectus abdominis, external oblique abdominis and the upper and lower fibres of the trapezius. Combining static work with the dynamic work of the legs causes facilitation of the trunk muscles based on the relative loads of each of the two types of work.

Realizing that different muscles of the body function differently, it is worth while studying the pattern of contractions of various muscle groups during different kinds and intensities of arm and leg work. Bigland-Ritchie and Woods (1974) found that the integrated EMG recorded over the quadriceps was linearly related to force and oxygen uptake (VO_2) during light dynamic work. Petrofsky (1979) observed an increase in the r.m.s. amplitude of the myoelectric signals from the quadriceps and an associated decrease in the centre frequency of the fast Fourier transformed power spectra of the EMG during fatiguing dynamic leg work. Komi and Tesch (1979) reported that in repeated knee extension at high angular velocities, the mean power frequency of the EMG of the vastus lateralis tended to decrease in persons with a higher percentage of fast-twitch fibres. In continuous concentric and eccentric elbow flexions, Hagberg (1981 b) observed that the development of muscle fatigue was well correlated with the changes of the r.m.s. EMG amplitude and the mean power frequency.

Since rhythmic dynamic work of given limbs may also induce isometric muscle contractions from other parts of the body, this study quantifies the muscular activities, to characterize the postural variations and their limits, in order to determine the ability of an individual to perform given different types of muscular work. In this study, the interplay of 22 muscle groups of the arms, the legs and the trunk were observed in order to compare the influences of postures such as standing, sitting with and without back support and reclining during arm and leg ergometry.

2. Ergometry and electromyography

The study included four motivated healthy young male volunteers (age: $23 \pm 3 \cdot 4$ years; body weight: $49 \pm 0 \cdot 9$ kg and stature: $160 \pm 2 \cdot 0$ cm). In none were the limb and trunk muscles specially trained. The subjects performed the following arm and leg work (Figure 1): (*a*) pedalling sitting (PS); (*b*) pedalling sitting with back support (PBS); (*c*) pedalling reclining (PR); (*d*) cranking sitting (CS); (*e*) cranking standing (CST); and (*f*) pedalling and cranking combined while sitting (PCC). The experiment was conducted on a modified mechanical ergometer (Nag 1984). The front wheel was driven by two transmission chains attached to the pedal and the crank wheel.

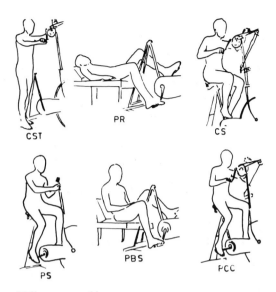

Figure 1.   Patterns of different arm and leg ergometry.

In the case of CST the subject stood adjacent to the ergometer seat and longitudinally at the centre of the driven wheel. The crank wheel was placed at about 10–15 cm above the heart level. The trunk and the upper limbs were apparently at rest in PR. In PBS a lumbar support was provided and the hands were kept immobile. The angle between the seat and the back support was about 120°. On the basis of the average $\dot{V}O_2$ max ($2\cdot2\pm0\cdot38$ l/min) yielded by the subjects, three braking loads equivalent to approximately $0\cdot5$, $0\cdot75$ and $1\cdot0$ times $\dot{V}O_2$ max were selected for each type of arm and leg work. The subjects performed the selected workloads randomly over 2 weeks. The rate of cranking and pedalling was 60 rpm for about 4 min. The myoelectric activity (EMG) of the following muscle groups of the right side of the body were recorded during the steady state of the arm and leg work:

1. Upper limb: flexor carpi radialis; brachio-radialis; biceps brachii (medial head); triceps brachii (medial head); deltoideus (acromial part).
2. Trunk: sternocleido-mastoidous; pectoralis major (clavicular head); rectus abdominis; external abdominal oblique; trapezius (upper and lower fibres); latissimus dorsi; erector spinae (L3–4).
3. Lower limb: gluteus medius; sartorius; rectus femoris; vastus medialis; semitendinosus; biceps femoris (long head); tibialis anterior; soleus and gastrocnemius.

The surface EMG was picked up by bipolar non-polarizable silver disc electrodes (5 mm dia.) placed in the direction of the muscle fibres. The locations of the electrodes were the same from day to day. The interelectrode distance was 25 mm. The myoelectric signals were processed through Beckman 9852A EMG averaging couplers (5·3–5000 Hz) and 9856A Universal couplers (5·3–1000 Hz) and the power amplifiers 412 series. In the EMG averaging coupler the signals were routed directly into the input transformer of the pre-amplifier. The signals were passed back into the coupler for further amplification by a single stage transistor circuit. The signals were coupled to a full-wave diode rectifier through a low pass filter. The signal turned to d.c. was then coupled to the power amplifier for recording. The area integration of the recorded signal, in a Beckman rectilinear R612 Dyno-graph, over 1 s duration was measured as IEMG (mV). Also, the amplified direct myoelectric signals were recorded on a Hewlett Packard 3964A instrumentation tape recorder (pass band: d.c. to 1·25 kHz) at a speed of 9·6 cm/s. The signals were displayed on a storage oscilloscope and further processed on an HP 85A computer to obtain the r.m.s. amplitude of the EMG.

3. EMG of upper-limb muscles

Since the size, number and orientation of the muscle fibres may not be the same in two different muscles, any comparison between the muscles was of no signifi-cance. Variations (intra- and interindividual) in the EMG were large, which demanded careful analysis of the myoelectric signals. With regard to the origin and insertion of the muscles, the activity behaviours of the individual muscles were compared under kinematic actions of the arms and legs in different postures. From the r.m.s. amplitude and IEMG of the myoelectric signals (as given in Table 1, and Figures 2–7), the amplitude of motor unit potentials and the number of active motor units in the muscle may be estimated.

The flexor carpi radialis (the radial flexor of the wrist and elbow, the abductor of the hand) was linearly active with increases in the resistances of arm work (CST, CS) and combined work (PCC). As noted from the r.m.s. amplitude and IEMG, the flexor carpi radialis was relatively more active in CST than in CS. This may perhaps be due to undesirable flexion and increased power of grip (Kamon 1966).

The brachio-radialis did not differ markedly in its activity between CS and CST at a workload equivalent to about 0·75 $\dot{V}O_2$ max. At a higher braking load, however, the muscle showed increased activity in standing arm work. The EMG of the brachio-radialis was similar between CS and PCC. Only a low-level sustained contraction of the muscle was noted during leg work.

The biceps brachii acted mainly across the long axis of the forearm, providing acceleration along the motion. The biceps brachii increased its strength of con-traction with the increase in braking loads and also with the posture of the arm work. Similarly to brachio-radialis, the biceps brachii also recorded higher EMG activity in the case of CST. Only moderate activity of the muscle was noted during PCC, when the braking load was almost equally shared by the arm and leg muscles.

The triceps brachii, a powerful extensor of the elbow joint, should be highly active in full extension of the arm. In none of the arm work (CS, CST and PCC) did the subjects have their arms sufficiently extended. Electrical inactivity of the triceps was sometimes observed in some subjects. In others, though the rhythmic rotation of the arm involved negative work for about half the period of the cycle, the durations of flexion and extension of the arm in different work modes were not identical (Figure 2). The triceps brachii had a shorter period of extension, especially in the case of CST. Obviously, the duration of bursts of activity of the biceps brachii was longer in CST compared with CS. By holding and grasping objects, the biceps and triceps brachii showed some static contractions during leg work.

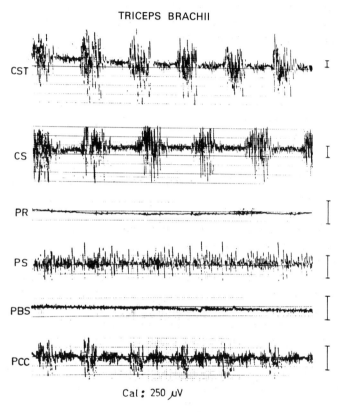

Figure 2. EMG of triceps brachii at a workload of 0·9 $\dot{V}O_2$ max in different ergometry.

The deltoideus (acromial part) and biceps brachii are synergistic to shoulder flexion and abduction movement (Hagberg 1981 a). In the cases of CST, CS and PCC the deltoideus stabilized during the first stage of the movement, i.e., as long as the resultant forces were below the sagittal axis of movement. This could be observed from the relatively low EMG activity in CS. For a given posture of work, however, there was virtually no difference in the r.m.s. amplitude and IEMG of the deltoideus against higher braking loads.

4.   EMG of trunk muscles

The sternocleido-mastoid (clavicular fleshy head) showed no marked activity in rhythmic arm and leg work. The pectoralis major (clavicular head) reflected a rhythmic component of activity in the cases of CS and CST. This activity of the pectoralis major was predominant only when the cranking was done against greater resistances (Figure 3). Because the centre of the arm movement is on the head of the humerus, the pectoralis major, in association with the deltoideus, provided the most favourable lever for movement, i.e., in flexing the arm at the shoulder joint and in medial rotation.

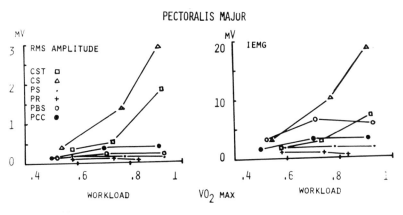

Figure 3.   R.M.S. amplitude and IEMG of the pectoralis major at different work intensities of arm and leg ergometry.

The latissimus dorsi also constituted a part of the unit to act on movement of the upper limb. Scheving and Pauly (1959) found that the latissimus dorsi is an abductor, a retractor and a medial rotator of the upper limb. The present study observed that the latissimus dorsi was more active in CS than in CST. In the case of PCC, due to the sharing of the braking load on the arm muscles, the latissimus dorsi was only moderately active. For leg work PR demanded higher activity of the latissimus dorsi.

The rectus abdominis showed a characteristic pattern of r.m.s. amplitude and IEMG (Table 1) in arm and leg work. In the case of CST there was significantly larger activity in rectus abdominis, which otherwise in a relaxed standing position picked up no activity. In the reclining posture (PR) too, the rectus abdominis was powerfully active against higher braking resistances. These indicate an increase in the general tone of the abdominal muscles. In contrast to reclining or standing work, bilateral leg work in a sitting position did not bring the rectus into significant activity. This observation was in line with that of Floyd and Silver (1950) who studied the activity of the abdominal muscles extensively in head-raising exercises. Floyd and Silver frequently observed differences between the right and left sides of the abdominal musculature, using EMG records taken by carefully matched electrodes.

The external abdominal oblique behaved differently. Due to the backwards inclination of the trunk, as in the case of PBS, the r.m.s. amplitude as well as the IEMG increased linearly with increased braking loads. In other arm and leg work the external abdominal oblique showed only slight activity. According to Partridge and Walters (1959) all portions of external oblique and rectus abdominis may be best activated by a lateral bend of the trunk, pelvic tilt, straight trunk curl and trunk curl executed with rotation.

The trapezius upper and lower fibres were more active in arm work with relative predominance of a rhythmic component (Figures 4 and 5). Since the upper fibres of trapezius undertake the function of suspending the shoulder and the lower fibres prevent downwards rotation of the scapula when the arm is elevated, the increased EMG of the muscle in CST indicated higher muscular load compared

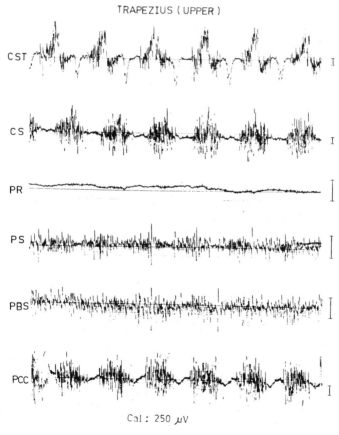

TRAPEZIUS (UPPER)

CST

CS

PR

PS

PBS

PCC

Cal : 250 µV

Figure 4.   EMG of the upper fibres of the trapezius at a workload of 0·9 V̇O$_2$ max in different ergometry.

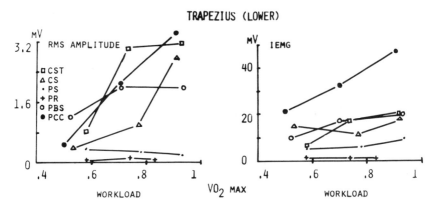

Figure 5. R.M.S. amplitude and IEMG of the lower fibres of the trapezius at different work intensities of arm and leg ergometry.

with that in the case of CS. Hagberg (1981 c) noted that the muscle load determined by EMG on the upper fibres of the trapezius was closely related to the external torque produced in the shoulder glenohumeral joint. Work tasks demanding repetitive arm flexion/elevation put a load on the trapezius fibres, leading to discomfort and pain in the neck (Onishi *et al.* 1976). The trapezius fibres are also significantly active in combined work. There was a linear increase in the r.m.s. amplitude and IEMG of the lower fibres of the trapezius in the case of PCC. This was largely the result of effort given by the arms. The trapezius fibres were less involved in the cases of PR and PS. During PBS, however, the lower fibres showed sustained contractions due to backward thrusts given by the trunk.

The erector spinae were minimally active in the case of back supported leg work, followed by CS, PS and CST. There was, however, a significant increase in EMG activity of the muscle during PR. In spite of a constant range of the r.m.s. amplitude, a larger IEMG of the erector spinae may be attributed to an increased number of motor units in action in PCC. There was a linear increase in IEMG with the increased braking load. The possible influence of PR in increased EMG is not clearly understood.

5.   EMG of lower-limb muscles

Because of the actions involved in ergometry, none of the adductors of the thigh were included in the study. Only the muscles that cross by the hip and knee were analysed. Most leg muscles were very much influenced not only by rhythmic leg ergometry but also by arm work in different postures. Because of their large size as well as close link between hip and knee regions, many of the muscles of the thigh have an important postural or stabilizing function.

The gluteus medius, though quiescent during relaxed standing, contributed towards fixation of the movement axis of the hip. The gluteus medius was active in

PBS and PR. The activity of the muscle was relatively less in PCC and PS. Houtz and Fischer (1959) also concluded that all the glutei were minimally active in bicycle pedalling. The fluctuations found in the EMG of the gluteus medius during arm and leg work were ascribed to positional changes in gravitional forces.

The sartorius had wide individual variability in its myoelectric activity. Since the muscle was supposed to be active during flexion of the hip and the knee joint or medial rotation of the tibia (Johnson *et al.* 1972), the sartorius was most active in PCC, followed by PBS and PS (Table 1). The muscle was least active for PR. Also in the case of standing arm work the sartorius was highly active with sustained contractions at higher braking loads, indicating the role of the muscle in postural stability.

Rectus femoris and vastus medialis, the two heads of quadriceps, though they did not act synchronously, had a similar activity pattern in all leg work. Rectus femoris was active to a higher degree in PBS and PR, and to a lesser degree in the cases of PCC and PS (Figure 6). Rectus femoris showed static contractions in arm work, with no marked difference between a sitting and a standing posture. The vastus medialis was highly active in the cases of PR and PBS (Table 1). Lieb and Perry (1971) observed that the vastus was active when the knee was held in extension. Any lag in full extension was a function of a large loss in mechanical advantage of the whole muscle. Therefore, the only function attributable to the vastus medialis was patellar alignment. However, in other leg work (PS and PCC) the vastus medialis was relatively less active. A low amplitude sustained contraction of the muscle during CST indicated some hyperextension of the knee in an upright position. Apart from the postural variations, the linear increase of the EMG of the quadriceps (rectus femoris and vastus medialis) with workload was evident. This supports the earlier observations of Bigland-Ritchie and Woods (1974) and Petrofsky (1979).

Hamstring (semitendinosus and biceps femoris) interplay was very similar in leg work (Table 1). Undoubtedly, however, the muscles were under high exertion during PBS and PR, compared with that in the cases of PS and PCC. The relative

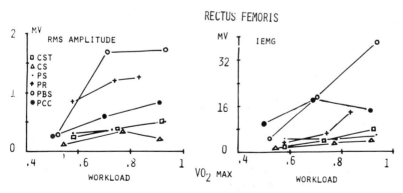

Figure 6. R.M.S. amplitude and IEMG of the rectus femoris at different work intensities of arm and leg ergometry.

Table 1. R.M.S. amplitude and IEMG of different muscles at a workload of $0.9\ \dot{V}O_2$ max in arm and leg ergometry.

	CST		CS		PR		PS		PBS		PCC	
	R.M.S.	IEMG	R.M.S.	IEMG	R.M.S.	IEMG	R.M.S.	IEMG	R.M.S.	IEMG	R.M.S.	IEMG
Flexor carpiradialis	1·43	22·2	0·52	11·3	0·02	0·4	0·08	1·3	0·05	1·0	0·64	20·4
Brachioradialis	1·93	23	1·30	18·5	0·03	0·4	0·10	1·2	0·09	0·8	0·59	16·8
Biceps brachii	2·11	25·8	1·79	18·4	0·30	3·8	0·15	2·0	0·12	2·8	0·81	11·8
Deltoideus	0·95	12·4	0·59	8·5	0·33	3·5	0·37	1·5	0·08	0·8	0·30	5·2
Sternocleidomastoid	—	2·1	—	0·86	—	0·2	—	0·27	—	0·6	—	0·3
Rectus abdominis	0·75	11·9	0·31	2·4	0·68	5·25	0·38	5·2	0·14	2·0	0·25	1·96
External abdominal oblique	0·40	3·0	0·49	3·76	0·42	6·3	0·13	4·0	1·20	15·7	0·43	5·9
Latissimus dorsi	0·53	4·25	0·86	12·7	0·22	4·5	0·07	2·17	0·03	0·8	0·07	2·0
Erector spinae	0·39	2·3	0·17	1·80	0·54	7·5	0·28	2·9	0·10	1·6	0·29	11·8
Gluteus medius	—	1·4	—	2·2	—	4·1	—	1·8	—	7·6	—	2·7
Sartorius	1·25	9·5	0·16	2·42	0·02	0·3	0·26	2·2	0·36	4·4	0·58	6·6
Vastus medialis	0·60	6·3	0·25	3·4	0·79	10·5	0·55	7·5	1·58	24·6	0·60	10·3
Biceps femoris	0·84	9·3	0·52	5·08	1·20	21·5	0·72	9·7	2·42	40·0	0·62	11·0
Tibialis anterior	—	0·8	—	3·5	—	1·1	—	4·9	—	1·9	—	3·2
Soleus	0·59	5·5	0·13	1·0	2·15	12·5	0·44	2·3	0·62	9·0	0·60	13·7
Gastrocnemius	0·75	6·6	0·11	0·9	0·9	17·5	0·35	5·33	0·86	19·4	0·49	7·2

predominance of the activity of the hamstrings in PBS and PR indicated greater postural influence on the muscles in knee extension after the hip was flexed (Basmajain 1974). Due to resisted extension of the knee, the semitendinosus was also more active in standing arm work than in CS (Figure 7).

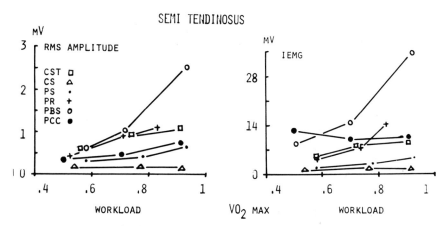

Figure 7. R.M.S. amplitude and IEMG of the semitendinosus at different work intensities of arm and leg ergometry.

The tibialis anterior showed pronounced activity during PS and PCC. As long as the long arches of the feet were unaltered (Jonsson and Rundgren 1971, Suzuki 1956) the tibialis anterior remained quiescent, as in CST. Also in PBS and PR the flat part of the toe was mostly supported on the pedal, and as a result the EMG of the tibialis anterior was less in PBS and PR.

The soleus was also under high tension in PR. In other leg work (PS, PBS and PCC) the EMG of the soleus did not vary markedly. Campbell *et al.* (1973) noted that the soleus was a strong mover of the foot on the leg and a stabilizer of the leg on the foot.

The gastrocnemius was active with the rhythmicity of leg work. The gastrocnemius also sustained a high level of activity in CST, due to body swings. The r.m.s. amplitude and IEMG of the muscle increased linearly with the workload only in PR and PBS. However, in PS and PCC, the activity levels remained the same. It is suggested that the gastrocnemius EMG may be sensitive to the conditions of body posture and speed of work (Herman and Bragin 1967) and not to the strength of contraction.

6. Discussion

Comparison of work patterns with muscle activities is helpful in establishing the skeleto-muscular links and the influence of body positions to enhance performance capability. Apparently, the factor limiting the amount of work is determined by the total mass of muscles that are mobilized to perform it. In reality, the force applications of particular body segments are transmitted through the whole body. This may essentially include continuous or intermittent static and dynamic work of the muscles of the upper and lower limbs.

A given amount of dynamic work, obtained by cranking or pedalling, also leads to isometric contractions in other parts of the body, depending upon the postural attitude of the person. Moreover, the trunk, which constitutes about 60% of the body weight, has a significant influence on different kinds of arm and leg work. In fact, participation of a muscle in rhythmicity or in sustained contraction characterizes the demand of the movement.

All arm muscles showed a low level of isometric contractions in holding objects and in supporting shoulder fixation during leg work. The act of cranking was an association of rhythmic flexion and extension with forward and backward abduction of the arm at the shoulder joint, when the hand was in a pronated position. Arm muscles in general were more strained in the standing posture. Flexor carpi radialis and brachio-radialis were more active in CST than in CS. The contraction and relaxation periods of both biceps and triceps brachii also indicated that there was a large tension on the flexor muscles in standing arm work. This study supports the author's own earlier view that the maximal power output for arm work while standing was less than that found in a sitting posture (Nag 1982, 1984).

Activity of the trunk muscles also varies with arm and leg work. Most of the trunk muscles contributed to facilitate arm movements. The rhythmic component of activity of the trunk muscles was predominant in arm work compared with that in leg work. The activity rhythms of the deltoideus, the pectoralis major, the latissimus dorsi and the trapezius upper fibres were in close association with one another during movements of the arm. However, there was no consistent difference in activity of the muscles in arm work in standing or sitting postures. While the latissimus dorsi and erector spinae (L3–4) were highly active in sitting arm work, the other trunk muscles under study (the deltoideus, rectus abdominis and the upper and lower fibres of the trapezius) were more active in standing arm work. Thus, the relative strain on the shoulder muscles was largely dependent on the working postures of the upper limbs (Herberts *et al.* 1980). The sustained isometric contraction of the latissimus dorsi, rectus abdominis, external abdominal oblique, upper fibres of trapezius and erector spinae in PR and PBS indicated that the trunk muscles were responsive to the control of the force of gravity acting on the shoulder and other joints when the legs were in rhythmic action.

The muscles of the thigh region have an important function in postural stability during arm and leg work. The gluteus medius, sartorius, vastus medialis and semitendinosus were found to be more active in CST than CS. These muscles showed static contractions during arm work. Among the leg muscles the quadriceps and hamstrings showed increased EMG activity in PBS, followed by PR. In addition to the increased activities of the trunk muscles, higher tension on the leg muscles yielded a larger energy debt in the cases of PR and PBS (Nag 1984). The contractions of the semitendinosus, biceps femoris, vastus medialis, rectus femoris and the gastrocnemius during flexion and extension form a closed kinematic chain of movements (Carlsöö and Molbech 1966) and were very much influenced by the posture of the whole body during leg work. Only in the case of upright sitting leg work, were these muscles relatively less strained. In combined arm and leg work the muscles of the lower limb were moderately active. In fact, the actions of most muscles of the body were in greater balance during PCC. This is due to a shared braking load between arm and leg muscles with the necessary rhythmic involvement of the trunk muscles.

However, these observations require to be studied further in depth; there is a lack of similar studies in the literature. The relative influence of the trunk muscles in arm and leg work may need to be explored on the basis of the EMG of trunk muscles, in persons of different body weights.

Acknowledgements

The author expresses his gratitude to Prof. B. B. Chatterejee, Director of the Institute, for his invaluable suggestions in the preparation of the manuscript. Thanks are due to Mr Ashish Goswami, Mr Sankar Ghosh, Mr S. P. Asthekar and Mr C. K. Pradhan for their kind help in the study.

Chapter 17
The Influence of Under- and Overstimulation on Sitting Posture

Tadeusz Marek and Czeslaw Noworol

Jagiellonian University, Cracow, Poland

1. Background

The paper presents results of empirical investigations which have been performed to build mathematical models of changes in operators' sitting posture, related to under- and overstimulation from the task.

It has been pointed out by Walsh (1964) that muscle tone and, connected with it, body posture, are controlled by many pathways of the nervous system. Figure 1 shows the main pathways descending to the spinal cord that can both increase and decrease muscle tone. The pathways causing a reduction of tone are: path 1, cortico-bulbo-reticular; path 2, caudato-spinal; path 3, cerebello-reticular; and path 4, reticulo-spinal. The pathways increasing the tone are: path 5, reticulo-spinal; and path 6, vestibulo-spinal. The reticular formation of the medulla (pathways 4 and 5) exerts important regulatory controls over muscle tone, body posture and the spinal motor mechanisms.

Magoun and Rhines (1946) and Rhines and Magoun (1946) demonstrated that the bulbar portion of the brain stem, superior to the inferior olivary nucleus, contained a number of facilitatory and inhibitory regions from which spinal reflex

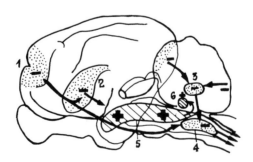

Figure 1. Pathways controlling muscle tone. From Walsh (1964). See text for explanation.

activity could be affected. Stimulation of the inhibitory region (pathway 4) reduced the basic spinal reflex, the motor response caused by cortical stimulation and the decerebrate rigidity.

A facilitatory area (pathway 5) is found in the reticular formation of the ventral diencephalon, midbrain tegmentum and pantine gray. Stimulation of these regions reinforces the excitatory inputs to the spinal motor neurons and facilitates reflex responses even in decorticate preparations. Then reticular influences act primarily to facilitate or inhibit spinal motor mechanisms and modify postural and phasic movements. Excitation of the reticular formation may effect complete postural adjustments such as contralateral extension of the limbs combined with ipsilateral contraction.

The concept of stimulation load is connected with Hebb's and Leuba's theories of optimal levels of arousal (Hebb 1965, Leuba 1965). The level of arousal depends on the level of stimulation and the level of stimulation depends on the variability, complexity and intensity of stimuli.

According to the point of view of Grandjean (1967), McFarland (1971), Cameron (1973) and many others, a high level of stimulation activates the facilitatory area of the reticular formation and when it is prolonged the inhibitory system begins to take over. A low level of stimulation activates the inhibitory system from the beginning (Marek and Noworol 1984 a). The level of arousal changes in the following way. The individual supplies himself with stimulation till he attains an optimal level of arousal. The individual who is excessively stimulated begins to act to reduce the stimulation. When the level of arousal is too low, stimulation is sought. Thus, from a theoretical point of view, muscle tone and body posture can be modified by the level of stimulation.

A group of VDT operators were examined in a preliminary investigation. Two kinds of 6 hour data-entry tasks were carried out; a low-stimulation and a high-stimulation task. The preliminary investigation focused on the four parameters of sitting posture at rest; that is, the angle of trunk inclination, the angle of head inclination, the vertebral column bend and the hunched-shoulders index. Sequential analysis of variance was applied (Marek and Noworol 1984 b). This analysis permitted us to minimize the number of operators examined (19 operators). The body posture was recorded on videotape during rest intervals after every 20 min of work. Having compared the values of the four parameters, and their changes, it was found that over- and understimulation imposed different characteristics of body posture during the first few hours ($p < 0.05$), especially concerning the angles of trunk and head inclination. In overstimulation conditions the angles of trunk and head inclination first decrease and then increase. In the same conditions the spinal bend and the hunched-shoulders index both decrease. In understimulation conditions both angles first increase and then decrease, and the other two parameters act similarly. The spinal bend and the hunched-shoulders index were not taken into account in our investigation because of statistically significant differences between variances.

The aim of the further research reported below was to find out whether changes in these two parameters of sitting posture at rest—angles of trunk and head inclination—under high and low levels of stimulation (over- and understimulation) will be different in terms of underlying mathematical models.

2. Method

Subjects

Twenty-four VDT operators were examined. The operators were 21–34 years old. Two kinds of data-entry task were carried out by the same female operators for 6 hour periods. These were a monotonous task (understimulation), where the data to be entered consisted of numbers only, and a relatively complex task (overstimulation), where the data to be entered consisted of alphanumeric signs, the operators converting a FORTRAN IV programme into a BASIC programme.

Procedure and technique

Operators were tested on the morning shift. The sitting posture at rest was recorded on videotape 25 times; that is, before work and then after every 15 min of work during the 6 hour session.

Before recording posture each subject was seated on a stool (without support) and asked to take up a posture at their leisure. Each recording was continued for 60 s. Two parameters of body posture were analysed in 2 s bands, the angle of trunk inclination and the angle of head inclination (Figure 2). Mean angles across subjects were calculated for each of the 25 recording periods.

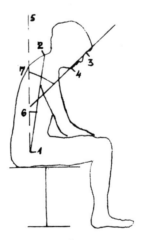

Figure 2. *The parameters of posture investigated.*
The anthropometric points: 1, trochanterion; 2, cervicale; 3, glabella; 4, pogonion. 6, the angle of trunk inclination (the angle between the vertical (5) and the trochanterion–cervicale line). 7, the angle of head inclination (the angle between the vertical and the glabella–pogonion line).

3.   Results

Empirical data

Tables 1 and 2 show the means of trunk- and head-inclination angles for over- and understimulation conditions, respectively. Standard deviations for each angle were less than 6° for trunk and less than 11° for head. Differences between standard deviations are statistically insignificant in every case.

Changes of angles were analysed as functions of time. Four empirical curves (two tasks by two measures) of angle changes were drawn for each person. Figures 3 and 4 show these empirical curves obtained for the means.

Models of changes in trunk and head inclination

All the empirical curves obtained (*A*, *B*, *C*, *D*) were approximated using the method of stick functions, a mathematical technique to deal with functions which change form on a continuum. Regression analysis for repeated measures was used.

Table 1.   *The mean angles (n = 24) of trunk and head inclinations—complex task (overstimulation).*

Work session	Angle of trunk inclination (degrees)	Angle of head inclination (degrees)
Before work	2·4	8·2
After work (min)		
15	1·1	7·0
30	0·3	6/3
45	0·5	5·9
60	0·9	5·8
75	0·5	5·7
90	0·8	6·0
105	0·0	6·1
120	1·6	6·6
135	2·0	7·9
150	2·7	9·2
165	3·6	9·8
180	3·2	10·5
195	4·9	12·0
210	5·7	12·9
225	7·0	15·1
240	6·8	16·2
255	8·5	16·9
270	9·4	17·5
285	10·0	17·9
300	9·5	18·0
315	10·2	18·2
330	9·5	18·1
345	10·0	18·4
360	10·0	18·3

The ergonomics of working postures

Table 2.   The mean angles ($n = 24$) of trunk and head inclinations—simple task (understimulation).

Work session	Angle of trunk inclination (degrees)	Angle of head inclination (degrees)
Before work	1·3	8·8
After work (min)		
15	2·2	12·2
30	2·8	13·9
45	4·4	13·8
60	4·9	13·1
75	3·0	13·0
90	0·2	8·8
105	−1·1	8·1
120	−0·8	8·9
135	0·2	10·1
150	2·3	12·0
165	1·9	11·8
180	3·5	14·4
195	4·0	13·8
210	4·9	16·2
225	5·9	16·0
240	7·6	18·8
255	8·1	18·2
270	9·2	20·2
285	9·5	20·5
300	9·8	20·5
315	9·9	20·6
330	10·0	20·0
345	10·2	21·1
360	10·1	21·0

Each empirical curve was divided into three parts, I, II and III at 2 and 4·5 hours. These partition points were identified by examination of the empirical data. Figures 5 and 6 show the models obtained. The model for changes in angles of trunk inclination is the same as for changes in angles of head inclination. It is shown in the model of Equations (1), for overstimulation:

$$a_I = \alpha + \beta t + \gamma t^2, \qquad 0\,h \leqslant t \leqslant 2\,h$$
$$a_{II} = \alpha' + \beta' t, \qquad 2\,h \leqslant t \leqslant 4\cdot 5\,h \qquad (1)$$
$$a_{III} = \alpha'' + \beta'' t + \gamma'' t^2, \qquad 4\cdot 5\,h \leqslant t \leqslant 6\,h$$

where a_I, a_{II} and a_{III} are the theoretical angles of inclination for curve parts I, II, and III, respectively; t is time; and the Greek letters are the equation coefficients. The coefficient γ is greater than zero and γ'' is less than zero.

Table 3 contains the regression coefficients obtained in the investigations, after

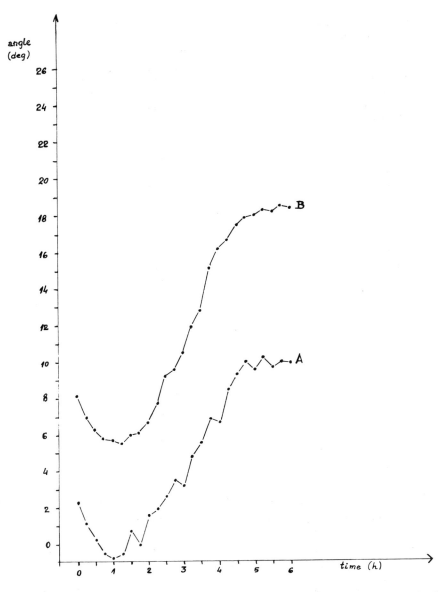

Figure 3. Changes in angles of trunk inclination (curve A) and head inclination (curve B) for the overstimulation condition.

the linear transformation of time given by Equation (2). The transformation was used for simplification only:

$$t = t/0.25 + 1 \qquad\qquad (2)$$

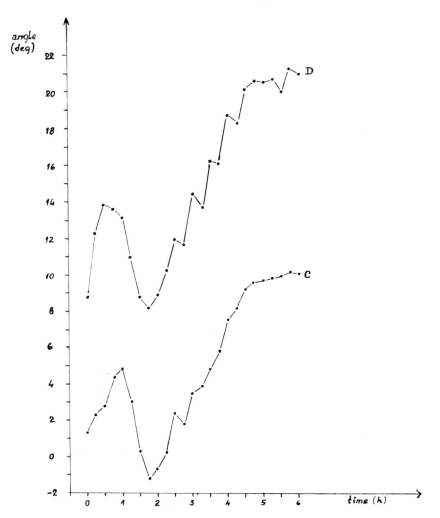

Figure 4. Changes in angles of trunk inclination (curve C) and head inclination (curve D) for the understimulation condition.

Table 3. Regression coefficients obtained in the investigations. Overstimulation, models under Equations (1).

	Trunk			Head		
	α	β	γ	α	β	γ
I	3·38	−1·30	0·12	9·1	−1·19	0·10
II	−7·14	0·81		−2·02	1·08	
III	9·47	0·13	−0·009	17·25	0·33	−0·025

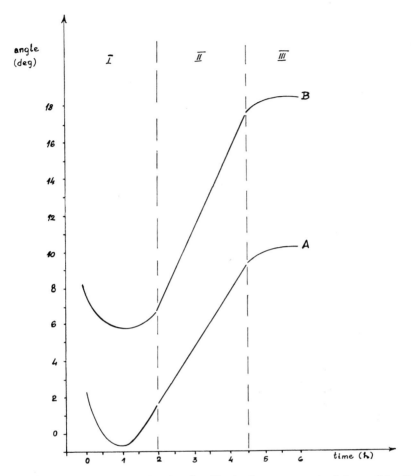

Figure 5.   Models of changes in angles of trunk and head inclinations; overstimulation condition. Parts I, II and III show theoretical curves a$_I$, a$_{II}$ and a$_{III}$ (Equations (1)).

Changes in angles of trunk and head inclination in understimulation conditions are described by the model of Equations (3):

$$a_I = \alpha + \beta t + \gamma t^3, \qquad \text{for } 0\,h \leqslant t \leqslant 2\,h$$

$$a_{II} = \alpha' + \beta' t, \qquad \text{for } 2\,h \leqslant t \leqslant 4\cdot5\,h \qquad (3)$$

$$a_{III} = \alpha'' + \beta'' t + \gamma'' t^2, \qquad \text{for } 4\cdot5\,h \leqslant t \leqslant 6\,h$$

Table 4 contains the regression coefficients obtained in the investigations, after transformation by Equation (2). Using analysis of variance for the regression lines it was shown that all the coefficients in Tables 3 and 4 are significant, with $\alpha = 0\cdot05$.

The regression straight lines for part II, and the second-degree polynomials of

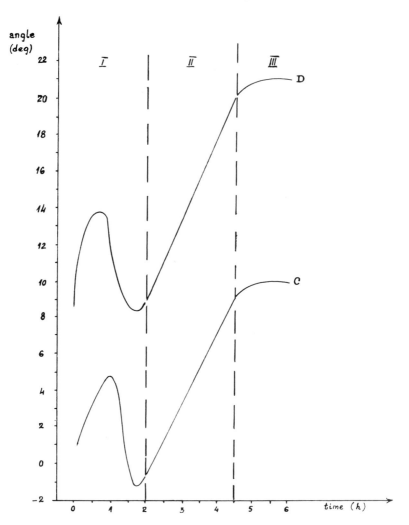

*Figure 6.   Models of changes in angles of trunk and head inclinations; understimulation condition.
Parts I, II and III show theoretical curves a_I, a_{II} and a_{III} (Equations (3)).*

*Table 4.   Regression coefficients obtained in the investigations. Understimulation,
models under Equations (3).*

	Trunk				Head			
	α	β	γ	δ	α	β	γ	δ
I	−2·76	4·16	−0·75	0·036	2·14	7·96	−1·59	0·087
II	−7·9	0·9			−9·75	1·18		
III	8·84	0·39	−0·03		20·41	−0·089	0·025	

part III were compared. There are no statistically significant differences between these pairs of lines for over- and understimulation, with $\alpha = 0.05$.

Figure 7 shows the four model curves *A*, *B*, *C* and *D* for over- and understimulation during the first 2 hours (part I). For understimulation the curves are polynomials of third degree and for overstimulation, of second degree.

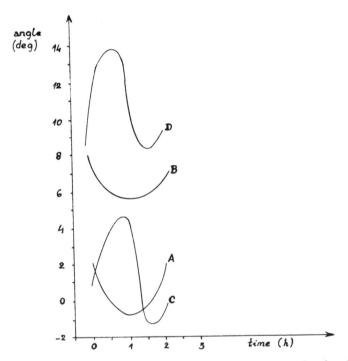

Figure 7.   Different models (for under- and overstimulation) of changes in angles of trunk and head inclinations in part I of the curves.

4.   Conclusion

The difference in changes of trunk and head inclination during the first 2 hours of work, between conditions of over- and understimulation, are caused by differences in the predominance of the inhibitory and facilitatory reticular systems during this time. The predominance of the facilitatory system follows overstimulation, and then muscle tone increases. As a result, trunk and head inclinations decrease. When overstimulation is prolonged the inhibitory system is activated, the muscle tone decreases and the inclinations of trunk and head increase.

In understimulation conditions on the other hand, the predominance of the inhibitory system follows understimulation from the beginning. Muscle tone decreases and angles of inclination increase in the interval 0–1 hour. The organism defends itself against excessive subactivation. As a result the facilitatory system begins to predominate (Fiske and Maddi 1967, Tranel 1962, Wachs 1977, Zuckerman 1979). Thus the muscle tone increases and angles of inclination decrease in the interval 1–2 hours. Finally, the inhibitory system begins to take over again, and the angles of inclination start to increase again at the end of the 2 hour period.

Thus the sitting posture at leisure is influenced differently by over- and under-stimulation. Changes of posture, using head and trunk inclinations as the two indicators, are different during the first 2 hours of work between the two kinds of stimulation. Following this period, changes in both cases are similar.

SECTION 4

MEASURES OF THE EFFECTS
OF POSTURE

Ecomomists, in their attempts to match work to people, seek methods which give a measure of the improvement in terms appreciable by the person. The 'mission', as it is often referred to in the United States, is the objective of a person's work, hence performance measures are essential. But if there are no human measures incorporated it could also be argued that there is no ergonomics.

The provision of standards which include ergonomic criteria is certainly one approach for the work investigator. For many problems the provision of standards is likely to be sufficient, they can stand in place of human measures for the necessary human-oriented decisions have already been taken. Thus, the shapes of figures and letters for minimum reading errors, the shape and spacing of knobs and most other control means are readily available from national and, in some cases, international standards. Noise and lighting levels are other areas where much can be determined without further studies.

When more complex problems arise, standards are not so easy to produce, nor are the data already published very easy to apply to a particular problem, although they may illuminate it and make the direction for its solution clearer. The lack of utility of standards in providing problem solutions is not surprising, since the effects on performance, and on people's experiences of the performance, can vary dramatically with their individual differences, and the relative degrees of importance of the various factors involved are all affecting the ultimate outcomes. As every experimenter knows, even if the subjects and the experiment stay the same, a change in motivation can ruin (or make!) a study.

So the student of working posture must have tools to measure the effects of the postures, and of the work activities, hose engaged in the tasks under study. Often the performance measures are not too difficult, but what is by no means sufficient are the measures of the effects. As earlier chapters in this book will have shown (e.g., the one by Westgaard), many of the long-term effects of posture are not firmly linked with clear-cut causes. Whilst it is argued that scientific rigour requires more investigations, and certainly the understanding of the aetiology of the widespread musculo-skeletal complaints suspected of arising from work situations must be pursued intently, it would be an inhuman stance to avoid taking actions until a high degree of relative certainty was achieved.

Thus it is that several of the methods in the following chapters use reductions in the loading of joints, muscles or skeletal components as indicators of improvement. This utilizes the current state of knowledge of the damage which can result from different levels of loading in order to give guidance on the levels which can be deemed appropriate. However, the degree of reduction, or the effects of increased exposure to lower levels of stress, on the subsequent development of disease is still a subject for wide discussion (see, for example, Hagberg 1982, Kilbom *et al.* 1983).

Rohmert and Mainzer's chapter provides a wide summary of the methods which can be employed, together with some applications for some of them. It will be noted that there are only a few methods which provide direct measures of the response of the body itself to loading. Heart-rate measures have a limited utility in static work, but various forms of analysis for myoelectrical activity of selected muscle groups is a major area of interest (see, for example, Kilbom *et al.* 1983).

A new method, using the long-recognized fact that the stature shortens during the day due to gravity and other loads upon the spine, appears to have a useful part to play in the study of whole-body postures. Its contribution is to the understanding of the behaviour of a specific part of the body, as with the use of EMG recordings, rather than the more integrating nature of heart-rate records.

Less precise, but very useful as measures of responses to load, are the subjective methods. Several of these are described in this section, and Chapter 23 (Manenica) proposes the use of subjective assessments to identify the maximum capacity for maintaining a posture. There is much investigation still required into subjective methods, using discomfort as a criterion, since discrepancies in findings from different but equally carefully conducted studies are notable. If we again refer to the paper by Kilbom *et al.* (1983) it is stated, as a result of her own studies, that "it seems clear that perceived pain intensity, especially if assessed at submaximal endurance time, would be a poor predictor of the static endurance capacity of the individual".

The paper by Montreuil and Laville brings in the totality of the work experience to aid in identifying the key points in jobs which require change. From discussions with those engaged on the work in question a number of beneficial features arise. The people involved understand better how work can damage them, and are thus better able to take proper precautions themselves; they are able to convey to the investigators a fuller understanding of the detail of the work and of their experiences of it, which in turn can lead to better solutions; and, finally, they can contribute to the solutions themselves, giving as a result a greater likelihood that the solutions will be adopted and developed as the work itself changes.

A major area of research in posture studies is in biomechanics, the calculation of the loads and forces within the body. Increasingly sophisticated models, linked to computer data collection, calculation and display, are bringing a new understanding of the causes of body damage. Early models, and many current models, are static, taking no account of the inertia forces arising from movement, but researchers are recognizing that static models of work situations are limiting. The introduction of inertia forces, however, reveals the weaknesses in the available body segment masses and centres of gravity data. Studies are under way in various centres to use, for example, *in vivo* data drawn from Nuclear Magnetic Resonator (NMR) equipment. Although time consuming, and using expensive equipment, NMR studies give the opportunity to identify the volumes of the various tissues in the different body segments, calculate their masses and distributions and derive the centres of gravity for the segments. A further important contribution from this safe imaging device is the opportunity to measure precisely the lever arms of many of the muscles which are the subject of biomechanic calculations.

It is interesting to note that, for certain of the uses of biomechanics, this increase in knowledge will not improve matters to any great extent. If absolute values are required, clearly the better data are essential, but where comparisons are being done and relative values are sought, it is unlikely that the situation will be changed. Indeed, if a separate 'anchor' can be established, the simpler current models may be difficult to improve upon. Thus, if a particular posture is found to load the spine by a certain amount as calculated from the rate and extent of collapse of the spinal discs, a biomechanic model for the posture in question, where co-efficients have been matched to the result, may give perfectly satisfactory answers to a wide range of practical problems.

Chapter 21, by Wangenheim and co-workers on the now well-known ARBAN procedure, illustrates a process, more complex than OWAS, which endeavours to draw a number of different factors into the evaluation of work. It is placed in this section as it is more extensive than just the body-posture record and its evaluation. The use of ARBAN involves judgements, and some equipment which might not sit easily on every factory floor. It also provides a sophisticated analysis which presents its data in a form useful for design decisions.

Chapter 18

Influence Parameters and Assessment Methods for Evaluating Body Postures

W. Rohmert and J. Mainzer

Institut für Arbeitswissenschaft, Technische Hochschule Darmstadt, Petersenstrasse 30, 6100 Darmstadt, F.R. Germany

1. Introduction

The aim of this chapter is to review methods for the recording, analysis and evaluation of body postures. The application of the principles presented and discussed is based on a systematic procedure within an ergonomic design. The topic covers a number of specific areas and problems, and requires a comprehensive discussion of methods, which will be supplemented by selected examples of practical applications. Many marginal areas of the topic cannot be presented within the scope of this article, since body postures in any given working task include a number of specific problems. Table 1 depicts the four basic types of working task

Table 1. Basic types of working tasks.

Type of work	Specific work content	Main strain of organs and capabilities	Ergonomics term	Examples
Mainly physical	Producing force	Muscles (heart and circulation)	Muscular work	Handling loads
	Coordination of motor and sensory organs	Muscles + sense organs	Sensorimotor work	Assembling, crane operating
Mainly non-physical	Converting information into reaction	Sense organs + muscles	Mainly non-muscular work	Controlling
	Transformation from input to output information	Sense organs + mental abilities		Programming, air-traffic controlling, book-keeping, translating
	Producing information	Mental abilities	Mental work (in the narrow sense of the word)	Dictation, designing, problem solving

Source: Rohmert (1975).

183

according to Rohmert (1975). In each of these types of working task—muscular, sensorimotor, mainly non-muscular and mental—body posture is an important problem area. The examples of work design selected have been classified according to the above-mentioned types of working task and will cover these systematically.

Because of the variety of ergonomic problems related to or influenced by body postures, a systematic review of the field of the analysis and assessment of body postures themselves is necessary. Problems in ergonomic methodology which concern body postures can be grouped into the following areas:

    1. Definition and recording of body postures.

    2. Identification of influence parameters in body postures.

    3. Criteria and methods of evaluation, based on the effects of body postures.

    4. Systematic application of the knowledge of body-posture optimization in work design.

These four areas form the principal contents of this chapter.

2.   *Definitions of body postures*

From the point of view of ergonomics, postures are defined as quasistatic biomechanic alignments, i.e., conditions of the body, and due to their three-dimensional character within the three-dimensional space they can be described by appropriate geometrical parameters. A geometrically reproducible posture can be described in three different ways (see Figure 1):

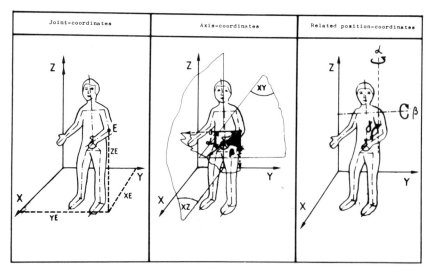

Figure 1. *Definition of body postures.*

1. Point-coordinates method. Recording of the three coordinates of individual joints.

2. Axis-coordinates method. Recording of the adjustment angles of the long axes of separate parts of the body in relation to the surrounding absolute space.

3. 'Anatomic' method. Recording of the adjustment levels of the long axes of separate parts of the body relative to the axis of the preceding body part.

The first two methods are limited to merely geometric data, without taking into consideration anatomic facts (e.g., limited range of movements). All three methods relate to the kinematic skeletal system; with their help all definable geometric parameters of any posture can be measured. The postures thereby become reproducible and quantifiable. Therefore these methods provide, on a *proportional* or *interval scale*, the data required for biomechanical models and analyses. Furthermore, it is significant that all possible mechanical rotational movements in each separate joint can be covered.

For practical applications, a number of methods have been developed, some of which aim at a simplification or even elimination of technical aids for posture recording (Figure 2) (Corlett and Manenica 1980). This does not necessarily lead to a reduction in accuracy or reproducibility of life parameters recorded. However, such a reduction is characteristic of some more simplified methods whereby a

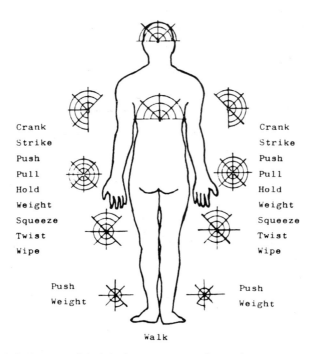

Figure 2.   A body diagram with limb displacement segments for recording posture. From Corlett and Manenica (1980).

limited number of the joints considered and their potential movements can be assessed in accordance with defined segments within the range of movement. Method modifications of this type are *ordinal-scale-oriented*.

On the level of the *nominal scale* one could place the description of postures based on posture typologies. A definition of a type of posture denotes a characteristic combination of the parameters considered. Such posture typologies generally relate to specific professions and are characteristic for them. As an example one could name the typology of dentists' postures, derived in an inquiry by Rohmert *et al.* (1984 b), and exemplified in Figure 3, which shows two postures from a total of eight types.

Definition and description of postures is related to specific research aims, and can be presented either generally or specifically (e.g., in the area of a given profession or a real work system). It is obvious that with a decreasing level of

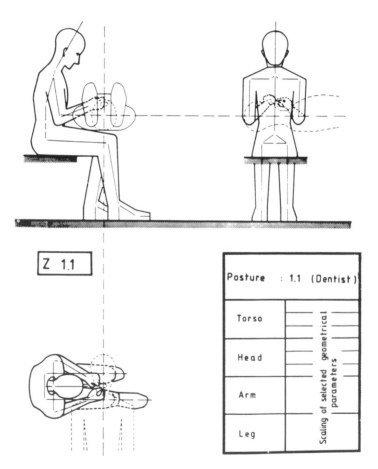

Figure 3.   Two types of dentist's postures. From Rohmert et al. (1984).

description and the use of different posture parameters, as well as a lesser degree of precision in the course of measurement of postures, the range of applicability of the methods decreases too.

The elimination of specific parameters in the description of postures does not necessarily result in each case in a disadvantage for the reproducibility of postures. In the instance of given space conditions within specific workplaces a number of posture parameters are predetermined. In specialized situations only specific postures (e.g., of the upper extremities or the bending of the torso) become of interest. In this context one could speak of a problem-oriented assessment derived from the basic statements.

It should be pointed out that in the evaluation of stress arising through posture, the holding time of the posture and its sequence also have to be considered.

3. Procedures for recording and analysis of postures and movements

Ever since the introduction of the 'cyclographic procedure' in motion study, a multitude of different methods of recording postures and a huge variety of technical devices have been developed. As a means for systematizing these procedures it is possible to find, through abstraction, the basic functional principles they all have in common. Following this path (Table 2) one can differentiate between: (*a*) mechanical or electromechanical; (*b*) optical; (*c*) acoustic; and (*d*) physiological principles.

The typical characteristics of these principles, as they occur in the course of the recording and analysis of postures, are derived from exemplary, commonly used systems. Non-instrumental technical procedures have also been taken into account—the so-called rating procedures, whereby the observations are verbally or graphically scaled on specially prepared rating forms. The typical characteristics of these procedures, as far as recording and analysis are concerned, are listed in Tables 3 and 4. The basic functional principles for the recording of postures show typical advantages and disadvantages which concern both their applicability and the level of feedback effects on the subjects rated.

Table 2. Systemization of methods for recording and analysing body postures.

Principle	Determination	Analysis
1. Technical principles	T	T
1.1. (Electro-) mechanical	A	A
1.2. Optical	B	B
1.3. Acoustic	L	L
1.4. Physiological	E	E
2. Rating methods	I	II
2.1. Verbal scaling		
2.2. Graphical scaling		

Table 3. Procedures of recording body postures.

Principle	Example	Function	Restrictions	Freedom of reaction
1. Technical principles				
1.1. (Electro-) mechanical				
Mechanical length-, angle- and time-measuring instruments	1.1.1. Tape measure, caliper, goniometer, stopwatch	Manual determination of certain dimensions	High effort	Bad
Three-dimensional length measurement	1.1.2. 'Dortmunder Würfel' (Lange, 1979); with distance measurements (three cords running from each body landmark)	Automatic determination of the space coordinates of selected points	Restricted space to move	Bad
1.2. Optical				
Usual photographic exposure	1.2.1. Camera, motion or video camera	Taking one or several pictures	Scarcely (depends on the instrument)	Good
Optical noting of selected points, passive markers	1.2.2. Coda - 3, Movement Techniques Limited, U.K.	Three thin planes of light are swept across the field of view and illuminate reflective markers (sampling rate: 300 Hz)	40° angle of view 10–60 m range, to 12 markers	Good; evaluation problems on account of the light planes
	1.2.3. Vicon, Oxford Medical Systems, U.K.	Up to 7 IR stroboscope-video cameras detect retroreflective markers (frequency 50/200 Hz)	To 7 m range	Good
Optical noting of selected points, active markers	1.2.4. Stereo-motography (Baum 1980)	IR light-traces are detected photographically in open-shutter technique	Cables lead to the markers at the subject	Rather good
	1.2.5. Selspot II, Selcom Inc., Sweden	Up to 16 video cameras detect up to 128 markers (frequency ca. 315 Hz)	To 5 m range	Rather good (cables at the patient)

1.3. Acoustic				
Ultrasonic location	Bundesanstalt für Arbeitsschutz und Unfallforschung, Dortmund	Time-delay measure	To present only for small parts, for example the hand	Rather good
Doppler's principle	1.3.2. Unopar (Nadler 1958)	Velocity dependent pitch vibration of ultrasonic vibration		
1.4. Physiological				
Measure of the muscle-action current	1.4.1. Grieve et al. 1975	Recording of the EMGs of relevant muscles		
Measure of forces and moments	1.4.2. Force plates, Kistler Instruments, Switzerland	Four multicomponent quartz transducers; subject stands on the platform	Leg position restricted (small area)	Good
Measure of acceleration	1.4.3. Multiple-axial accelerometer	Inertial mass in reciprocation with piezo- or inductive transducer		Rather good
2. Rating methods				
2.1 Verbal	2.1.1. AET (Rohmert 1979)	Observer interview		
	2.1.2. ARBAN (Holzmann 1983)	Detection, coding, computerizing, valuation		
	2.1.3. OWAS (Karhu 1977)	Classification by use of film material		
	2.2.1. Benesh Notation (Benesh 1955)	Notation system by use of symbols		
	2.2.2. Corlett 1980	Scaling in a standardized body diagram		

Table 4. Procedures of analysis of body postures.

No.	Example	Evaluation	Results—parameters and representation
1.1.1	Tape measure, caliper, goniometer, stopwatch	Manual or by computer (depends on expenditure)	According to selection and expenditure; analog resp. digital
1.1.2	'Dortmunder Würfel' (Lange 1979; Pynsent 1983		Space coordinates of selected points, for slow movements also: direction, time and speed; analog resp. digital
1.2.1	Camera, motion or video camera	Manual resp. with the help of: Vanguard Analyzer or coding by computer (Rohmert 1970) or OWAS (see below)	
1.2.2	Coda-3, Movement Techniques Limited, U.K.	Automatically (microcomputer)	Space coordinates of selected points—time; digital; graphical representation of trajectories—phase diagrams, stick-figures
1.2.3	Vicon, Oxford Medical Systems, U.K.	Interactive (marker identification) and automatic (microcomputer)	Of selected points: space coordinates, direction, speed, acceleration; digital; graphic similar to Coda-3
1.2.4	Stereo-motography (Baum 1980)	Manual	According to selection and expenditure
1.2.5	Selspot II, Selcom Inc., Sweden	Automatic (microcomputer)	Of selected points: space coordinates, time; digital
1.3.1	Bundesanstalt für Arbeitsschutz und Unfallforschung, Dortmund	Automatic	
1.3.2	Unopar (Nadler 1958)		
1.4.1	Grieve et al. 1975	According to expenditure	According to selection and expenditure, in general only classification of the posture of single parts of the body
1.4.2	Force plates, Kistler Instruments,	In general by computer	Three force-components, evaluation: direction and speed of the centre of gravity of the body
1.4.3	Multiple-axial accelerometer	According to expenditure	Of selected points: acceleration in given direction of selected distances, angle to the vertical
2.1.1	AET (Rohmert 1979)	By computer	Classified types of body posture
2.1.2	ARBAN (Holzmann 1983)	14 'functional units' of the body are examined by computer under the aspect of postural strain	Under the aspect of strain-valued parts of the body
2.1.3	OWAS (Karhu 1977)	Manual, by experts	Division in four classes concerning the necessity of a variation of the working conditions
2.2.1	Benesh Notation (Benesh 1955)	Manual	Qualitative body posture—description, features
2.2.2	Corlett 1980	Manual	Like Benesh

The mechanical or electromechanical principles have been proven to be neither easily applicable nor free of feedback effects. Their advantage lies in the quick and inexpensive installation of simple instruments for length, angle and time measurement.

Most of the measuring systems work according to optical principles, using photogrammetric calculation procedures. Their strength lies in the fast and precise listing of virtually unlimited numbers of items, which may also be selected and specified as required. Flexibility with regard to specific measuring problems can be achieved through diversity of lens, filter and monitoring systems. Problems of feedback could arise due to lighting condition requirements or because of active and passive markers attached to the subject.

Acoustic procedures seem to have come into use only for the assessment of movements in a smaller range (e.g., hand movements). This pertains also to physiological procedures, like the EMG analysis procedure or electro-oculography.

With force plates or accelerometers, specifically selected parameters (e.g., only arm postures) are being assessed individually. Also, aspects of body postures and body movement are being assessed, the impact of which can be seen in a specific formulation of a problem or as a supplement to other procedures.

In principle, all of these procedures allow for an assessment of macro—as well as micro—situations. However, the technical principles of the instruments limit the movement range of the subject. Rating procedures are not influenced by this limitation in so far as they are independant of technical equipment (e.g., AET according to Rohmert and Landau 1979). The advantage of the rating procedures lies in the very rapid qualitative and quantitative recording, analysis and, in certain cases, also evaluation of postures (see later).

The nature of the analysis of the data collected depends on the presentation of the basic data as well as on the specific problem in question. According to the number of marked points in space and the sampling frequency, quite a large quantity of calculations may become necessary, to such a degree that they can only be handled by automatic data processing. Facility of evaluation (be it automatic, interactive or manual evaluation) and the number and presentation of the event parameters prove themselves to be dependent not on principles, but on system-specific characteristics. Often, the most comfortable systems are based on optical principles. In practice, however, computer-monitored systems contain weak points in details, which can make their application considerably more difficult.

Additional and more specific statements on the precision and reliability of individual procedures would go beyond the scope of this paper. It should, however, be recognized that the technically oriented procedures show advantages which depend to a considerable degree on the form of their actual realization. In summary, only by means of a more precise formulation of the questions concerning the choice and application of a specific system can a path be found leading through the huge multitude and diversity of procedures.

It must first of all be decided whether a precise biomechanical analysis or an evaluation and classification of workload is necessary.

4. Evaluation of body postures

The evaluation of postures is based on their different short- and long-term endogenic and exogenic effects. The methods for posture evaluation (in general dealing with ergonomic problems) can be systematized according to Table 5. They can be classified as to whether they are objectively or subjectively oriented. Objectively oriented methods of evaluation relate to quantifiable parameters on a higher-level scale (proportional and interval scales). The phenomena evaluated within the objective approach can be classified into exogenic (postures as external mechanical activities) and endogenic ones (short- and long-term reactions, i.e., morphological changes in organs). The subjectively oriented approach can also be classified into methods relating to externally influenced aspects of work (e.g., workplace design) and those relating to the subjectively experienced strain.

As characteristic methods of objective evaluation of external activities one could mention biomechanics, activity analyses (e.g., analyses of change of postures) and work studies (work performance-oriented studies). Physiology provides the classic objective means of evaluating reactions of organs with respect to postures; epidemiology on the other hand deals with long-term morphological changes.

The subjective evaluation of external influence factors (in this context also stress factors) is the domain of psychophysics. The subjectively experienced strain can be investigated with methods of introspection. Below, the different approaches for evaluation, in accordance with the classification in Table 5, are characterized and clarified through selected examples.

Objective methods of evaluation

External activities
BIOMECHANICS
By means of biomechanical models of the human body different problems may be investigated, including analysis of the mechanical strain (forces, torques, pressures)

Table 5. *Classification of methods for evaluating body postures.*

	Objective		Subjective	
Phenomenon	External activities	Reactions of organs	External factors	Experienced strain
Methods (examples)	Biomechanics	Physiology (short-term reactions)	Psychophysics	Introspection
	Activity analyses (adopted postures)	Epidemiology (long-term damages)		
	Work studies (performance)			

scale level/objectivity
High ―――――――――▶ Low

of the movement and support apparatus (ligaments, tendons, joints, muscles). Furthermore, examination of the mechanical equilibrium of the body (e.g., during exertion of body forces in different body postures) as well as mechanical analyses of static and dynamic muscle work are possible. From the multitude of investigations within these areas two relevant practical studies will be presented here.

Jäger *et al.* (1984) investigated strain on the spine during the transport of dustbins (considered further elsewhere in this volume). Based on a biomechanical model (see Figure 4), the pressure forces in respect of the torques effective at the

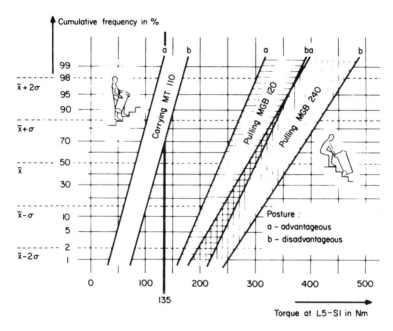

Figure 4. The load on the spine during the transport of dustbins. From Jäger et al. (1984).

lumbo-sacral point L5/S1, in different situations, were calculated. By taking into account the variable frequency of distribution of weights in dustbins the cumulative frequency of torques in the lumbo-sacral area was estimated. The differing postures lead to different strains or load. The biomechanical evaluation of the results according to Tichauer (1978) classifies muscular work into four categories depending on the difficulty of the accompanying circumstances (the most difficult class covers those tasks in which moments of more than 135 N m are reached). Based on these results, recommendations for the manual handling of dustbins in relation to postures can be described.

In a basic research study horizontal arm movements were analysed by Jenik (1972). The entire working task energy expenditure per minute, calculated with the help of indirect calorimetry and gas analysis, comprises the energy expenditure for:

1. Maintaining a sitting posture (without additional strain).
2. Holding a weight in the hand.
3. Developing the dynamic work necessary for acceleration and deceleration.
4. Developing the reaction forces and momentums for fixation of the shoulder joint and the elbow joint, which are necessary and arise due to the arm movement.
5. The static load (point 4) as an additional load on the torso and the legs in a sitting position.

The first two components of the energy expenditure due to the working task are to be regarded as constant and correspond to the 'static' component s (cal/mov.) in Figure 5. The other three expenditure components are conditioned by the movement itself, and when summed they correspond to the 'dynamic' component d (cal/mov.). Through a biomechanical analysis of over 50 movements it was found that the value of the specific dynamic component depends on the frequency of movement to the third power (Jenik and Rohmert 1972).

Figure 5 shows that the specific energy expenditure e (cal/mov.) has its minimum at the point where the static part amounts to two-thirds of the entire energy expenditure. Furthermore the optimum frequency of movement can then be estimated. With the help of this example it becomes clear that the bio-mechanical analysis of postures plays an important role in the dynamic working tasks. The problem of the static equilibrium (stability) of postures is not considered in detail here. We can only refer to the investigations of Grieve (1979) as well as Mainzer and Rohmert (1983).

In summary, one can state that biomechanical models concentrate on different quantitative parameters in the course of mechanical activities, i.e., they virtually do not cover the effects of repeated processes. Thus, biomechanics is primarily *practicability* oriented.

With the help of knowledge of functional relationships between bio-mechanical parameters and the kind and frequency of morphological changes (i.e., illnesses), stress limits can be derived. An important advantage of biomechanical models lies in the possibility of evaluating strain on passive tissue.

Improvements are needed in the application of mechanical models. For example:

1. A great number of simplifications and assumptions (e.g., concerning the direction of force vectors, the corresponding mass parameters) makes it evident that the significance of quantitative results is to be regarded critically.
2. Most models are static and related to the sagittal plane. Torques of the torso, frontal bends and movements still create difficulties.
3. The quantitative limits, e.g., the intradiscal pressure, cannot be calculated and applied due to subjective conditions. The significance of the limits values in relation to interindividual cumulative frequency curves or individual dispositions is difficult to estimate.

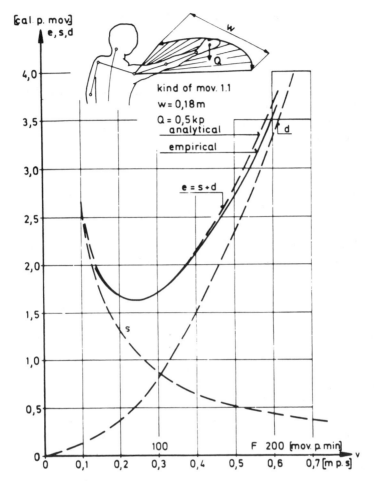

Figure 5.   Relationship between the static (s) and dynamic (d) components of the specific energy expenditure (e) (cal/mov.) during arm movements.
From Jenik (1972).

WORK STUDIES AND ACTIVITY ANALYSES

External work activities can be classified into the areas of work performance and change of postures. The applied disciplines of these areas are work studies and activity analyses.

Figure 6 shows an example by Tichauer (1975), which puts forward the relationship between parameters of postures (abduction angles, elbow height) and work performance as well as the work metabolism. Optimal work postures can be recommended with regard to these parameters.

In general, Sämann (1970) has presented the differences between favourable and unfavourable postures. Figure 7 qualitatively presents the relationship between physical load and the costs per output unit in a sparsely mechanized workplace.

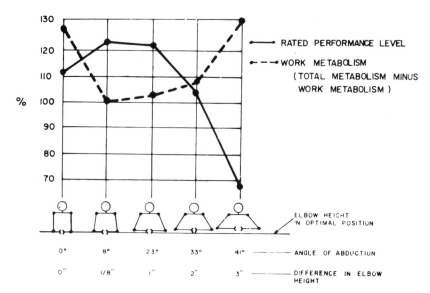

Figure 6.   Effect of the angle of abduction on physiological, as well as economic working efficiency in food packing.
From Tichauer (1975).

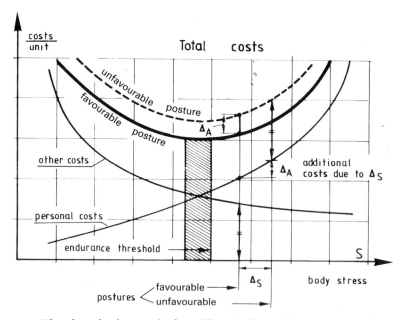

Figure 7.   The relationship between the favourable and unfavourable postures due to physical costs and the costs per output unit in a sparsely mechanized workplace.
After Sämann (1970).

The curve of the total costs, which was derived from the two opposing tendencies of the portion of cost for production and for manpower depending on the physical load, shows an economically optimal area. According to Sämann this is to be expected close to the endurance threshold. With additional physical load through more unfavourable postures the cost for manpower rises, whereas the production cost stays stable (given performance). The total cost rises (broken line), and the optimal economic area cannot be reached.

Stier (1959) showed that, depending upon the arm's position within the reaching area, there are favourable movement directions of the arm due to maximum velocity (see Figure 8). Within the circle with differently shaded sections the heavy lines show the favourable directions of movement. This example, from the area of movement studies, exemplifies that posture has a direct influence on efficiency of movement. Influences on the precision of movement can be shown as well. (Note also that the empirical results of the relationship between postures and performance can as a rule be analysed biomechanically.)

Among the activity analyses for postures, there are many results available for changes within sitting postures. The research recently carried out by Grandjean *et*

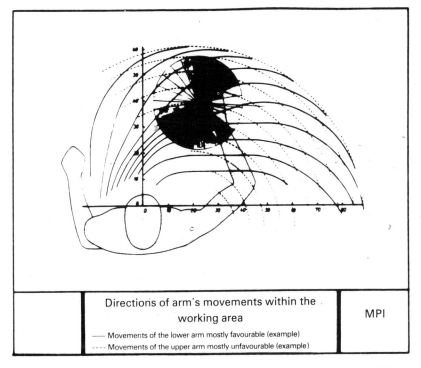

Figure 8. *Favourable movement directions of the arm (depending upon the arm's position) due to the maximum velocity.*
From Stier (1959).

al. (1984) illustrates that in the course of work with VDT devices considerably different inter- and intra-individual postures are being assumed.

According to an activity analysis, those postures which are preferred do not allow for a direct statement as to what degree these postures can be graded as favourable or unfavourable. A further evaluation, as well as an assessment of the influence factors, are necessary for this purpose.

In order to evaluate the desired posture changes during working activities in a given workplace, a workplace simulator was developed in our department. Figure 9 shows that with a number of motor-driven supports an adaptation of the anthro-pometric parameters of the work place to the instantaneous posture (operated by the subject or externally) is possible during work. Additionally, with the help of the videosomatographic method (Martin 1981), using drawings and overlapping video recordings of the real workplace, a number of variables important for the design of the workplace can be examined.

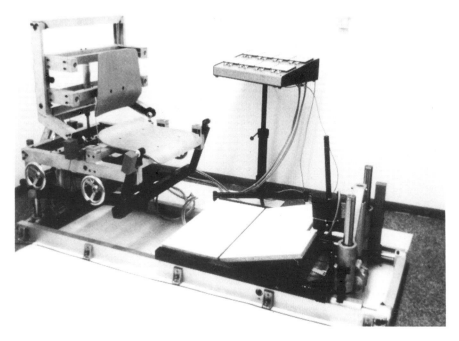

Figure 9.   Motor-driven workplace simulator.

In addition to this, with the help of a so-called *harmonogramme*, the simultaneity and sequence of manual and visual activities can be monitored. As a rule these are also connected with a change of body postures.

Research on the practical application of possible adjustments within a workplace, together with the parallel investigation of strain parameters, can be carried out with the help of the workplace simulator. Concrete design recommendations related to the as yet undefined problem of 'dynamic' postures can be derived.

It can be stated that *activity analyses* lead to statements about *practicability*. However, an additional interpretation of the activities investigated is needed.

Reactions of organs

The objective workload during external activities must be evaluated with regard to its further effects—the subjective strain. The stress/strain concept (see a concise presentation of the development of this concept in Rohmert (1984)) can be utilized for this purpose.

In doing so, relationships are being investigated between stresses caused by postures and their effects on the organs concerned. With the help of physiological measuring and evaluation methods, changes within the state of the affected organs—especially of the cardiopneumonal system and the musculature—can be measured as time functions. Heart-rate frequency and myoelectrical activity may be regarded as typical measurement parameters.

The physiological approach for evaluating body postures is based on *short-term reactions* of the affected organs as a time function. The physiological evaluation methods are arranged on the evaluation level of *tolerability*. Table 6 (according to Sämann (1970)) shows a relative comparison of three types of sitting postures based on physiological measurement values. This type of presentation can be characterized as an integrative comparison. Based on the time-related changes of a physiological strain value (myoelectric activity) (Figure 10), it is possible to derive conclusions about maximum possible work duration.

The physiological evaluation methods are especially well suited to dynamic changes of strain during phases of varying stress levels in real work. The physiologically 'dynamic', i.e., time-dependent, measurement values correspond to the dynamic character of real work. Furthermore, a specific advantage lies in the objectivity of such procedures, which allows for the establishment of causal relations between the stress factors (in relationship to the postures) and their short-term physiological effects.

An evaluation of passive structures (tissue, fibre) with the help of physiological evaluation methods with regard to muscle strain is impossible, a functional 'bottleneck analysis' is necessary for this purpose.

The *evaluation criteria*, i.e., *limiting values*, have been derived with regard to short-term changes; their relevance concerning long-term influences (physical complaints, morphological changes) has to be additionally validated. Also, as opposed to biomechanical evaluation approaches, the physiological evaluation criteria have an *integrating character* and are less suitable for an evaluation of isolated elementary activities.

It is *epidemiology* that deals with the *long-term objective morphological changes* with respect to the damage of organs (passive as well as active tissues). Epidemiology aims to show a statistical correlation between exposition (body postures) and

Table 6.   Overall assessment of sitting body postures.

Assessment criterion	Bodily posture		
	Normal sitting	Sitting bent	Sitting upright and arms above head
Energy requirement compared with resting position (kcal/min)	0·06	0·15	0·16
Increase of pulse rate compared with resting rate (beats/min)	7	13	13
EMG findings without loading (points)	1	6 Particularly back muscles	11 Particularly back and shoulder muscles
Particularly bad features	Little force can be applied within restricted working area Superficial circulation in buttocks and backs of thighs is impaired in the long term		
		Respiration and digestion impaired by abdominal compression	Long additional recovery time is necessary
Particularly good features	The provision of an armrest provides for a rapid transition between working and resting postures Relieves the supporting tissue of the legs Little stabilizing work necessary (good for precision motor activity) Circulatory and energy demands low Favourable working posture		
Range of application	Whenever conditions permit	Unjustifiable (work design)	Only if working point is only accessible from below
Overall assessment (rank)	2	4	10

Source: Sämann (1970).

damage indicators (i.e., physical complaints, morphological changes, absenteeism, early retirement, the area of physical complaints here is ranked together with the subjective evaluation methods).

As a criterion for calculating limiting values, 'absence of organ damage' can be used. Thereby epidemiology also becomes *tolerability oriented* and *integrating*. Epidemiology offers a potential basis for the evaluation of different postures. However, it is rather difficult to establish a cause-and-effect relationship between postures assumed in working life and specific morphological changes; the

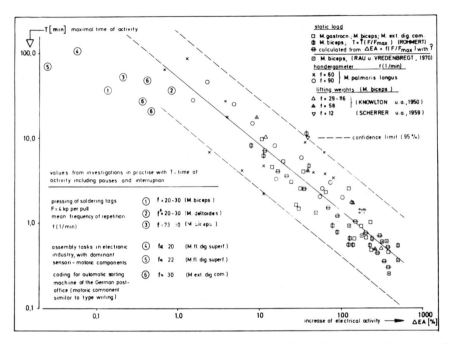

Figure 10.   Maximal time of activity for different kinds of muscular work in relation to the increase of electromyographic activity.
From Laurig (1973).

corresponding factor analyses presume an adequate description of both. This, however, has only been carried through in rare cases.

One of the most recent publications in this area is by Westgaard and Aarås (1984), which aims at ranking muscle strain arising as a result of postures as a causal factor in the development of musculo-skeletal diseases. Figure 11 shows the differences between two tasks having different postures (Figure 11 (*a*) indicates the setting up of telephone switch systems, (*b*) indicates the production of cables), with regard to musculo-skeletal diseases. Detailed analysis of the type, duration and localization of the diseases shows specific differences between the two work systems. A comparison of the respective postures which changed during the observation period, together with the development of new workplaces, can only be accomplished through the help of a far-reaching analysis.

In general, it can be stated that the epidemiological approach and the suggestions derived thereby, can only be reasonably realized with the help of related cause analyses (e.g., concerning load, with the acceptable intra-abdominal and intradiscal pressure, respectively). A parallel, reproducible observation and analysis sufficient for the aims of realization is necessary for this purpose, both for the parameters of strain (i.e., postures) and for their effects.

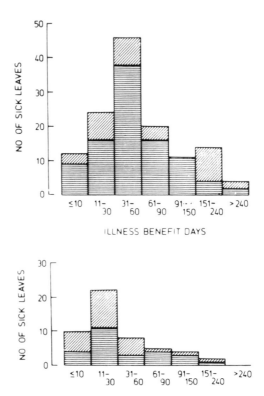

Figure 11. Duration of sick leaves due to musculo-skeletal illness in different work systems. From Westgaard and Aarås (1984).

Subjectively oriented evaluation approaches of body postures

A *subjective evaluation* of postures can be oriented to the external influence factors (design characteristics of the workplace, different stress factors); however, it is principally based on the subjectively experienced strain. The division of this area (as shown in Table 5) is relevant for pragmatic reasons. Evaluation of the external influence factors related to stress enables us to point out the design characteristics directly, whereas the experienced strain has an integrative subject-oriented character.

With the help of the *psychophysical method*, limiting values can be estimated for specific *stress parameters* in relation to postures. Figure 12, according to Snook, shows the distribution frequency of acceptable loads which were estimated with the help of the psychophysical method for different handling and reaching areas.

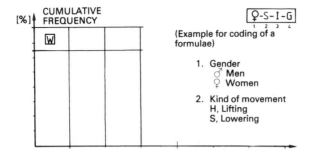

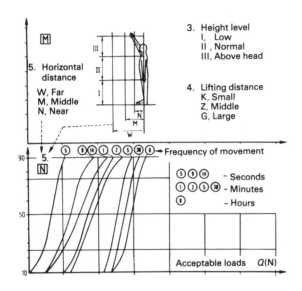

Figure 12.   Acceptable loads for manual handling.
From Snook (1978).

Reversed conclusions about acceptability of body postures at a given load thus become possible.

A classic *evaluation of experienced strain* can be carried through with the RPE (Rated Perceived Exertion) method. The so-called Borg scale (Borg 1982) leads to an integrative evaluation of subjectively experienced strain due to different postures. Figure 13 shows a result from the area of evaluating different dentists' postures, derived by means of the Borg scale. The different types of static postures assumed were subjectively evaluated in constant periods. The time-dependent rise of ratings differed according to postures and allowed their relative evaluation. The *subjective complaints* due to different postures (as a result of longer-lasting exposure)

The ergonomics of working postures

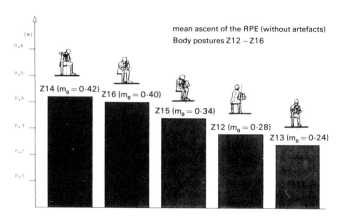

Figure 13. *Ranging of different dentist's body postures, based on subjective ratings by means of the Borg scale.*
From Rohmert et al. (1984).

have to be coordinated to this area as well. In this context the already-quoted publication of Westgaard and Aarås (1984) can be again cited.

The advantages of the subjective-evaluation approaches lie in their *integrative character* and they are significant with regard to the possible analysis of unrecognized bottlenecks. The evaluation levels of *tolerability* and *acceptability* are being reached through subjective evaluation.

Comprehensive view

The different approaches for posture evaluation concentrate on different parameters and use different evaluation criteria. The choice of a given method in any context depends on the specific aims as well as its relevance.

The following examples from investigations carried out by our department concerning postures at different types of work tasks clarify this approach.

Maximal duration of exertion of static forces in different body postures (muscular work)

The maximal holding time during static exertion of body forces depends on the force needed, as a percentage of maximal voluntary contraction (MVC), due to the respective body posture. Figure 14 summarizes the results of completed investigations (Rohmert *et al.* 1985). Depending on different distributions of load on actively (muscle) and passively (joints, ligaments) supporting elements, different maximal holding times are obtained. In bent postures, especially, there is a prolongation of holding times compared with cases where the static forces are being exerted with the upper extremities, which are primarily characterized by muscle strain. Furthermore, in order to evaluate the postures investigated, electromyographic investigations of exposed muscles and a subjective evaluation with the

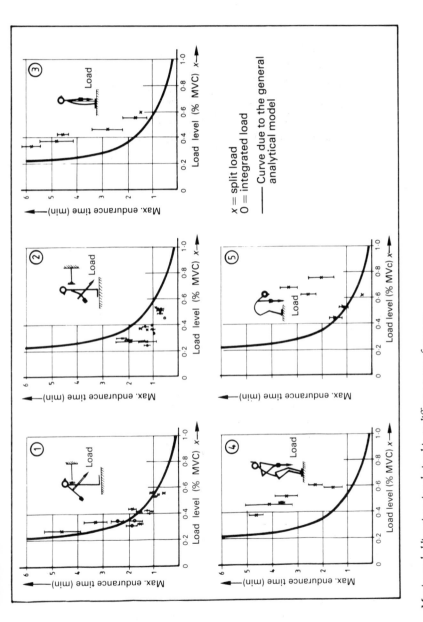

Figure 14. Maximum holding times in relationship to different types of postures. From Rohmert et al. (1985 b).

help of the Borg scale were carried out. Based on these results the maximal holding times for different postures can be evaluated (divided into those with less or more participation of the passive transposition elements of forces) as can the 'bottle-neck' muscle groups and the reporting of complaints by the subjects. This example from the area of *muscular work* covers objective and subjective evaluation approaches to body postures.

Mounting electronic parts onto printed-circuit boards (PCB) (sensorimotor work)

In the area of sensorimotor work an assembly workplace in the electronic industries was investigated and redesigned (see Rohmert *et al.* 1979). The work task consisted of placing approximately 70 different electronic parts onto a horizontally fixed car radio's printed circuit board (PCB) (see Figure 15). The work method was characterized by two-handed, simultaneous, intermittent placing of the parts (cycle periods in the realm of seconds).

The containers with the parts were placed behind the PCB. The points of placement were often covered by the working person's hands. This sight impediment was confirmed in the course of an objective investigation of the working person's external activities. The sight impediment became especially crucial where, together with the respective change of postures, the right-oriented PCB half (in relation to the sagittal plane) had to be worked on with the left hand and vice versa. Additional frontal and sagittal bending as well as rotation of the

Figure 15. Body posture while mounting electronic parts onto printed circuit boards (PCBs). From Rohmert et al. *(1979).*

entire trunk and the neck area were necessitated. A suggestion for a logical distribution of the assembly parts between the right and left hand resulted in a reduction of this activity. To a large extent posture is predetermined by *manual activity*, location of work elements and *sight geometry*.

Here, the unfavourable, almost horizontal, arrangement of the PCBs had an especially negative influence on bent postures. This horizontal arrangement and the steep angle of incidence of the lines of vision forced the working person to bend forward constantly.

In order to evaluate the resulting physiological strain in the course of several shifts, the electrical activity of selected muscles (m. trapezius, m. deltoideus) was estimated; the results re-affirmed the disadvantages mentioned.

In redesigning the workplace a systematic procedure was required, which could regulate itself according to the logical relationship of the most important problem areas (sight geometry, anthropometry, work movements and posture). First (see Figure 16) the optimal sight geometry was determined (sight distance, sight inclination angle), and the PCB positioned in such a way as to make it manually accessible. The series of movements was determined by the angular positioning of the PCBs relative to the working person as well as the part container relative to the PCB. The next step allowed the posture to be optimized. Comparative physiological measurements, activity observations as well as subjective evaluations by the working person, lead to a better evaluation of the new workplaces with their optimized postures. Those postures ranked as favourable had to be adjusted for the entire range of (female) body heights. From this resulted the necessary adjustment ranges for supports, and the respective measurement of the work location.

In this case of *sensorimotor work* the objective as well as the subjective evaluation approaches for postures were taken into consideration.

Visual inspection of PCBs (mainly non-muscular, reactive work)

The visual inspection of PCBs with the help of microscopes can be used as an example for reactive work. Figure 17 shows that during existing conditions microscope use is carried out for the entire time in an unfavourable bent posture. Measurable fatigue in the areas of the neck, shoulder and back could be traced objectively with electromyography. When compared with the average of other work locations, subjective ratings showed an occurrence of cases of work in capacity due to ailments three times higher.

The use of unsuitable microscopes (too little horizontal and vertical distance between the eye piece and the lens) only allows for a corrective adjustment within the work location. The microscopes were not developed for the type of work situation in which they were being used. Figure 18 shows that it was possible to improve postures by positioning the microscope higher, and with a special support for both arms.

This example from the area of *reactive work* demonstrates the combined use of several problem-oriented evaluation approaches for postures.

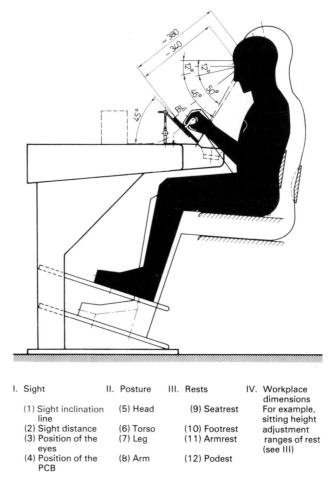

I. Sight	II. Posture	III. Rests	IV. Workplace dimensions
(1) Sight inclination line	(5) Head	(9) Seatrest	For example, sitting height
(2) Sight distance	(6) Torso	(10) Footrest	adjustment
(3) Position of the eyes	(7) Leg	(11) Armrest	ranges of rest (see III)
(4) Position of the PCB	(8) Arm	(12) Podest	

Figure 16.   A systematically redesigned workplace for mounting PCBs.
From Mainzer and Rohmert (1983).

Police command-control centre (mainly non-muscular, combinatory work)

Ergonomic measurements were conducted in a police command-control centre (Haider et al. 1982) on the use of computer-controlled text processing systems with three alternative input devices: function keys, a light pen and a touch panel. A group of 14 subjects operating these three devices was examined. The investigations reported here are restricted to the tasks of dispatching and reporting.

The work activities were recorded as a time series, through electric impulses generated by every functional action of the keyboard, light pen and touch panel. Horizontal and vertical electro-oculograms (EOG) were recorded in order to get information about the frequency of eye movements during the use of the devices.

Stress reactions were expected primarily in areas of muscular workload (the

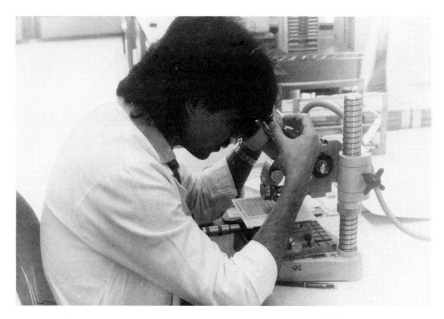

Figure 17. *An unfavourable posture during the visual control of PCBs with the help of microscopes. From Rohmert* et al. *(1984 a).*

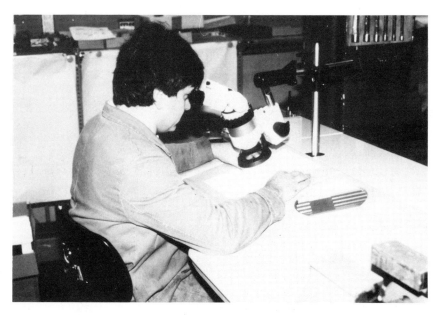

Figure 18. *An improved posture, due to the redesigned workplace for the visual control of PCBs. From Rohmert* et al. *(1985 a).*

holding of fingers over the alphanumeric keys, holding the light pen far away from the body, posture of the trunk). Therefore, electromyograms (EMG) of the m. deltoideus and the m. erector trunci were taken in order to assess local muscular strain. The difference in electrical activity of the muscles during operation of the touch panel and the light pen was not statistically significant, and it was concluded that the operation of the touch panel and the light pen led to increased strain of the m. erector trunci because of fatiguing postures. The operators had to bend and twist their trunks in order to reach the display (see Figure 19).

When averaged over 14 subjects, the advantage of the light pen was shown through the lower values for frequency and amplitude of electro-oculographical activity (OA). The performance activities with the light pen were significantly lower than those with the function keys. The mean values of motor strain as shown electrophysiologically point to different qualities of layout design. These results, plus the fatiguing postures, mean that although the dynamic motor stress produced when working with the light pen was low, this device caused high values of static muscular strain.

Polygraphic measurements made it possible to relate specific design deficiencies of the different devices to physiological measurements. The concept of measurements used in this investigation enabled us to find certain cause and effect relationships between workplace deficiencies and physiological reactions.

With increased frequency in touching the display, the EMG of the m. deltoideus showed a higher ranking of the physiological strain measurement. The

Figure 19.   Body posture of the VDT operator in a police command-control centre. From Haider et al. (1982).

location of the monitors was beyond the optimal reaching distance. Consequently, the operators had to work with their arms extended, which resulted in static muscular strain (see Figure 19). In order to reduce the strain, the subjects tended to bend and twist their trunk, as registered by the EMG for the trunk muscles. This indication of physiological strain was evident both for the light pen and the touch panel, since both devices required contact with the surface of the monitor.

The results revealed intolerable motor strains, and pointed out the necessity of an anthropometrically and physiologically based work-system design. The firm involved in the development of the command-control centre decided to redesign the prototype. The newly designed workplace is shown in Figure 20.

5. *Vocational comparative analysis of job demands concerning body postures*

Postures arising within a job can be highly variable and they are connected with a number of influence factors and their effects. A detailed ergonomic analysis of separate parameters of postures (in the sense of a measurement of available geometrical individual parameters), as well as their individual evaluation, would be too extensive. For this reason more integrative analytical methods are called upon which have made possible analytical comparisons of job-related postures (generally on the level of the ordinal scale and to formulate a mutual coordination of the tasks and their relative evaluation).

According to the aim of the methods of assessment and evaluation the postures are assessed in a more or less detailed fashion (e.g., through codification by a person in the course of observing the activity or through a recording with different

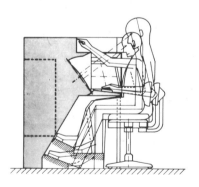

M=1:10

Figure 20.   A redesigned VDT workplace in a police command-control centre.
From Haider et al. (1982). Side elevation of recommended workplace, for users of heights 1900 mm and 1500 mm.

technical devices). The relevant methods were mentioned earlier. For illustration, an analysis based on the AET (Rohmert and Landau 1979) will be presented here.

A data sample based on a new technique for job analysis (AET), which was applied for each one of the selected 485 tasks, was analysed with regard to different postures (Landau and Reus 1979). In this study the dependence of postures on a given task, on a given type of work, as well as on the sex of the worker was proved quantitatively.

Altogether there were 14 types of postures taken in account (Landau *et al.* 1975). Amongst those were:

    1 = standing normally
    2 = standing bent (bend of torso up to approximately 45°)
    3 = standing bent (bend of torso over 45°)
    4 = standing, arms above head
    10 = sitting normally
    11 = sitting bent
    12 = sitting, arms above head

Other postures were related to kneeling, squatting and lying.

For each work system, body postures had to be assessed according to the time of exposure within a shift. For each activity, variables for the classification of the working task had to be set up. For further research, relevant variables had to be assessed from the area of the work system and the demand analyses.

Figure 21 summarizes the average distribution of the most frequently arising specific postures for different tasks in the course of a shift. As a comparison with the

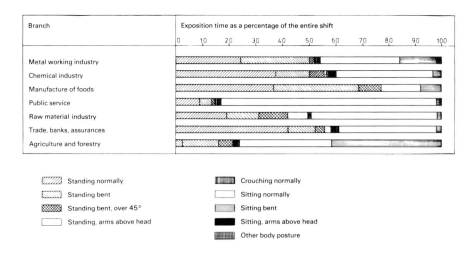

Figure 21.   Average distribution of the most frequently arising specific postures for different tasks in the course of a shift in different branches.
From Landau and Reus (1979).

activity demands related to the AET characteristics of postures (the 14 afore-mentioned posture types were applied here), the procedures of cluster analysis are highly suitable (see Figure 22). The cluster analysis for postures was carried out with the application of the method combination 'group-average'/W-distance, according to Skarabis (1970).

By analysing the graphical presentation (dendrogram) of the results of the cluster analysis we can determine different clusters concerning the characteristics of postures. The statistical distance between the clusters obtained is due to the scale of distance, i.e., similarity, which is plotted next to the dendrogram. Figure 22 shows a cluster analysis of 30 activities by the area of tasks concerning planting, cultivating, breaking of stones, which are clustered into four main groups.

1. Different modes of standing. With 15 activities (characteristic for these are mining activities and activities in the area of farming and forestry) the main work postures lie within the range of standing normally, standing bent, standing bent over and standing with arms over the head.

2. Sitting normally. With five activities the employees during almost all of the shift time have a normal sitting posture (driving jobs in the area of mining and forestry).

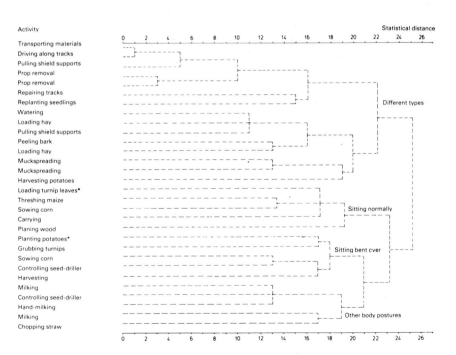

*Figure 22.   Dendogram resulting from cluster analysis of postures. Thirty activities analysed by 14 measures of body posture. * indicates a group of jobs.*
From Landau and Reus (1979).

3. Sitting bent. If the driving activity for harvesting natural products contains especially high information-flow components—depending on the position or the nature of the work object—bending of the torso in a seated task can be discerned.

4. Special forms of work postures. Agricultural areas with few mechanized activities; e.g., squatting or kneeling.

The analytical assessment of postures with the help of AET allows a comparison of the demands of postures within different working systems.

6. Design aspects

The design of work systems and workplaces is connected with the problem of postures in many ways. According to the activity model in each type of work, the problem of postures must be considered in the course of *acquiring visual information* and *mechanical activity* (see earlier examples).

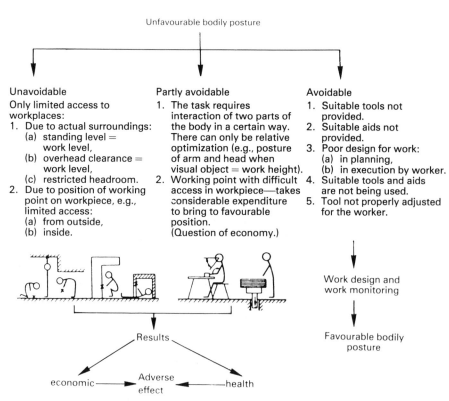

Unfavourable bodily posture

Unavoidable

Only limited access to workplaces:
1. Due to actual surroundings:
   (a) standing level = work level,
   (b) overhead clearance = work level,
   (c) restricted headroom.
2. Due to position of working point on workpiece, e.g., limited access:
   (a) from outside,
   (b) inside.

Partly avoidable

1. The task requires interaction of two parts of the body in a certain way. There can only be relative optimization (e.g., posture of arm and head when visual object = work height).
2. Working point with difficult access in workpiece—takes considerable expenditure to bring to favourable position. (Question of economy.)

Avoidable

1. Suitable tools not provided.
2. Suitable aids not provided.
3. Poor design for work:
   (a) in planning,
   (b) in execution by worker.
4. Suitable tools and aids are not being used.
5. Tool not properly adjusted for the worker.

Work design and work monitoring

Favourable bodily posture

Results

economic — Adverse effect — health

Figure 23.   Classification of body postures into three groups.
From Sämann (1970).

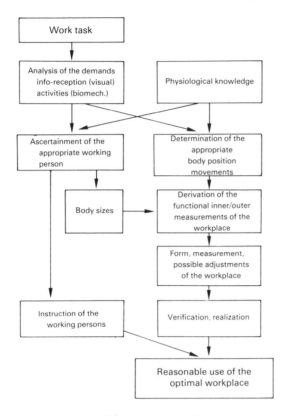

Figure 24.   A systematic design approach for optimization of body postures.

In order to find ergonomically reasonable solutions a systematic procedure during the process of design is necessary although conditions often necessitate compromise. An ergonomically favourable posture cannot be reached in each case. Allowing for real-world conditions, Sämann (1970) classifies unfavourable body postures into those which are unavoidable, partly avoidable and avoidable. Figure 23 shows the criteria used for classifying posture into one of these three groups. The classification is made by recording unfavourable postures in the actual working conditions. The author points out the detrimental effect of unfavourable postures on health, and also from an economic point of view.

In a systematic procedure, however, it is possible to consider causal influence factors of unfavourable postures or to exclude them. Figure 24 shows that an analysis of posture demands which arise within a given working task from the areas of *information* flow (type and arrangement of information sources, sight geometry, lighting, etc.) as well as the *mechanical activity* (the type of working tools, work objects and work surfaces and the resulting necessary movements, forces and individual postures), present a basis for optimization. Consideration of the natural physiological characteristics and limitations is a prerequisite in the resulting design.

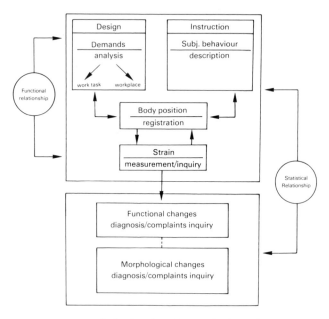

Figure 25. A systematic approach for describing functional relationships in the causal context of effects of body postures.

An ergonomically 'optimal body posture' is thus always adapted to a given activity and/or relevant demands, and has the characteristics that it considers equally the aims of *optimal strain* and *optimal efficiency*, it is not forced, it is generous and can be adapted or changed such that the three-dimensional visual and manual relationships in the work activity stay favourable.

A number of specific evaluation methods and criteria of postures were presented in Section 4. The illustrative examples show that a posture is the result of a causal chain of influence factors and subjective aspects. *Adequate training* of the working person (e.g., in handling loads or a reasonable use of given adjustment possibilities at the workplace) is therefore as important and necessary as the design of the work system itself.

7.   Conclusions

The ergonomic methods for the evaluation of postures, as the previously discussed systematization clarifies, are based on differing approaches. Figure 25 shows (Mainzer 1984) that a systematic approach for describing functional relationships in the context of the causal chain of demands and resulting strain, makes possible an evaluation of short-term effects of postures and their influence factors, and thereby offers a basis for a reasonable design.

As opposed to this, the correlation between functional relationships and possible morphological changes is burdened with a number of undetermined factors. The coordination of the two columns presented in Figure 25 can be achieved on the basis of statistical methods. The relationships that can be assumed between short-term measurable muscle strain, subjective complaints and morphological changes can be shown.

Comprehensive evaluation of body postures on the basis of a single methodological approach is not possible. The ergonomic evaluation approach is systematically oriented to an analysis of relationships between causes of body postures and their effects.

In the synthesis of the knowledge obtained two design approaches are necessary; through the design of both the work activity and work place, objectively necessary postures can be positively influenced and furthermore, subjective behaviour can be optimized.

On this basis ergonomics offers an instrument for a systematic optimization of postures.

Chapter 19
Torque of the Dorsal Erectors and the Compressive Load on the L5/S1 Disc at Inclined Postures and When Lifting

Janko Susnik and Tatjana Gazvoda

Health Centre of Carinthia, Department of Occupational Medicine, Ob Suhi 11, 62390 Ravne na Korosken, Yugoslavia

1. Introduction

"Avoid lifting heavy goods" or "avoid lifting loads exceeding 10 kg" are recommendations made by individual physicians and medical boards. When making such recommendations, it is unlikely that authorities think of the fact that 30 kg of feathers is a bulkier load than 30 kg of lead: such an amount of feathers forms a voluminous package that can be held only with stretched arms, whereas lead can be lifted close to the body. To use the language of mechanics: the lead load is on a short lever-arm, the feather load on a long one, and, consequently, the torque caused by the gravity of lead is different from that caused by the gravity of feathers.

This paper presents a simple and quite applicable method for computing the torque and the compressive load on the L5/S1 disc in bent positions and when lifting loads. The results are compared with tolerance limits: for continuous work allowable loads are those not exceeding 30% of such extreme loads, and for peak periods the allowable loads are those not exceeding 60% of the extreme load.

2. Experimental methods

The sample tested consisted of 29 young male volunteers, selected at random; 14 involved in testing with the load on a short lever-arm and 15 in testing with the load on a long lever-arm. Statistically, the two groups did not show any significant anthropometric differences, especially not in the weight of the head, neck, trunk and upper extremities above the fulcrum in the hip ($t = 0.808$) and above the fulcrum in the L5/S1 spine ($t = 0.909$).

The torques of the dorsal erectors and the compressive loads on the L5/S1 disc were studied by using a series of symmetrical two-dimensional biomechanical models. The limits of the body segments of the subjects were marked with perforated markers and the volumes of the body segments were computed by the immersion method. The mass of each of the segments was determined by multiplying the volume by the specific weight. The L5 spina was located with the fingers and a lateral marker was stuck on it. Then the subjects were instructed to take positions with trunk inclinations of 0–15, 30, 60 and 90°. The angles were adjusted with a protractor between the vertical line and the line between the two anthropological points; the acromion and the trochanterion. The knees were kept erect. The subjects bent without external load and with loads of 10, 20, 30 and 40 kg. The load with a short lever-arm was represented by a cylindrical container containing different amounts of lead shot, and the load with a long lever-arm was represented by a wooden case filled with different amounts of lead plates.

The so-called 'back-bent–knee-erect' type of lifting was chosen because in this case the torque of the dorsal erectors is maximal, but it does not differ very much from the 'back-bent–knee-bent' type seen very often in practice. In the 'back-erect–knee-bent' type, on the other hand, the torque of the dorsal erectors is minimal.

The subjects were photographed laterally with a 50 mm lens from a distance of 6 m. Colour slides were projected in the ratio 1:10 on a graphpaper and 'contour-grams' were drawn (Figure 1). By using a geometrical method, the centres of

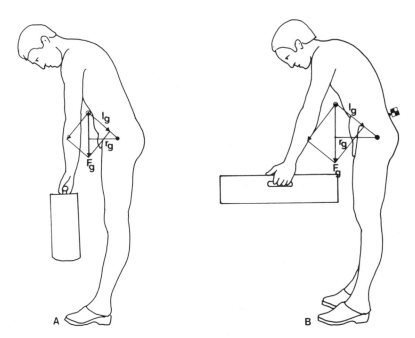

Figure 1.   'Contourgrams' of two different subjects. Subject A holds the load with a short lever-arm and subject B with a long lever-arm. F_g is the gravitational force of the masses of the head, neck, trunk, upper extremities and external load, l is the lever and r is the lever-arm of this force. The marker on the right-hand figure shows the L5 spina.

gravity of the segments and the common centres of masses above the fulcrums in the hips and in the lumbo-sacral spine were located. (Dempster 1961, Donskoi 1975). To calculate the torques, the hips were taken as fulcrums in all positions, and to calculate the pressures on the L5/S1 disc this was taken as the fulcrum, although from the upright position to 27° man bends in the lumbo-sacral spine, and from 27° onwards two-thirds in the hips and one-third in the spine (Ayoub and McDaniel 1971).

Torque is calculated according to the rules of mechanics. The line between the fulcrum and the centre of the masses of the head, neck, trunk, upper extremities and external load represents the lever (l_g), its horizontal projection representing the lever-arm (r_g). The product of the lever-arm and the gravitational force of the masses above the fulcrum (F_g) is their torque (M_g). In the state of equilibrium this torque is the same as the muscle torque (M_m). Muscle torque is the product of the force of the dorsal erectors (F_m) and their lever-arm (r_m) estimated by most authors at 0·06 m (Troup and Chapman 1969). That is,

$$M_g = F_g r_g \qquad (1)$$

$$M_m = F_m r_m \qquad (2)$$

and in equilibrium,

$$F_g r_g = F_m r_m \qquad (3)$$

The maximum voluntary isometric muscular strength was measured in positions analogous to the experiments (Warwick et al. 1980). The tested person stood upright with erect knees (7·5°), leaning against a wall or bending at angles of 30, 60 and 90°, respectively, and actuated a dynamometer according to the standard procedure (Kroemer 1977). This strength was needed as a capability limit.

To calculate pressures on the L5/S1 disc, the formulae published by Roozbazar (1975) were supplemented. The advantage of these equations is that they need no functional radiographs.

The supplemented equations read as follows:

$$F_m = \frac{[(-p \times 483 r_p)/760] + [F_{LS} r_{TLS}) + (F_B r_B)]}{r_m} \ (kg) \qquad (4)$$

$$F_p = (p \times 483)/760 \ kg \qquad (5)$$

$$F_c = F_m - F_p + (F_{TLS} \cos \alpha) + (F_B \cos \alpha) \ (kg) \qquad (6)$$

where

F_m = extension force of the dorsal erector (kg);
F_p = force of intra-abdominal pressure (kg);
F_c = resultant compressive force on the L5/S1 disc (kg);
p = intra-abdominal pressure (mmHg);
483 cm^2 = average cross-section of abdomen (Eie and When 1962);
r_p = lever arm of intra-abdominal pressure F_p. Morris et al. (1961) propose a

length of 11 cm but we prefer the equation we obtained on the basis of Chaffin's (1969) data and which takes into consideration the change of length of lever-arm in inclination of trunk:

$$r_p = 15 \cdot 07 - 0 \cdot 09 \, \theta \, (\text{cm}) \tag{7}$$

where θ is the angle between the axis of the trunk and the horizontal line: consequently, $\theta = 90 - \alpha$, where α is the angle of inclination of the trunk:

F_{TLS} = gravitational force of the body mass above the fulcrum L5/S1 (kg);
r_{TLS} = force arm F_{TLS} (cm);
F_B = gravitational force of the mass of external load (kg);
r_B = force arm F_B (cm);
r_m = force arm F_m (6 cm; Troup and Chapman 1969);
α = inclination of trunk in degrees.

Instead of using an average cross-section of abdomen of 483 cm² it is preferable to calculate an approximation (Schultz *et al.* 1982 b):

$$S_{\text{abd}} = \frac{WA \times DA}{100} \, 46 \, (\text{cm}^2) \tag{8}$$

where WA is the width of the abdomen (laterolateral diameter) (cm) and DA is the depth of the abdomen (ventrodorsal diameter) (cm). (In the original, Equations (4), (5) and (6) relate to the pressure on L4/L5 disc.)

3. Results

Tables 1–4 show arithmetic means and standard deviations of empirically obtained torques or pressures on the L5/S1 disc.

Individually empirically obtained data were processed by using step-by-step regression analysis. After several tests, four equations were chosen which corresponded most to biomechanical functions.

When lifting a load with a short lever-arm:

$$\text{torque } M_m = -112 \cdot 1631 + 2 \cdot 386 \, \alpha + 1 \cdot 480 F_{TK} + 3 \cdot 295 F_B$$
$$+ 7 \cdot 153 \log F_B \, (\text{N m}) \tag{9}$$

Significance of the equation	95%
Determination coefficient	0·9215
Multicorrelation coefficient	0·9599
Standard error of dependent variable	26·3153
Error for 95% significance ($\pm 1 \cdot 96$ SE)	51·5779

When lifting a load with a long lever-arm:

torque $M_m = -119 \cdot 3624 + 2 \cdot 374 \, \alpha - 0 \cdot 012\alpha^2 + 2 \cdot 501 F_{TK}$

$$+ 10 \cdot 171 F_B - 57 \cdot 864 \log F_B \text{ (N m)} \qquad (10)$$

Significance of the equation	95%
Determination coefficient	0·8287
Multicorrelation coefficient	0·9103
Standard error of dependent variable	51·0358
Error for 95% significance (±1·96 SE)	100·0303

When lifting a load with a short lever-arm the pressure on the L5/S1 disc is as follows:

$$F_c = 10(-331 \cdot 1756 + 5 \cdot 529 - 0 \cdot 054\alpha^2 + 7 \cdot 182 F_{TLS}$$

$$+ 6 \cdot 898 F_B - 0 \cdot 006 F_B^2) \text{ (N)} \qquad (11)$$

Significance of the equation	95%
Determination coefficient	0·9216
Multicorrelation coefficient	0·9553
Standard error of dependent variable	33·8825
Error for 95% significance (±1·96 SE)	65·8825

When lifting a load with a long lever-arm the pressure on the L5/S1 disc is as follows:

$$F_c = 10(-255 \cdot 6384 + 3 \cdot 055 \, \alpha - 0 \cdot 04\alpha^2 + 7 \cdot 086 F_{TLS} + 9 \cdot 429 F_B) \text{ (N)} \qquad (12)$$

Significance of the equation	95%
Determination coefficient	0·9506
Multicorrelation coefficient	0·9750

Table 1. *Torque (N m) at inclinations of 0–15, 30, 60 and 90° when lifting external loads of 0, 100, 200, 300 and 400 N with a short lever-arm.*

Angle (°)	F_B (N)				
	0	100	200	300	400
0–15	−4·8[a]	22·7	41·7	60·6	79·4
	(11·1)	(9·1)	(10·6)	(11·8)	(14·8)
30	59·6	87·8	116·8	144·1	173·5
	(12·2)	(15·2)	(17·2)	(21·3)	(23·7)
60	101·8	144·6	186·0	228·5	267·6
	(20·5)	(18·3)	(21·7)	(35·4)	(32·7)
90	141·9	189·1	239·4	288·9	336·4
	(19·3)	(16·2)	(19·2)	(22·7)	(28·2)

[a] Arithmetic means; the figures in parentheses are the standard deviations.

Table 2.   Torque (N m) at inclinations of 0–15, 30, 60 and 90° when lifting external loads of 0, 100, 200, 300 and 400 N with a long lever-arm.

Angle (°)	F_B (N)				
	0	100	200	300	400
0–15	6·7[a]	77·2	137·0	189·5	241·7
	(7·8)	(15·2)	(19·6)	(13·7)	(21·1)
30	64·9	131·1	186·1	225·3	271·9
	(12·1)	(16·2)	(25·6)	(17·4)	(14·6)
60	111·3	176·1	224·4	273·0	315·1
	(24·3)	(14·1)	(22·0)	(24·2)	(19·8)
90	146·6	206·9	255·8	302·0	359·9
	(21·5)	(22·0)	(22·0)	(29·5)	(24·0)

[a]Arithmetic means; the figures in parentheses are the standard deviations.

Table 3.   Pressure on the L5/S1 disc (N) at inclinations of 0–15, 30, 60 and 90° and when lifting external loads of 0, 100, 200, 300 and 400 N with a short lever-arm.

Angle (°)	F_B(N)				
	0	100	200	300	400
0–15	611·3[a]	1209·0	1698·9	2183·5	2664·3
	(199·4)	(286·5)	(308·0)	(335·9)	(367·5)
30	1256·0	1845·9	2393·2	3035·2	3628·8
	(397·6)	(417·9)	(517·1)	(509·8)	(562·2)
60	1152·7	1893·3	2644·4	3437·0	4134·0
	(346·0)	(383·8)	(438·0)	(516·5)	(578·8)
90	174·6	1062·3	1866·7	2624·7	3468·4
	(375·6)	(406·3)	(447·2)	(419·5)	(498·7)

[a]Arithmetic means; the figures in parentheses are the standard deviations.

Table 4.   Pressure on the L5/S1 disc (N) at inclinations of 0–15, 30, 60 and 90° when lifting external loads of 0, 100, 200, 300 and 400 N with a long lever-arm.

Angle (°)	F_B(N)				
	0	100	200	300	400
0–15	637·3[a]	1900·9	2954·0	3972·4	4863·3
	(268·9)	(331·1)	(414·4)	(436·5)	(594·4)
30	1229·4	2387·3	3304·5	4132·7	4979·6
	(405·4)	(464·5)	(442·3)	(501·8)	(369·5)
60	1108·9	2236·1	3164·2	3952·9	4662·4
	(472·7)	(419·7)	(496·9)	(517·7)	(532·0)
90	265·4	1329·5	2209·6	3221·0	3728·5
	(465·6)	(596·3)	(546·0)	(894·1)	(632·2)

[a]Arithmetic means; figures in parentheses are standard deviations.

Standard error of dependent variable   31·8056
Error for 95% significance (±1·96 SE)   62·3389

Symbols in these equations have the same meaning and dimensions as in Equations (4)–(8).

Figures 2–5 show arithmetic means and standard deviations of experimentally obtained data. Also shown are theoretical curves calculated from Equations

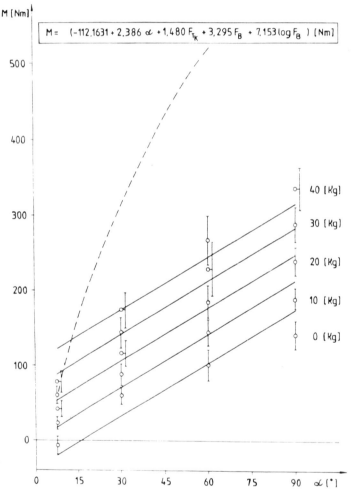

Figure 2. Extend loads with a short lever-arm. The abscissa shows the angle of inclination of the trunk in degrees, the ordinate presents the torque in Newton metres. The small circles represent arithmetic means of torques in holding loads of 0, 10, 20, 30 and 40 kg with a short lever-arm. Standard deviations close to them are shown. The equation at the top of the diagram, computed from experimental data, was the basis for the computation of the theoretical curves shown on the diagram. The dashed-line curve represents the average maximum torques of the test group.

(9)–(12) in which arithmetic means of the weights of respective segments of the tested persons were taken as F_{TK} and F_{TLS}.

4. Discussion

In the tests we analysed a series of static events although lifting in reality is a dynamic action. Smith *et al.* (1982) believe that, by ignoring acceleration, loads are

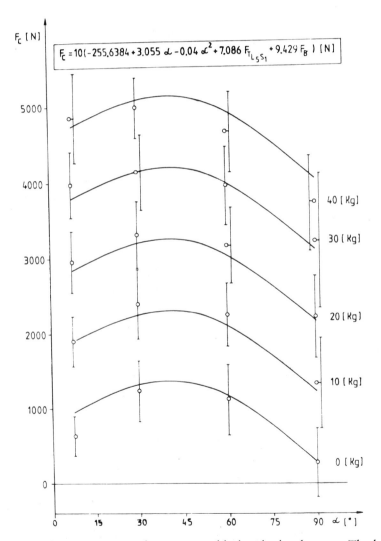

Figure 3. Analogous to Figure 2, relating to external loads with a long lever-arm. The theoretical curves for the loads of 30 and 40 kg show exaggerated torque and are therefore inapplicable. The equation is applicable only to loads not exceeding 25 kg.

either overestimated or underestimated. According to the experience of Andersson *et al.* (1976), increases in speed and acceleration augment the intrinsic pressure on the disc. In contrast, Chaffin (1969) and Martin and Chaffin (1972) claim that acceleration in load lifting is negligible and there are no significant differences if lifting is studied as a series of static events.

We studied the so-called 'back-bent'–knee-erect' type of lifting. The torque

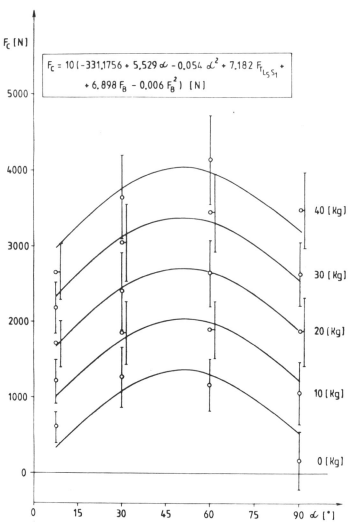

Figure 4.   Extend loads with a short lever-arm. The abscissa shows the angle of inclination of the trunk in degrees, the ordinate presents the pressure on the L5/S1 disc in Newtons. The small circles represent arithmetic means of pressures in holding loads of 0, 10, 20, 30 and 40 kg with a short lever-arm. Standard deviations close to them are shown. The equation at the top of the diagram, computed from experimental data, was the basis for the computation of the theoretical curves shown on the diagram.

while lifting 'back-bent–knee-erect' or 'back-bent–knee-bent' is rather similar, whereas it is lower in the 'back-erect–knee-bent' type due to the shortening of the lever-arms r_T and r_B.

In spite of that, the 'back-erect–knee-bent' type of lifting is not a better alternative. The knees are subjected to a greater load (Carlsöö 1980) and a greater mass of lower extremities has to be lifted (Leskinen *et al.* 1983). If lifting is done from a 45° inclination to an upright position, the intrinsic pressure on the disc is the same as in the case of lifting with erect or bent knees (Andersson *et al.* 1976). In the 'back-erect–knee-bent' type, a proportionally lower torque of the dorsal erectors entails also an essentially higher torque in the shoulders (Yates *et al.* 1980). Even the torque in the wrist joint depends on the method of lifting (Garg *et al.* 1980).

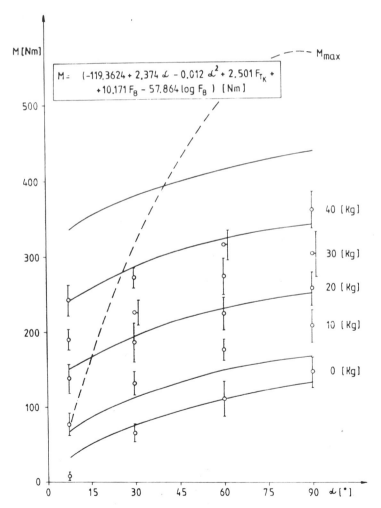

Figure 5.   Analogous to Figure 4, relating to external loads with a long lever-arm.

Joints are protected by well-developed and active muscles. In other words, joints are more vulnerable if the muscles are weaker or their activity is reduced, as in the case of fatigue. Fatigue occurs after long-lasting hard dynamic work and after isometric strain when it exceeds 15–20% of the maximum muscular power (Rohmert 1960 a).

Equations (9) and (10) are used to compute the torque of the muscles maintaining external equilibrium. First, the lever-arm of the external load r_B is defined. The data of the inclination angle (α), the F_{T_K} force and the F_B force are applied in the equations. F_{T_K} is the gravitational force of the mass of the upper extremities, head, neck and trunk, above the fulcrum in the hips, or 65% of the body weight T_T in kg, respectively. F_B is the gravitational force of the external load (kg).

The computed torque is compared with the maximum voluntary torque, which is individually measured in an analogous position or obtained from the equation for average maximum voluntary isometric torque (M_{max}):

$$M_{max} = -24 \cdot 332 + 13 \cdot 414\,\alpha - 0 \cdot 076\alpha^2 \text{ (N m)} \tag{13}$$

Figure 2 shows curves representing the torques of loads with a short lever-arm. The curvatures are hardly noticeable, although they are distinctly indicated by the experimental points. With an inclination of 0–15° and with an external load of 40 kg, the torque computed from this equation is overestimated, but with an inclination of 90° it is underestimated. It is also overestimated in the case of an inclination of 90° without an external load.

Analogously, Figure 3 shows experimentally obtained torques and curves (computed from Equation (10)) for loads with a long lever-arm. The last term of the equation causes the difference between experimental and theoretical data for external loads of 30 and 40 kg. The difference is so big that we have limited the applicability of the equation to $F_B = 0$–25 kg.

For torques with a short lever-arm and long lever-arm loads, the authors have published polynomials which are more precise, but they can only be applied to compute torques relating to the average weight of segments (T_K) of the persons involved in the experiment, and cannot be used to determine torques relating to the individual weight of segments. These polynomials have an advantage over Equations (9) and (10) when torques are determined in a workplace analysis (Susnik and Gazvoda 1983, 1984).

Compressive loads on the L4/L5 disc were calculated by using the equations published by Roozbazar (1975) and modified by us. They have been studied by many other authors, e.g., Debevec (1973) and Chaffin (1969), but direct measurement such as described by Nachemson and Elfström (1970) or Andersson *et al.* (1976) has been used here. Measurements were carried out intravitally with a piezo-electric pressure transducer.

In the literature, data concerning pressures on discs of the lumbo-sacral spine differ considerably. Our results are in agreement with the pressures directly measured by the above-mentioned Swedish researchers and with the data given by Strait *et al.* in Debevec (1973). By the biomechanical method described by Chaffin

(1969), for which a functional radiograph of the spine is required, the pressure in situations analogous to ours is about 30% higher.

The curves calculated from Equations (11) and (12) are in good agreement with the experimental data. Both equations are perfectly applicable. In these formulae also, α is the angle of inclination, F_B is the gravitational force of external load and $F_{T_{LS}}$ is the gravitational force of the body segments above the fulcrum in L5/S1, or 61% of body weight (T_T).

Results obtained in this way have been compared with the pressures reported for discs in other work. For people between 40 and 50 years of age Perey (1957) states 5003–10 791 N; Sonoda (in Junghans 1979) 15 000 N in the case of axial pressure and 3940 N in the case of axial tension; Nachemson (1966) 4900–7530 N, Troup (in Chaffin 1969) 5270 N. For older people this limit has been reported as 2845–5199 N (Perey 1957), 1560–5800 N (Lin *et al.* 1978) and 3261 N (Ranu *et al.* 1979). Based on these data we determined the values in Table 5 as critical pressures. In our evaluations we accept the usual physiological criterion that in the case of a continuous 8 hour load, 30% of the maximum load is allowable, and in the case of a load occurring from time to time, 60% of the maximum load is permitted. The maximum load means the maximum voluntary torque or the critical pressure on the lumbo-sacral discs.

Table 5. *Critical pressure on the lumbo-sacral discs (N).*

Sex	Age	
	Up to 50 years	Over 50 years
Male	8000	3700
Female	6000	2800

5. *Summary*

The torque of the dorsal erectors and the pressure on the L5/S1 disc have been investigated testing a series of two-dimensional, symmetrical, static events, using the biomechanical method of contourgrams, and applying a geometrical construction of centres of gravity, i.e., mechanical vectors. Subjects were tested in upright and bent positions, with the trunk inclined 30, 60 and 90°, without external load or holding, in both hands, a load of 10, 20, 30 or 40 kg and with a short or a long lever-arm. Four regression equations were constructed from the experimentally obtained data and these were adapted for practical use. They can be used for satisfactory, valid and discriminative calculation of the torques of the dorsal erectors and of pressures on the L5/S1 disc for lifting loads in characteristic bent positions.

The user will choose the equation according to his evaluation of the lever-arm of the external load. The equation will be applied to the data on the weight of the

Table 6.   Torque of the dorsal erectors (M_m) *and the compressive load on L5/S1 disc* (F_C) *in a bent position and when lifting loads.*

1.0.   Type of lifting (position) (Method: observation)
    1.1. Back bent, knee erect   1.2. Back bent, knee bent   1.3. Back erect, knee bent

    Note: The equations do not hold for the back-erect–knee-bent type of lifting.

2.0.   Angle of inclination (α) (Method: protractor)

3.0.   Lever-arm $F_B(r_B)$ (Method: observation)
    3.1. Short   3.2. Long

4.0.   External load (F_B) (Method: weighing)
    0–40 kg except Equations (6.3) and (6.4)

5.0.   Body weight (T_T) (Method: weighing)
    Kilograms

6.0.   Equations
    6.1. Torque: short lever-arm (α up to 27°)
$$M_m = -112.1631 + 2.386\alpha + 1.480(61T_T/100) + 3.295F_B + 7.153 \log F_B \text{ (N m)}$$
    6.2. Torque: short lever-arm ($\alpha = 28$–90°)
$$M_m = -112.1631 + 2.386\alpha + 1.480(63.7T_T/100 + 3.295F_B + 7.153 \log F_B \text{ (N m)}$$
    6.3. Torque: lever-arm (α up to 27°)
$$M_m = -119.3624 + 2.374\alpha - 0.012\alpha^2$$
$$+ 2.501(61T_T/100) + 10.171F_B - 57.864 \log F_B \text{ (N m)}$$
    6.4. Torque: long lever-arm ($\alpha = 28$–90°)
$$M_m = -119.3624 + 2.374\alpha - 0.012\alpha^2$$
$$+ 2.501(63.7T_T/100) + 10.171F_B - 57.864 \log F_B \text{ (N m)}$$
    Note: Equations (6.3) and (6.4) hold for $F_B = 0$–25 kg
    6.5. Pressure on the L5/S1 disc: short lever-arm (α up to 27°)
$$F_c = 10(-331.1756 + 5.529\alpha - 0.054\alpha^2 + 7.182(61T_T/100) + 6.898F_B - 0.006F_B^2) \text{ (N)}$$
    6.6. Pressure on the L5/S1 disc: short lever-arm ($\alpha = 28$–90°)
$$F_c = 10(-331.1756 + 5.529\alpha - 0.054\alpha^2 + 7.182(63.7T_T/100) + 6.898F_B - 0.006F_B^2) \text{ (N)}$$
    6.7. Pressure on the L5/S1 disc: long lever-arm (α up to 27°)
$$F_c = 10(-225.6384 + 3.055\alpha - 0.04\alpha^2 + 7.086(61T_T/100) + 9.429) \text{ (N)}$$
    6.8. Pressure on the L5/S1 disc: long lever-arm ($\alpha = 28$–90°)
$$F_c = 10(-225.6384 + 3.055\alpha - 0.04\alpha^2 + 7.086(63.7T_T/100) + 9.429F_B) \text{ (N)}$$

7.0.   Limits and tolerances
    7.1. Torque
$$M_{max} = -24.332 + 13.414\alpha - 0.076\alpha^2 \text{ (N m)}$$
    Age 25 years: $M_{max} = 100\%$; age 40 years: 90%; age 50 years: 85%
    Women: 70% of men's M_{max}
    Continuous load limit: ($M_{max} \times 30$)/100 (N m)
    Discontinuous load (not more than twice in an hour) limit: ($M_{max} \times 60$)/100 (N m)

Table 6 (contd).

7.2. Pressure on the L5/S1 disc:
Determine the highest F_c which rises up to 50° (when with a short lever-arm), rises up to 40° and falls again up to 90° (when with a long lever-arm).
F_c (critical pressure) (causing damage to disc)
Up to 50 years: 8000 N (male); 6000 N (female)
Over 50 years: 3700 N (male) 2800 N (female)
Continuous load limit: 30%
Discontinuous load limit: 60% of the limit.

EXAMPLE
A 47 year old man: weighing 76·4 kg is lifting a case (long lever-arm) weighing ($F_B=$) 24 kg from a table. The angle of inclination varies from 65° to about 0° (upright position):

Equation (6.4)
$M_m = -119·3624 + 2·374 \times 65 - 0·012 \times 65^2 + 2·501 (63·7 \times 76·4)/100 + 10·171 \times 24 - 57·864 \log 24$
$= 270·20$
M_{max} from Equation (7.1) $= 526·48$ (N m), age correction 87% $= 458·04$ (N m)
The highest torque in lifting is at 65°, 270·20 (N m), which is 59·0% of M_{max} after age correction. Consequently, lifting is permitted more than twice an hour but less than continuously.

Equation (6.8)
$F_c = 10(-225·6384 + 3·055 \times 65 - 0·04 \times 65^2 + 7·086(63·7 \times 76·4)/100 + 9·429 \times 24$
$F_c = 3449·4$ N
But in lifting from an inclination of 65° to the upright position the movement passes the inclination of 40° where F_c is the highest (see 7·2). Analogously computed F_c for $\alpha = 40°$ is 3450·9 N— the difference is negligible.
The pressure amounts to 43% of the limit of 8000 N. Consequently, lifting is permitted more than twice an hour but less than continuously.
Note: Lifting, particularly lifting of loads with a long lever-arm, is often limited by the torques in the shoulders and knees

---

load, the body weight of the tested person and the angle of inclination of his trunk. The result is compared with the maximum allowable torque or pressure on the disc. Detailed instructions are given for practical procedures and evaluation in Table 6.

Acknowledgements

We thank the Communal Research Communities of Carinthia, the Slovenian Ironworks and the Health Centre of Carinthia for financing this research study.

Chapter 20
Change of Stature as an Indicator of Loads on the Spine

E. N. Corlett

Department of Production Engineering and Production Management, University of Nottingham, Nottingham NG7 2RD, U.K.

and J. A. E. Eklund

Division of Industrial Ergonomics, University of Linköping, S-581 83 Linköping, Sweden

1. Introduction

Methods which measure the effect of work on the individual are particularly interesting for ergonomists. Such methods, together with subjective assessments by the people doing the work, form a more solid ground for conclusions than expert judgements. The control of spinal load is important for defining good working conditions, but existing methods for measurement have limitations regarding equipment, personnel costs and field applicability. Eklund and Corlett (1984) proposed that the precise measurement of body-height shrinkage could be used to assess the effect of loads imposed on the spine.

Intervertebral discs have been shown to respond visco-elastically to loading (Hirsch 1955, Marholf 1972). Increased time under load and increased load will increase the disc height loss, and also affect the stiffness of the disc. The process is due to fluid loss and also, to some extent, structural deformation (Adams and Hutton 1983 a, Armstrong 1958). The reverse process takes place when the discs are unloaded (Krämer 1973). It is known that decreased disc height not only affects the mechanical properties of the joint but also decreases the space for the nerve paths and thereby causes increased risk of pain (Adams and Hutton 1983 a, Krämer and Gritz 1980).

As a result of the disc height change, the body height decreases during the day and recovers when lying in bed at night. The difference is approximately 15 mm (Bencke 1897, De Puky 1935, Tyrrell et al. 1985). There are reports that astronauts who spent 2 weeks in space without gravitational forces increased 5 cm in height.

Increased back load also increases the rate of body-height loss (Fitzgerald 1972, Forssberg 1899, Krämer and Gritz 1980). Eklund and Corlett (1984) showed, by using equipment and procedures sensitive enough to measure body height to within 1 mm, that stature changes are related to biomechanical loads. They also demonstrated different stature changes due to different loadings during 0·5–1·5 hours, using small groups of subjects. It has also been shown that stature changes are related to perceived discomfort and to exertion (Troup et al. 1985).

The method is suitable for the evaluation of workloads, postures and designs. It has been used in studies of chair design (Eklund and Corlett 1985, Williamson 1984). Advantages of the technique are that it has field applicability, the equipment is relatively inexpensive and does not demand highly trained and educated personnel to collect data. Further, the method is non-invasive and does not disturb people during the performance of their tasks.

Assessment of intradiscal fluid influx and efflux is possible with the method, which is of particular interest in relation to discussions about disc nutrition and its possible connection with disc degeneration (Holm and Nachemson 1982).

The purpose of this paper is to describe the apparatus and procedure currently used for height measurements, with the different sources of error, and to demonstrate the relation between body-height changes and spinal loads.

2. Methods and equipment

Apparatus and procedures

The original apparatus for measuring stature (Eklund and Corlett 1984) has been modified in many technical aspects to achieve greater control over the measurements. Variations of the apparatus are in use in three places in the United Kingdom and one in Sweden, and experience from these other centres has been incorporated with that of the authors to indicate the current position. The variations between the various pieces of apparatus, all of which operate on the same broad principle, are intended to explore different methods for achieving precision in, and control over, measurements. The Linköping version is illustrated in Figure 1 and described further in the text.

The basic principles of the method are to place a standing subject, leaning slightly backwards, on a platform. The weight distribution between the heels and soles is controlled, as is the position of the subject's back at a number of points. A measuring device has its contact surface lowered onto the top of the subject's head to measure stature. The changes in height are obtained by subtracting successive readings, the absolute value of the stature is not measured. By instruction and practice the subjects are taught to adopt a consistent posture from trial to trial. Several readings are taken at each trial.

The version of the apparatus shown in Figure 1 consisted of a stiff rectangular platform 90 cm × 60 cm, on which a 210 cm high aluminium pipe was attached at right angles. It can be removed for transport. Backward tilt of the whole apparatus was obtained, between 0 and 20°, by adjustment of the two front legs. It was found that a small backward tilt permitted better muscular relaxation but kept substantially the same compressive load on the spine. Positions which unloaded the spine gave a quick recovery which disturbed the measurements. Angles of tilt between 5 and 15° have been used and found to be satisfactory. Increased tilt was perceived to be more uncomfortable and it also proved to be more strenuous to get into and out of the measurement position. An acceptable value of 15° was chosen for the study reported later in this chapter.

Figure 1.   The Linköping apparatus and a subject in position for measurement.

An alternative form of the apparatus is shown in Figure 2. It incorporates a broader back board and provides adjustments for a backward lean of some 30°. The measuring head, described later, differs from that shown in Figure 1 but otherwise the systems for measurement and control are identical. On the right of Figure 2 is the mirror used by the subject to align the spectacles used for head position control (see Figure 3).

A wooden plate positioned the feet at a 20° included angle, with the heels 4 cm apart. The soles were supported on a weighing scale, with its top surface on the same level as the fixed surface for the heels. This arrangement allowed accurate repeated foot positions and measurement of the weight distribution between heels and soles. The weight on the soles was allowed to vary ±2·5 kg from each subject's preferred or average reading during training trials. The importance of this factor for the variability of the measurements is still uncertain.

Knee instability, and as a consequence muscle tension in order to keep the knees extended occurred with some subjects. The wooden plate for the heels was then moved slightly forwards, which avoided the problem.

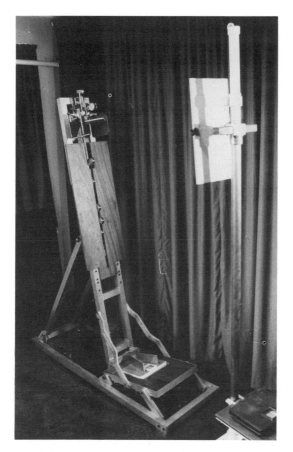

Figure 2. The Nottingham apparatus, with the different form of height measurement and broader back support.

The subject leaned against six supports, individually adjustable for height and protrusion. They were designed for the sacrum, the midlumbar spine, the lower thoracic spine, the midthoracic spine, the cervical spine and the head, respectively, and could accommodate most individuals' back profiles. Scales marked the positions of the supports.

For the instrument shown in Figure 1 the sacral support consisted of a 9 cm high, 5 cm wide and 1·3 cm thick plate attached to a 20 cm high and 40 cm wide wooden plate. The larger plate was incorporated in order to aid getting into position and to prevent instability when standing against the supports. Without it, the feeling was of leaning against a narrow vertical support, and muscular activity arose in some subjects to avoid rotation about the vertical axis.

The four back supports were 2 cm × 5 cm, except for the midthoracic support, which was 2 cm × 10 cm. They were all rounded in the sagittal plane. One development from the original apparatus was that the lower thoracic support was added, placed approximately between the midthoracic and lumbar supports. Comparative tests showed that adding this fourth back support gave a tendency for less variability in the measurements. It seemed as if the posture of the cervical spine caused more variability than any other factor. There were also indications that the lower thoracic support could prevent some variability, due to rib-cage tilt.

Thin flexible brass metal sheets were partly electrically insulated, and mounted with a small air gap on the four back supports. These worked as microswitches in order to indicate contact and very small forces were needed for this. When all four switches made contact at the same time, a light was illuminated, only visible to the experimenter. Contact indicators ensured a reproducible posture in the sense that contact with all four back supports was obtained, but there still remained another source of error. Increased pressure on one support meant a small change in posture, which could not be prevented. This was especially critical in the cervical region where there was some contact with soft muscle tissue.

The head support consisted of two wooden plates, 12 cm × 10 cm, mounted vertically in a V, with 90° between the plates (see Figures 3 and 4). This design of the head support increased the precision of the positioning of the head in repeated trials in both the lateral and anterior-posterior directions.

A spectacle frame with markers (bead, front sight), to be aligned by the subjects when looking to the front in a mirror, controlled variability of head rotation and flexion/extension. The head support and the spectacles position the head, or rather the point to which body height is measured, with an accuracy better than ±2 mm in the horizontal plane.

The height measurements are taken by lowering a measurement head, connected to a linear transducer, on to the top of the subject's head. The transducer measures within a range of 140 mm and can be positioned on the main frame every 10 cm, in order to measure subjects between 145 and 195 cm height. Instead of using a plate, which could cause some variability due to the amount of hair which can be trapped beneath it, the measurement head shown in Figure 3 was constructed. It consisted of a 90 g weight, under which five cylindrical pins, diameter 1 mm and length 17 mm, were attached. Four pins formed the corners of a 5·5 mm square, and the fifth was placed in the centre. This construction was found to penetrate the hair and compress the skin on the skull in a reproducible manner. It did not cause any feelings of pressure or discomfort on the head of the subject.

Another version of the measuring head (Figure 4) uses a counterbalanced flat plate mounted on a precision linear system of rollers. The plate and roller unit can be repositioned on the main frame at 2·5 cm intervals to cover a wide range of the adult population. The measuring head can use either an engineer's precision clock gauge, reading to 0·01 mm, or a linear transducer. The counterbalance system controls the pressure on the scalp. The subjects are required to fold their arms, which decreases the movement of the shoulders and is one means of creating a more reproducible posture. It also seemed that less variability arose if the subjects were instructed to inhale and stretch themselves a little or imagined that they were taller when they 'rolled' into position and made contact with the back supports.

The following procedure was adopted when making the height measure-

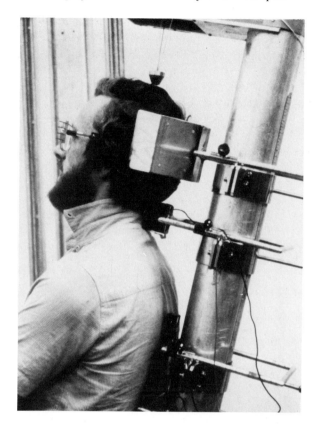

Figure 3. *The measurement head, the spectacle frame with markers, the head support and the cercival support during a measurement.*

ments. The subject first positioned the feet, and then the sacrum made contact with the sacrum support. Subjects then folded their arms over their chest and 'rolled' into contact with the back supports, going from bottom to top of the spine. The head angle was then adjusted, subjects exhaled to relaxed level, relaxed muscle tensions and said when they were ready for measurement. The measuring head was lowered onto a subject's head during 1 s, and the reading was taken after another second. This whole procedure was repeated five times, the subject stepping off and back onto the apparatus each time, which gave a set of five measurements. It was essential that the subjects repositioned themselves totally between measurements, in order not to introduce any systematic error. If a subject considered that a measurement felt strange or different, a new measurement was taken. With only a little practice subjects were able to adopt the measuring posture on each occasion very consistently. With some subjects it has been found helpful to get them to inhale, stretch slightly upwards or 'think tall' as they fold their arms and position themselves on the instrument. This is just prior to the exhalation and relaxation phases, which then follow prior to measurement.

The experimenter noted the height readings, the weight on the scales and the signal for contact with all four back supports. If the weight on the scales was outside the approved interval or if contact was not indicated for even one back support, the measurement had to be repeated.

Two measurements were taken before each set of five, for checking the support adjustments, the procedure and the equipment. In spite of the fact that the positions of the supports were noted for each subject and set prior to the arrival of the subject, small final adjustments were often found necessary. This was carried out just before the start of the experimental session.

Training and instruction for 20–60 min were given to each subject a few days before his first measurement session. The subjects were taught to perform the procedure without any commands by the experimenter. The positions of the supports on this occasion were noted for future sessions with that subject.

All measurements in the study to be described started 75–90 min after the subjects rose from bed in the morning. Their sleeping hours, morning activities and travel to the laboratory were close to their normal pattern, and comparable every experimental day, according to instructions given in advance. Measurements were taken before and after the 45 min experimental session.

3.   The experiment

Eight subjects performed four experimental sessions each. Shoulder loads of 0, 10, 20 and 25 kg were administered by using a waistcoat with pockets for lead weights, which were distributed equally on the front and back. The subjects stood upright, but were free to move around a few steps when they wished. The procedure described above was followed.

Subjects

Six male and two female subjects participated. Their mean age was 30 years (range 18–34 years) and mean height 178 cm (range 166–189 cm).

Results

The stature loss was closely related to the shoulder load and therefore also to the biomechanical load. The regression line, marked in Figure 5, was described by the equation $y = -0.093x - 1.68$. The standard deviation for the regression line was calculated to be 0.17 mm. The differences between the four loading conditions were all significant at the 5% level (*t*-test).

The standard deviation for the height measurements, based on all sets of five height measurements and all subjects, was 0.62 mm. Individual differences in the variability of height measurements can be described by their standard deviations, which ranged from 0.43 to 0.73 mm. These values are based on all the sets of five height measurements for each subject. The average stature change from the four

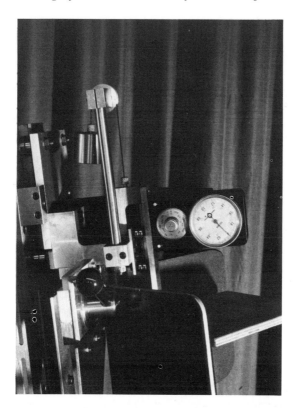

Figure 4.   The measuring head of the Nottingham version. The counterbalanced head contact plate runs on a four-roller precision track. The angled plates to locate the head laterally are adjustable horizontally and the entire measuring head can be raised or lowered in 2·5 cm steps.

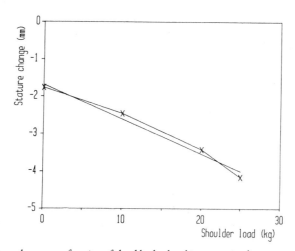

Figure 5.   Stature change as a function of shoulder load and its regression line.

loading conditions plotted as a function of the subjects' ages showed a decrease of stature loss with increasing age. The equation of the regression line was calculated to be $y = 0 \cdot 17x - 8 \cdot 0$. Further, considering the variability of each subject's average stature change, individual differences can be clearly noted in Figure 6.

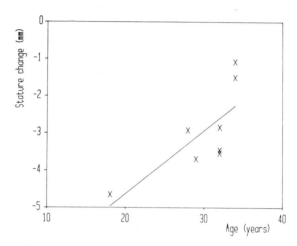

Figure 6.   Average stature loss as a function of the subjects' ages and its regression line.

By comparing the heights measured on each of the four mornings at the start of each experiment, an indication of the consistency of the control of sleeping time and morning activity was obtained. It must be remembered that just because the same heights were found on each day cannot lead to the conclusion that the previous load–time course was the same for each of the subjects. However, it is a control measure in the sense that it suggests that the degree of shrinkage experienced by each subject's spine was closely similar for each of the experimental days. Thus one aspect of the state of the subject's activities and loading was identifiable. It can be seen in Table 1 that the range of morning height differences was $3 \cdot 34$ mm on average and the standard deviation $1 \cdot 56$ mm. This variation for each subject showed some agreement with the standard deviations of height measurements within all sets of five readings. Any variation which could have occurred as a result of having to adjust the position of the supports is included in the results from these morning height measures.

Table 1.   The range and standard deviation of the eight subjects' stature in the four mornings.

	Subject								
	1	2	3	4	5	6	7	8	Average
Range (mm)	$2 \cdot 2$	$3 \cdot 9$	$2 \cdot 5$	$2 \cdot 5$	$2 \cdot 4$	$3 \cdot 5$	$5 \cdot 3$	$4 \cdot 4$	$3 \cdot 34$
Standard deviation	$1 \cdot 04$	$1 \cdot 59$	$1 \cdot 22$	$1 \cdot 06$	$1 \cdot 27$	$1 \cdot 51$	$2 \cdot 32$	$2 \cdot 01$	$1 \cdot 56$

The experiment involved a total of 64 sets of height measurements. The fifth measurement was lower than the first in 50 of these, and the opposite occurred in the other 14. On average, the fifth measurement was 0·71 mm lower than the first. The differences were statistically significant at the 5% level.

The standard deviation of the 64 sets of measurements was compared for the first, second, third and fourth experimental session. No significant differences were found, but there was a tendency for a decrease in the standard deviation to occur with an increase in the number of experimental sessions. There was also an indication that the standard deviation could increase if a longer time elapsed since the previous experimental session.

4. Discussion

The method can be seen to have several advantages. It has been demonstrated to be easy to use, it demands no very expensive equipment and no particularly specialized or trained personnel. No restrictions are imposed on the subjects, as, for example, being attached to or having to carry equipment.

The method is also non-invasive and has been shown to be sensitive in identifying changes in body loads. Of particular interest is the fact that the method is a measure of the *effect* on the person, an aspect which has a particular attraction for ergonomists. It is suitable for comparative ergonomic evaluations of tasks, workplace and equipment designs, when the subjects are used as their own controls.

The standard deviation of the height measurements obtained with the present apparatus and procedure indicates sufficient precision to identify statistically differences in imposed conditions. In spite of the technique having been in use since 1982, there is no doubt that many features can still be improved on the height measurement apparatus and also that the procedure for measurement control can be improved further. The results from this experiment, and of others now in the literature, point to the importance of effective training of the subjects as a means of further decreasing the variability of the measurements. The effect of additional training sessions is one possibility which requires investigation.

The systematic difference between the first and the fifth measurement was noticed not only in this experiment but also in earlier experiments during the development of the method. This reflects the spinal loading during the period of the measurements, which normally took up to 5 min. There might also be other factors in subject behaviour causing measurements of shorter body height towards the last measurements in the set of five. The results emphasize, however, the importance of performing the measurements consistently, quickly and without interruptions and disturbances. It also appears important that the subject is not stressed or feels that there is a shortage of time. Motivation towards the experiment is also important.

The results have demonstrated clear differences between individuals regarding the variability of their measurements. The characteristics of 'good' and 'bad' performers have not yet been determined. However, the effects of individual differences, as well as any characteristic effects of sex, age, work experience, etc., are all factors which have yet to be extensively studied.

The properties of the discs depend on their loading history. Proper control and experimental design must ensure that errors are not introduced due to differences in disc conditioning. Important influences are sleeping time, time between getting up in the morning and the start of the experiment, and also any activities giving rise to back loadings before the experiment. In many cases, loadings and activities of a subject are difficult to control during a day. It has therefore been the practice wherever possible, and was decided for this study, to perform the experiments in the morning, when the discs are more pliable and responsive to loadings. These influences cannot be totally controlled, but it has been proposed in this paper that the variation of the subjects' heights at the start of the experiment can be used as a control measure of the initial back loading. It is also essential to take notes about actual sleeping time, getting up time, pre-experiment activities and the time of the start of the experiment.

The choice of 45 min experimental sessions was a compromise between time availability, subject acceptability and experiment efficiency.

The experimental design is most important, a fact which is emphasized by the substantial age effect and individual differences demonstrated here. Using the subjects as their own controls makes the method very suitable for comparisons between different loading situations. It was also demonstrated that the method permits the assessment of individual responses to spinal loading, a fact which may open new possibilities in future back-pain research.

Chapter 21
ARBAN—A Force Ergonomic Analysis Method

Michael Wangenheim, Björn Samuelson† and Henrik Wos

The Research Foundation for Occupational Safety and Health in the Swedish Construction Industry, Box 94, S-182 11 Danderyd, Sweden and

†Department of Clinical Physiology, Akademiska Sjukhuset, Uppsala University, Uppsala, Sweden

1. Introduction

ARBAN is a method for systematic force ergonomic analysis of work. The project was started in 1979, reports of the first stage appearing in Holzmann (1982) and Holzmann and Wangenheim (1983). Since then, thanks to numerous case studies and further research, the method has developed to its present form.

ARBAN, in its present state, is a 'person-adjusted' ergonomic method, forming a bridge between physical measurements, psychophysiological measurement methods and statistics. The purpose of ARBAN is to be a simple and reliable method of studying work situations, from a force ergonomic aspect. Experienced ergonomists and persons with limited knowledge of ergonomics can carry out qualified ergonomic analyses, after participating in an ARBAN course. The work studied is not disturbed or affected by the analysis process. The results obtained are easily evaluated, giving a clear picture of the problems and thus forming a basis for corrective measures in the future.

An ARBAN analysis comprises: (a) time-study leading to the choice of representative cycles; (b) video filming the chosen cycles; (c) 'freezing' the film at short intervals; (d) observational analysis and coding of the loads due to posture, force, vibration, and static loads with the help of guidelines, reinforced by certain measured data; (e) running a computer programme; and (f) combining the results of the cycles selected into an overall picture.

The method is intended for use in identifying, defining and documenting problems at workplaces, planning and optimizing production lines and workplaces and developing and designing machines and products. Thus it relates to routine production engineering and produces results that can be interpreted by engineers, designers and health-care personnel.

Readers familiar with the original form of ARBAN will find that particular developments are:

1. A greater degree of systemization in the assessment of load or effort, through the use of laboratory-tested guidelines.

2. The use of laboratory-tested data routines to analyse static load, recovery and force requirements (in place of simple addition of means over time).

3. Expansion from 6 to 14 Functional Units (or body parts).

This paper provides a description of the updated method and of its applicability based on case studies.

2.   *The ARBAN method's structure*

The *planning* of an ARBAN study includes a comprehensive time-study. The work is broken down into phases and elements contained in each phase. In this way one becomes familiar with the applied work technique and can minimize the later filming. Even if the tasks are repeated throughout the working day, it is of importance to note the variations in the task during the day, as well as the breaks, so that the compilation of the different cycles in the computer program will represent the whole task or working day. The planning phase usually requires the participation of one person at the workplace during the whole day, and this is essential for the next phase, which is the documentation of representative sequences for analysis.

Documentation at the workplace without disturbing the worker is effected with a two-camera video system. The video system produces a split-image picture, from two angles (allowing reliable analysis of rotated postures) with a timer (100th of a second) within the picture-frame. Various information is recorded on the sound track, including the subject's anthropometric measurements. The information is used later to assess the subject's ability to develop force. Other information recorded includes production data such as quantity or area per time unit, measurements of the size and weight of the materials handled and production costs (people and machinery). Furthermore, an interview with the subject is recorded while viewing the film. In this interview the worker comments spontaneously about perceived muscle effort while handling external loads, as well as how hard different aspects of the job appear to him. He may also comment on his general state of health, as well as on possible improvements based upon his experience on the job.

The *analysis* begins with a study of the particular work phase, first at ordinary speed and then in slow motion in order to become familiar with the dynamic elements of the work. An interval between each 'frozen' frame to be analysed is chosen. The interval varies between $\frac{1}{2}$ s for high tempo tasks and 2 s for slower tasks.

The *postural load* is assessed with the aid of a guide (Carlsöö *et al.* 1985). The body is divided into 14 parts, so-called Functional Unit (FU) (Figure 1), and stress associated with the following factors is assessed: (*a*) posture; (*b*) generation of force whilst handling external loads; (*c*) vibration; and (*d*) static load (for the duration of the above stresses) and the remaining stress under recovery from the static load. For each FU, a number of pictures presenting a recommended evaluation of different possible real-life work postures is presented in eight position charts: head and back plus three for hand/arm extremities and three for foot/leg extremities (extremity

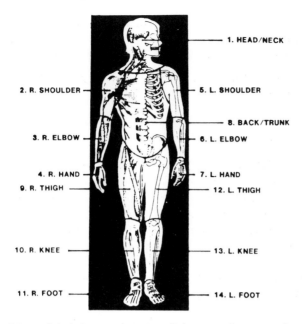

Figure 1. Breakdown of the body into regions, so-called Functional Units. Each extremity has been broken down into three parts. Each part covers one of the three major joints as well as the distal part.

charts are valid for right and left sides of the body). The recommended evaluation of the postural load is presented with the body in an upright position except for the pictures covering back/trunk postures.

There are two types of position charts. In the first, circular, type (Figure 2), the normal position of rest is situated in the middle. The circle rated as 3 indicates the limit of the 'comfort zone', regarded as a moderate load. By studying numerous films from workplaces of all sorts, we estimate that around 80% of postures range within the 'comfort zone'. The extreme voluntary position in each direction is shown on the external circumference. The second type of position chart (Figure 3) has no special form but simply tries to illuminate work postures in a pedagogical manner.

The recommended assessments in each of the pictures were based on an anatomical and physiological line of reasoning that takes into consideration the biomechanics of the postures as well as, in some cases, clinical experience. The numerical values 0–10 in the guide relate to verbally formulated stress categories, as on Borg's scale (Table 1) (Borg 1982). Before recommending the assessment for the different postures, tests were conducted with 28 workers from the building industry, 33 final year physiotherapist students and 6 ergonomists. Each one of the over 100 postures was repeatedly assessed on Borg's scale. The results of these tests are to be published in the future.

The *force load* is also assessed in numerical values of 0–10, relating to the formulated stress categories on Borg's scale. The anthropometric and technical data

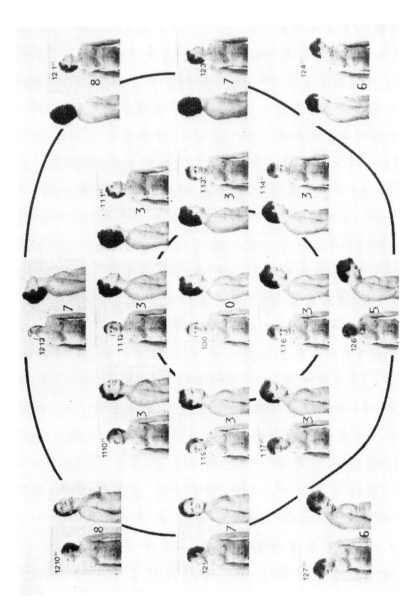

Figure 2. The position chart for the head/neck. Each position is represented by a split picture showing the head from the front and from the side. Each picture has a code number (small numbers), this way the position can be identified in the guide book, where an explanation of the 'load' number (large number in the middle of each picture) is presented.

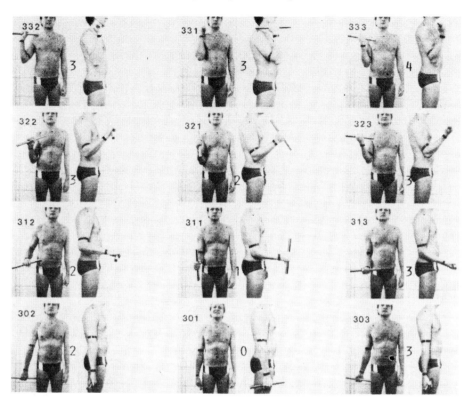

Figure 3.   The position chart for the elbow/forearm. The structure is the same as in Figure 2, except for the movement pattern. Here the left column of pictures describes pronation and the right column supination. In each horizontal series of pictures the angle between the upper arm and forearm is the same. In the top series the arm is flexed maximally, while in the bottom series it is fully extended.

gathered during the documentation phase are used to estimate the force exerted in each 'frozen' sequence. If it is possible, measurement of the subject's maximal force in the relevant directions is performed during the documentation phase with a field dynamometer. By comparing the load handled with the maximum force an assessment (0–10) is obtained.

If it is impossible to measure the forces, the subject's ratings of the applied forces recorded on the video voice track provide the observer with the necessary background data to accomplish an observational assessment on Borg's scale. Based on three series of experiments relating subjective ratings (of the person working) and observed ratings on Borg's scale, the observer ratings can be converted to estimates of the worker's own evaluation with the aid of mathematical formulae in the software (Holzmann and Wangenheim 1983). In the three series a panel of observers rated different ergonomic strains while at the same time the worker in question rated the individual strain himself. The first series involved static work

Table 1. *Borg's category scale with ratio properties that relate verbally formulated stress categories to mathematically defined stress ratios. The stress ratios have a mathematical correlation with physiological stress.*

Borg's scale	Verbally formulated stress category
0	Nothing at all
0·5	Extremely weak (just noticeable)
1	Very weak
2	Weak (light)
3	Moderate
4	Somewhat strong
5	Strong (heavy)
6	
7	Very strong
8	Extremely strong (almost maximum)
9	
10	Maximal

procedures, in which a load was held in a fixed posture, and two dynamic manual handling tasks. The second series consisted of lifting boxes of different weights. The third comprised combined stress in assembly-line work during a full working day. The studies showed that observers were able to discriminate between the different loads in all stress situations over the full range of loads. It was also shown that the relationship between observer ratings and self-ratings approximates a logarithmic function whereby the observers tend to estimate the low stresses greater than do the workers, while there is closer agreement in the rating of high stresses (Figure 4).

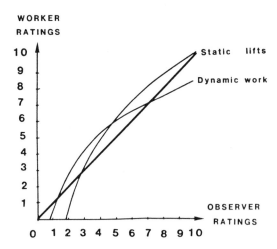

Figure 4. Correlation between ratings by the panel and by the worker himself, on Borg's scale. Summary of three experimental series. Two different curves have been plotted: one applicable to static lifting and one applicable to dynamic work.

Vibration is rated in a procedure similar to that for the force load. Both the subjective (worker's) assessment (on Borg's scale) of different situations and notes made of the tools handled form the basis for the rating. A guide covering common hand-held tools is now being prepared. Measurements of the tools' frequency and acceleration are presented. In some cases recording of shock forces and damping effects on the body surface by double pulse holography (Wangenheim and Holzmann 1984) contribute to the assessment of the vibrational load. By identifying the tool in the guide a basic evaluation is made. If the tool cannot be identified, the worker's rating on the video voice track, converted through a recent study determining a stimulus–response relationship for hand vibration, is used. This study showed that self-rating can be used to assess vibrational intensity for common hand-held tools. The perception of increases in the amplitude of vibrations at a certain frequency is progressive. We found a significant rating pattern of a negatively accelerating power function as an expression of the stimulus–response relationship.

The *static load* as well as the recovery phase are assessed with the aid of modified models based on Rohmert's (1962) studies. In tests that have been going on in collaboration with Professor Rohmert, maximum endurance time in static force exertions was used as a fundamental parameter for the assessment of five different working postures. Three male subjects participated in tests exerting static postural forces on different load levels (25, 35, 50 and 75% of their own maximal force) until exhaustion.

As a result of the tests, the model has been adjusted for more complicated postures such as stooping forward or hanging passively without muscle activity. In addition, more specific models for the FUs in the postural guide are now being programmed. A second series of tests, in which the same postures are being tested in order to obtain more specific models for the recovery phase, was completed during 1985.

The static load model is activated for each FU at each 'frozen' sequence when the FU is nearly motionless, or in a situation where the muscles are contracted continuously.

The *total analysis* is presented in 15 curves and one table. One curve shows the integrated stress on the whole body over time and the other 14 show the integrated stress for each FU (Figure 5). The values of the four loads assessed, as well as different remarks, are noted above each point analysed in time on all the 15 curves. The table sums up the whole work analysis (Table 2).

For each FU at each stop, the values of the four loads assessed are transformed into one value, presented on the curves. The transformation is to add the loads, each raised to the fourth power, dividing the total by four and extracting the fourth root.

This way, the heavy loads are not disproportionately reduced as in a usual calculation of the mean value. The theory behind this mathematical transformation is that the body adapts, by integrating the different loads at each instant, the heavy loads being given a larger amount of consideration. The model forming the total stress for the whole body is built mainly on statistics concerning injuries in

250 *The ergonomics of working postures*

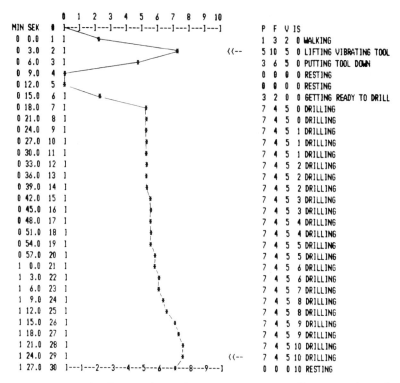

Figure 5. An example of a typical stress/time ARBAN curve. All 15 curves have the same structure. The stop number relates to each situation studied, the time taken from the timer on the video film. Above each stop the four loads assessed as well as different remarks are noted. Sometimes an arrow appears above a stop; this indicates that one of the four loads has reached a dangerous level.

different professions. Before the analysis is started, a coefficient for each FU is fed into the computer. Finally, every load for each FU is integrated over the whole analysis time and presented in one table (Table 2). The ARBAN software is written in Pascal and runs on a personal computer.

3. Applicability of ARBAN based on case studies

The ARBAN method has been developed for the purpose of analysing force-related ergonomic problems. It can be used for studying both men and women. The method aims primarily at analysing heavy mobile work situations, but has been shown to be useful in analysing sitting work thanks to further development of the postural guide. It is not yet adjusted for precision work. One should bear in mind that the ARBAN method produces a 'trend' analysis and not an absolute measurement of the work situation. By using a computerized technique, the results are presented in a manner allowing all decision-making personnel to take appropriate steps.

Table 2. ARBAN's summarizing table showing the distribution of ergonomic stress over the functional units.

Functional unit	Posture	Force	Vibration	Static	Total
1. Head/neck	6·1	5·5	2·3	1·6	4·9
2. Right shoulder	5·0	5·0	2·6	0·3	4·2
3. Right elbow	6·4	5·0	2·3	1·4	4·9
4. Right wrist/hand	5·2	5·2	3·4	0·7	4·5
5. Left shoulder/arm	5·6	5·0	2·3	0·6	4·5
6. Left elbow/arm	4·9	5·2	3·5	0·3	4·4
7. Left wrist/hand	5·2	5·0	2·3	0·4	4·3
8. Back/trunk	5·5	5·8	2·3	0·9	4·7
9 Right thigh	6·7	5·8	3·1	3·3	5·4
10: Right knee	5·9	5·2	2·6	0·9	4·7
11. Right foot	6·3	5·7	2·8	2·4	5·1
12. Left thigh	6·9	5·4	3·0	3·5	5·4
13. Left foot	5·9	5·4	2·3	0·9	4·8
14. Left knee	6·9	5·9	2·8	3·5	5·5
The whole body					4·8

The method can be used for:

1. Comparative studies of alternative work methods, products or equipment.
2. Identifying optimal man–machine combinations.
3. Guidelines in product development.

Below is a presentation of two recent studies demonstrating the first two points above. In these studies the results are presented in different forms deriving from ARBAN analysis data.

The first study: comparing alternative scaffolding methods

Statistics show that scaffold builders are the most vulnerable group in construction work. Their work is very heavy; several tons of materials are handled manually every day, often in difficult postures and at high levels.

This study has concentrated on an ergonomic comparison of the two most common types of light scaffolding, Burton and Haki. Both types are built by two workers. The scaffolding was built around identical three-storey houses, the ground and other conditions being the same. The work was performed in both cases with one worker mounting the scaffold and the other remaining on the ground as a helper. The Burton scaffold consists of plain steel pipes which are locked with screw clamps. The Haki scaffold is fitted into place by a system of welded hooks that are inserted into sleeves. The Haki scaffold is considered to be easier to build than the Burton scaffold.

The ergonomic study of the two types of scaffold shows that the workload is high, in some cases very high, with both of them. Quite unexpectedly, the Haki scaffold appeared to involve a higher load than the Burton for all parts of the body

(Figure 6). This can be explained by the fact that the Haki scaffold is more sensitive to careless treatment than the Burton scaffold and is therefore damaged after a certain period of use and thus becomes more difficult to mount. Therefore, a recommendation to reinforce the hook/sleeve system was reported to the manufacturer.

Figure 6 demonstrates how a comparison of the two scaffold types can be made with the help of the ARBAN data. Such comparisons can be made for the different work sequences as well as for the different relevant load factors, such as force, etc. In this example the total force load for all work sequences combined is shown for each part of the body.

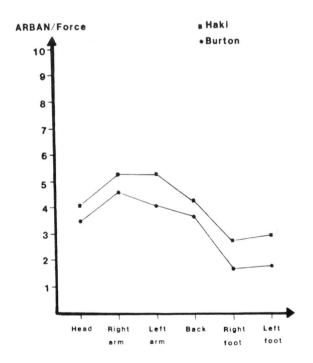

Figure 6.   The ARBAN force load is presented for a few FUs. The figures for the two scaffold types are taken from the summarizing tables.

Figures 7 and 8 show the total load for different steps in the work sequence which are representative of these types of scaffold. Such figures are quite usual when presenting an ARBAN study. They are based on the time-study performed in the planning phase, showing the distribution of the total time for each step in the work sequence during a whole day. Each work sequence is, of course, repeated many times during a working day.

ERGONOMIC LOAD
ON WHOLE BODY

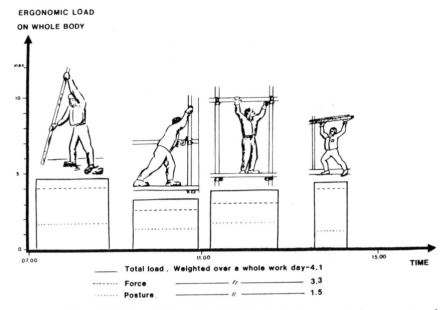

Figure 7. The ARBAN total load on the whole body for the four steps in the construction of a Burton scaffold. Each step is presented horizontally, describing how long the worker is engaged in it out of the whole working day. Inside the block for each step, the mean values for the posture and force loads are given.

ERGONOMIC LOAD
ON WHOLE BODY

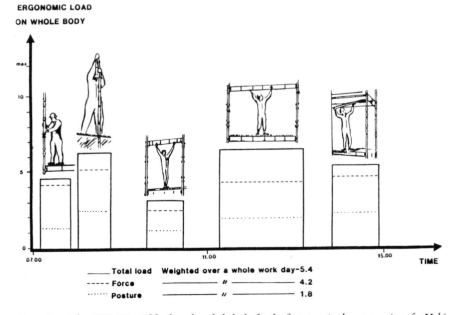

Figure 8. The ARBAN total load on the whole body for the five steps in the construction of a Haki scaffold. The structure is the same as in Figure 7.

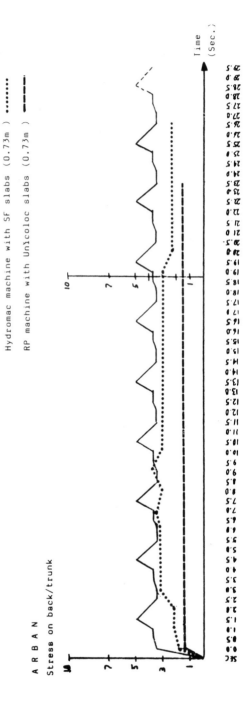

Figure 9. A typical ARBAN curve (such as in Figure 5) presenting the load whilst laying concrete floor slabs. The helper's load on the back is compared in three combinations of technique/slab.

The second study: finding the optimal man–machine combination in laying concrete slabs

A study was made of different methods of laying concrete slabs. The aim was to indicate which combination of slab/technique was optimal from an ergonomic point of view. All combinations were used at the same time at the same workplace; therefore, all relevant conditions were exactly the same. The methods studied were manual laying, machine laying with the Hydromac machine and machine laying with the RP machine. In each case two types of slabs were tested. All methods require two workers; one slab-layer or machine operator and one helper. Both workers were studied.

In every case the most heavily loaded part of the body appeared to be the back and trunk. Figure 9 shows three basic ARBAN curves presenting the load on the back/trunk as a function of time for three combinations. One can clearly see which slab/technique combination is easier for the back/trunk. Figure 10 shows the load for the back and trunk in relation to the production capacity for each method.

In this case, with the aid of the ARBAN method, we have been able to demonstrate clearly the advantages of a certain combination: the RP machine with the Unicoloc slab. Productivity was higher, costs lower and, last but certainly not least, the ergonomic load was fully acceptable.

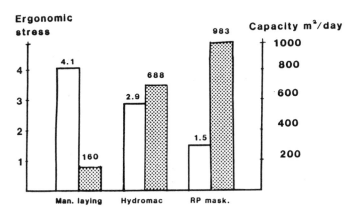

Figure 10. ARBAN stress on back/trunk and production capacity (square metres per workday), in the three combinations of technique/slab.

Chapter 22
A Postural Load Evaluation Technique: Based on Energy Expenditure, Heart Rate and Subjective Appreciation

Kamiel Vanwonterghem

Instituut voor Mijnhygiene, Havermarkt 22, B-3500 Hasselt, Belgium

1. Introduction

The technique presented here is not a cut and dried solution to the problem of postural load evaluation, but is a practical technique developed in the field.

From the ergonomics points of view, decision-making about improvements in work conditions and worker behavioural patterns is of paramount importance. Therefore, there is a need for convincing information for both industrial partners, management and workforce. For management purposes, this information should by preference be more oriented to concrete and objective elements, as, for example, production quality and quantity, absenteeism, labour accidents, environmental conditions and social complaints. The work force, on the other hand, is more concerned about personal and subjective aspects, the degree of discomfort, annoyance, fatigue, etc. Hence, any ergonomic study of 'man–machine–environment' systems should take these dual considerations into account.

Following Cazamian (1973), our ergonomic studies are based upon the simultaneous collection of objective measurements of external load, such as environmental conditions and task load (work rhythm, hours of work, workplace arrangement, etc.), and of internal load, particularly respiration, energy expenditure and heart rate. Finally, studies are completed by means of more subjective parameters, collected through (in this case) the CERGO Subjective Workload Index questionnaire. This explores fatigue, interest, risk, task requirements, work rhythm, the effect of postural and environmental constraints and so on.

These theoretical and philosophical considerations characterize our approach to any ergonomic problem, and therefore are relevant for the 'postural load' problem, which is a complex problem because of its anthropometric, biomechanical and energy expenditure aspects. These are so complex that no techniques are available for an evaluation of the whole bodyload. The interrelational difficulties of those problem elements, interfering with the inter- and intra-individual differences of the operators, are also an obstacle in communication between both industrial partners because of the objective *and* subjective character of the information.

To compensate the lack of direct information, an indirect evaluation technique for 'postural load' has been developed. This method is based on a physiological phenomenon, the discrepancy between the energy and circulatory systems which normally change in a closely similar way. To give a context to these objective data, the operator himself (herself) expresses his (her) own experiences, as mentioned above.

Hypotheses

Physiological studies on energy expenditure and its influence on the circulatory system show a high level of correspondence between oxygen uptake and heart rate (e.g., Åstrand and Rodahl 1970, Cerretelli 1973, Lehmann 1962). This correspondence is shown in Figure 1 for a standardized cyclo-ergometer test ($r = 0.89$). However, the correlation increases when the test has a more dynamic character, as, for example, during a full dynamic step test ($r = 0.95$) (Figure 2). The difference between these two fundamental tests may be explained by the 'sitting position' on the cyclo-ergometer, which therefore includes some static muscle load. Inter- and intra-individual differences do not influence the concordance, although slight slope modifications may occur.

During a study of work procedures in which the physical work contains more static than dynamic components, we have observed the following phenomena:

1. Correlation is at a lower level, which varies with the kind of muscular work.

2. Oxygen consumption and heart rate can change in opposite directions. This phenomenon drastically disturbs any conformity.

In the next study we tested these hypotheses in activities which could be characterized as having miscellaneous muscle loads of static and dynamic components.

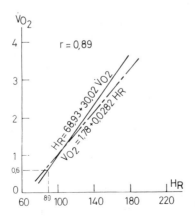

Figure 1. Correlation and regression between oxygen consumption and heart rate during a cyclo-ergometer test.

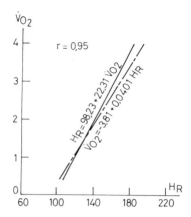

Figure 2.   Correlation and regression between oxygen consumption and heart rate during a step test.

2.   The study

A series of ergonomic studies, sponsored by an insurance company, covered work accidents in a metalworking industrial plant, the main activities of which consisted of manufacturing and assembling combine harvesters.

The activities studied

Two kinds of assembly job were chosen for the postural study. The first concerned assembling the rotor-threshers and included several tasks, of which two were selected:

1. Assembling the rotor-thresher, by mechanic I, which had more static characteristics.

2. Assembling the main disc drive, by mechanic II, which had more dynamic characteristics.

The second job, assembling the steering cab onto the frame by mechanic III, was of mixed form, including both static and dynamic aspects of postural load.

The individuals studied

The three mechanics all had at least 10 years experience in their jobs. Their base values of oxygen uptake and heart rate are shown in Table 1. This shows a difference between the correlation coefficient of the first two mechanics (I and II) and of the third mechanic (III). This is due to a difference in the test conditions. The first two subjects performed a cyclo-ergometer test (5 min at 50 W, followed by a 3 min rest, then increasing the load in steps of 50 W to 150 W), giving a certain static component (sitting). The third subject performed a full dynamic step test, 5 min on a bench of 30 cm height at a rhythm of 30 rises per minute. The different

Table 1. *The physiological characteristics of the mechanics—basic values.*

	Mechanic I	Mechanic II	Mechanic III
$\dot{V}O_2$ max (l/min)	2·60	3·06	2·74
$\dot{V}O_2$ 8 hours (l/min)	0·87	1.202	0·91
HR max (beats/min)	168	189	182
HR 8 hours (beats/min)	136	131	140·7
r^2 ($\dot{V}O_2$ and HR)	0·79	0·83	0·90
r ($\dot{V}O_2$ and HR)	0·89	0·91	0·95
$\dot{V}O_2$/HR regression coefficients			
a	68·93	93·56	98·23
b	30·0164	29·6172	22·31
HR/$\dot{V}O_2$ regression coefficients			
a	−1·78	−2·26	−3·81
b	0·0282	0·0266	0·0401
$\dot{V}O_2$ kg/body weight (ml)	31·7	34·0	30·6

test conditions were because of equipment availability problems. These base values serve as a reference for the results obtained during the actual performance of the jobs.

Following the investigation principles discussed earlier, the external load, the internal load and the subjective findings were measured and collected by appropriate apparatus.

The environmental conditions comprised a neutral climate (19 °C and a relative humidity of about 68%), a luminance of 70 cd/m² (which is satisfactory for this kind of assembling task) and a variable noise level between 87 and 95 dB(A) at 250 and 2000 Hz frequencies with peaks of 105 dB(A) during the mechanical fixing of axles.

The task description

The mean assembly time allowed was 30 min. The work rhythm was free and the remaining time was used for recovery (rest) and informal contacts.

Mechanic I (see Figure 3)

	Minutes	Seconds
Setting up the rotor beater	2	30
Assembling and stretching of pulleys	6	30
Setting up protection plates	3	
Setting up the belt guide	3	
Setting up the belt protection	2	30
Greasing and paint corrections	2	30
Preparing the belts stretcher	2	
Total	22	

Figure 3.   View of mechanic I.

This task contained periods with limbs in a static position, sometimes holding heavy weights (up to 35 kg).

Mechanic II (see Figure 4)

	Minutes	Seconds
Helping setting up, and sealing the rotor beater	2	35
Setting up the main disc drive	7	25
Preparation of next disc drive	2	
Setting up the threshing drum	4	
Setting up the tension pulley	3	
Tightening the tension pulley	2	30
Preparing the tension pulley	1	30
Total	23	

Mechanic III (see Figure 5)

The task of mechanic III consisted of positioning and fixing the steering cab onto the frame. A co-operator collaborated and was positioned in the cab. They had to fix and to fasten several small screws and mechanic III also had to keep the screws on the right spot while his colleague fastened them. The structure of the frame and the wheels hindered the operator and caused a significant postural load. Besides this static aspect there was also an important dynamic load, circulations around the machine, climbing steps and ladders.

Figure 4.   View of mechanic II.

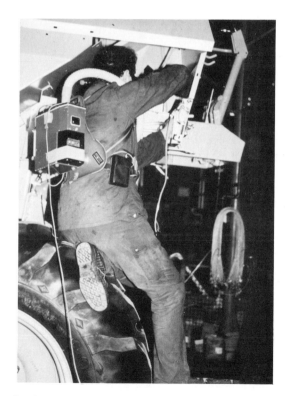

Figure 5.   View of mechanic III.

The physiological measurements

Basic physical conditions for each operator were tested before the work began so that they were already used to the different apparatus. The following have been calculated from the measurements taken during the basic tests:

1. The maximum oxygen uptake ($\dot{V}O_2$ max), calculated by means of the von Dobeln formula:

$$\dot{V}O_2 \text{ max} = 1 \cdot 29 \frac{L}{HR - 60} e^{-0 \cdot 00884T}$$

where L is the workload in kilogram metres per minute, HR is the heart rate (beats per minute) at the end of the effort, T is the age in years and $\dot{V}O_2$ max is the maximum oxygen uptake capacity in litres per minute.

2. The maximum allowable oxygen consumption for an 8 hour workday (one-third of the $\dot{V}O_2$ max).

3. The absolute maximum heart rate (220 − age in years).

4. The maximum allowable heart rate for an 8 hour workday;

$$HR(\text{max}) \text{ 8 hours} = \frac{220 - \text{age}}{3} + HR(\text{rest})$$

5. The correlation coefficient and the coefficient of determination, calculated by the usual formula.

The linear formula has been chosen after a preliminary study where the logarithmic, the exponential and the linear formula have been compared. This study showed the best correlation for the linear formula ($r = 0 \cdot 95$) compared with the logarithmic ($r = 0 \cdot 87$) and the exponential ($r = 0 \cdot 84$) formulae.

The measurements above, and those taken during the working cycles, have been made by means of the following apparatus:

1. The energy consumption (oxygen uptake) was measured by means of Mijnhardt's portable Oxycon. This consists of a volume-meter which is worn by the worker and is provided with a polargraphic cell for the Oxycon analysis. The results are transmitted to a stationary device where the following parameters are calculated and displayed: (*a*) the expired volume, in litres per minute; (*b*) the respiratory frequence, in breaths per minute; (*c*) the difference in oxygen concentration between inhaled and expired air, as a percentage of O_2; and (*d*) the oxygen consumed, in litres per minute.

2. The heart rate was measured by means of Novo's Memolog apparatus. This is a very light (370 g) two-channel instrument and allows the simultaneous capture of two physiological or physical parameters, including, for example, heart rate, body temperature (rectal, skin, oral and tympanic temperatures), electrical skin resistance and acceleration. The information collected is stored directly. Besides these values, the sample number of the time base (from 15 s up to 4 min) is also displayed on a liquid crystal display and allows permanent quality control of the

measurements. Connected to a microprocessor (RS 232 C interface) the information stored can be treated in digital or graphic forms.

The sample time for the energy expenditure was 1 min, and for the heart rate the time-base was set at 15 s. Mean values for every operation have been calculated over several assembling cycles.

The subjective workload

The subjective workload was determined by means of the Subjective Workload Index (SWI) questionnaire. This contains eight elements, six of which are estimated as strengthening: fatigue, risks, work rhythm, attention required, task complexity and responsibility. These are quoted positively. The other two, rather favourable elements and therefore quoted negatively, are of a more motivational kind: interest in the job and independence. The values are obtained by drawing a mark on a 0–10 scale. The sum of the values divided by the number of parameters taken into account gives the SWI index, which is then compared with the following nuisance scale: 0, no nuisance at all; 1, light nuisance; 2, nuisance; 3, bothering; 4, very bothering; and 5, intolerable, painful.

3. Results

Assembling the rotor-thresher

The physiological results for mechanic I are illustrated in Figure 6. The SWI was 2·24 which indicates some nuisance. Noteworthy are the scores 9 for 'interest in the

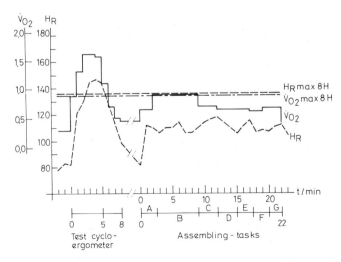

Figure 6. Energy and circulatory profiles for the cyclo-ergometer test and the assembling cycle for mechanic I.

job' and 8 for 'independence', two values which being scored negatively explain the relatively low overall score. There is a high 'fatigue' score for the following operations:

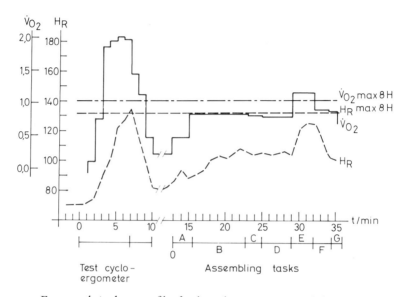

Figure 7. Energy and circulatory profiles for the cyclo-ergometer test and the assembling cycle for mechanic II.

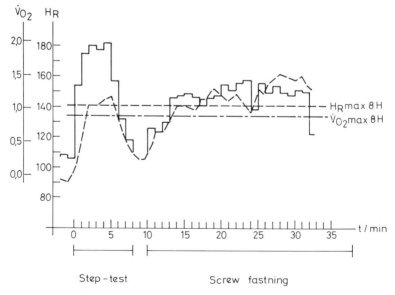

Figure 8. Energy and circulatory profiles for the step test and the assembling cycle for mechanic III.

1. Setting the rotor-beater (score of 8) with complaints of pain in the wrists and forearm.

2. Setting up the protection plates (score of 6) with pain in the knees, feet and ankles.

The contradiction between the values for the fatigue and motivational variables has been explained by a more intensive interview, which revealed a hypermotivated 52 year old person.

The physiological results for mechanic II are illustrated in Figure 7. The SWI index was 2·75, indicating again only some nuisance. A high fatigue score (8) was reported for setting up and stretching the tension pulley. This operation causes an important muscle strain because of over-shoulderline movements. A particular complaint was about vibrations and noise.

Considering the physiological measures for the second mechanic the two parameters seem to evolve normally; an increasing energy expenditure is coupled with a proportional increase in the heart rate. For the first mechanic, on the contrary, the rise in heart rate is:

1. More important for certain operations, for example the operation of setting the rotor-beater. This gives an oxygen consumption of 0·6 litre O_2/min. According to the dynamic cyclo-ergometer test (see Figure 9) for 0·6 litre O_2/min a heart rate of 86·9 could be expected. The measured heart rate of 113 beats/min is, however, 30% above this dynamic level.

2. Another discrepancy is observed for the operation of setting the protection plates, in which the muscle effort is rather low, proved by the 0·68$\dot{V}O_2$ level which is lower than for the preceding operation. Where an associated decrease of the heart rate may also be expected, this parameter evolves in the opposite direction. This phenomenon may be explained as a postural load caused by a squat position held for about 3 min. This effort is also subjectively experienced as 'bothering' while the operator complains about pain in the legs. The overall load seems to be acceptable, neither the 1–3$\dot{V}O_2$ max nor the heart rate for the 8 hour limit being exceeded.

Assembling the steering cab

The physiological results for mechanic III are shown in Figure 8. The SWI was 3·56 which means bothersome, and which is in full agreement with the objective measurements. Some complaints of pain in the quadriceps and the muscles of the shoulders have been quoted as severe (8 and 9·5).

From the physiological profile we see the same phenomena as for the previous activities although the physical intensity of this operation is significantly higher, and exceeds the 8 hour limits for oxygen uptake and heart rate at several periods. The higher level could be explained by the heavier dynamic muscle load. The SWI seems to be more influenced by energy expenditure although the complaints expressed refer to more static aspects. For this task job design corrective measurements are necessary.

Comparison of the three assembling tasks

The results obtained from these three assembling tasks are summarized in Tables 2–4 and in Figures 9–11. These permit a comparison between:

1. The values measured during the basic test: (*a*) for mechanics I and II a cyclo-ergometer test and (*b*) for mechanic III a step test.
2. The values for the daily jobs.
3. The difference between these two values, expressed as a percentage.

These data show the breakdown in correspondence between oxygen uptake and heart rate found in a dynamic type of muscular work (Figure 12). It is conjectured that the weight of the object handled (mechanic I) is of major importance, although postural load (mechanic III) seems also to be important.

Table 2. *Results of the basic tests and for the assembling task of mechanic I.*

	Basic values of the cyclo-ergometer test	Values of the assembling task	Difference (%)
r^2	0·79	0·24	−69·2
r	0·89	0·49	−44·9
Regression coefficients			
$\dot{V}O_2$/HR			
a	68·93	69·41	
b	30·02	50·80	
HR/$\dot{V}O_2$			
a	−1·78	+0·15	
b	+0·0282	+0·0048	

Table 3. *Results of the basic tests and for the assembling task of mechanic II.*

	Basic values of the cyclo-ergometer test	Values of the assembling task	Difference (%)
r^2	0·83	0·52	−37·3
r	0·91	0·72	−20·9
Regression coefficients			
$\dot{V}O_2$/HR			
a	93·56	68·99	
b	29·62	31·05	
HR/$\dot{V}O_2$			
a	−2·26	+1·25	
b	+0·02666	+0·0209	

Table 4. Results of the basic tests and for the assembling task of mechanic III.

	Basic values of the step-test	Values of the assembling task	Difference (%)
r^2	0·90	0·37	−58·9
r	0·95	0·61	−35·8
Regression coefficients			
$\dot{V}O_2/HR$			
a	98·23	111·54	
b	22·31	22·83	
$HR/\dot{V}O_2$			
a	−3·81	+1·04	
b	+0·0401	+0·016	

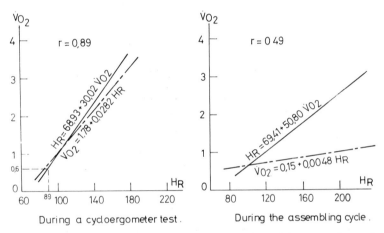

Figure 9. Correlation and regression between oxygen uptake and heart rate during the cyclo-ergometer test and the assembling cycle for mechanic I.

4. Conclusions

From the statistical point of view, the greater the static muscular (postural) load, the more the correlation between the physiological parameters of oxygen consumption and heart rate is broken down.

The phenomena observed should be confirmed by other studies. If changes in correlation between heart rate and oxygen consumption could be linked, in a systematic way, to the changes in the composition of work (static and dynamic components), the actual measuring techniques and methods could be used to estimate the whole body load. In such a research programme, the influence of other environmental parameters on oxygen consumption, heart rate and the correspondence between these two, such as, for example, some climatic conditions, should also be taken into account.

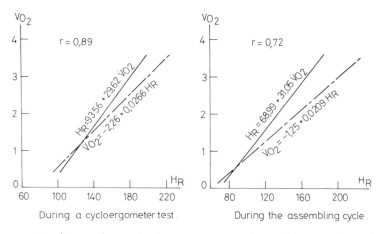

Figure 10. Correlation and regression between oxygen uptake and heart rate during the cycloergometer test and the assembling cycle for mechanic II.

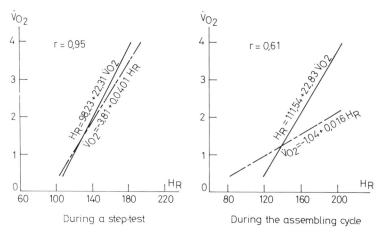

Figure 11. Correlation and regression between oxygen uptake and heart rate during the cycloergometer test and the assembling cycle for mechanic III.

The subjective appreciation of the nuisance caused by body postures is very important too. If there is an agreement between the objective and the subjective findings (as, for example, in the cases of mechanics II and III) one objective, namely to find information which could convince management and operators, is satisfied, and this results in concrete actions in the field of technical, organizational and behavioural applications. If there is no agreement between the SWI and the objective measures (the case of mechanic I) there are other aspects, perhaps of a more sociocultural kind, which should be analysed.

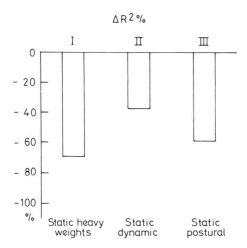

Figure 12. *Destruction of the correlation between oxygen uptake and heart rate for 'static heavy weights', for 'static and dynamic' loads and for 'static and postural' loads.*

Chapter 23
A Technique for Postural Load Assessment

I. Manenica

Department of Psychology, Faculty of Science and Arts, Zadar, Yugoslavia

1. Introduction

Research interest in working postures and postural load during recent years has, no doubt, been initiated and maintained by the realization of the importance of this problem from both the productivity and human suffering points of view. Evidence in the form of complaints from operators working in different industries throughout the world has been strong enough to attract the attention of ergonomists and other specialists in search of an answer to the problem.

Westgaard and Aarås (1984) pointed out that about 30% of the time lost through sick leave in Norway was due to the effects of inappropriate postures on musculo-skeletal systems. A comparison of industrial standards in Norway with standards in some other countries may suggest that this percentage could be much higher elsewhere.

Stubbs *et al.* (1983) have reported the results of their investigation in the nursing profession. The results showed that in about 16% of cases back pain was directly initiated by a patient-handling incident. The authors advocated an ergonomic assessment of the tasks performed in the nursing profession, with the hope of reducing back-pain problems.

It is recognized that most injuries and/or back complaints are due to bad equipment design, inappropriate handling methods and workplace design. It seems, though, that a significant proportion of skeleto-muscular complaints could be avoided by avoiding 'bad' postures, which are not always due to the load induced by the task itself. It is therefore important to find techniques for postural load assessment which could be of use on the shop floor and in workspace and equipment design.

Corlett *et al.* (1979) proposed a posture recording technique which included separate position recordings of ten major body parts in terms of their angular deviations from a 'standard position'. The technique was tested in industrial and laboratory conditions and it was found sufficiently discriminative and reliable. A major problem connected with the technique was data analysis and interpretation.

A further step would imply finding a technique for quick and reliable postural load assessment. An attempt was made by Corlett and Manenica (1980) to establish the relationship between the posture holding time and the muscular force applied, expressed as a percentage of the maximal muscle force. At the start, their work went along the lines of some previous work, such as that of Monod and Scherrer (1965) and Rohmert (1973). The authors then went a step further and successfully established, *inter alia*, the relationship between the holding time and the muscle force of those muscles which, although they could not be identified, were the limiting factor in maintaining the posture. The relationship, which was logarithmic, could be expressed in linear form of the type:

$$y = a - bx$$

where y is the log of the posture holding time, x is the muscular exertion (%), a is the intercept on the y-axis and b is the regression slope.

Manenica (1980) stated that the holding time in any posture depended on the load of the relatively most loaded muscle involved in holding the posture. Taking this as a starting point, on the basis of the known regression line between the holding time (log values) and the force the author put forward a model to establish the magnitude of postural load on the back muscles, on the basis of posture holding time only.

The postural load was expressed in terms of the percentage of muscle strength used to maintain the body in a particular position without any additional load. In five observed postures, which were defined on the basis of work height and work distance (Corlett and Manenica 1980), it was found that these impaired the back muscle working capacity by from 27·5 to almost 60%. This meant that the residual working capacity ranged from 72·5 to about 40% only, compared with the body working in an erect position.

The difficulty with the model seems to be the apparent diversity of force–time relationships for different muscle groups (Figure 1). Its application seems to require a separate regression line for each muscle group, which would make the model almost inapplicable.

2.   Re-analysis of the results

The data on which the regression lines (log of holding time and the muscle force) were based were re-analysed. Two-way analysis of variance showed significant effects of the muscle force applied on holding time ($F = 40·10$, df $= 4/16$, $p < 0·01$), but it also showed that a significant amount of the variance was due to the effects of the five different muscle groups studied ($F = 3·27$, df $= 4/16$, $p < 0·05$). This could mean that various muscles have different force–time relationships. There was, however, a striking similarity amongst the regression slopes, which suggested that the obtained differences amongst the muscle groups might be due to an experimental error.

Re-analysis of the data shown in Table 1 showed significant differences in holding times at the 100% load level (the maximal muscle force). This could mean

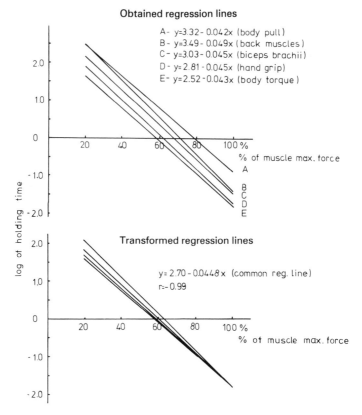

Obtained regression lines

A- y=3.32- 0.042x (body pull)
B- y=3.49- 0.049x (back muscles)
C- y=3.03- 0.045x (biceps brachii)
D- y= 2.81 -0.045x (hand grip)
E- y=2.52 -0.043x (body torque)

Transformed regression lines

y= 2.70 - 0.0448x (common reg. line)
r=-0.99

Figure 1. Obtained and transformed regression lines.

Table 1. Obtained data.

Muscles involved	Force (%)				
	20	40	60	80	100
Obtained posture holding time (min)					
Hand grip	6·75	2·74	1·12	0·45	0·18
Biceps brachii	8·41	3·42	1·39	0·57	0·23
Body pull	11·94	5·16	2·23	0·96	0·41
Body torque	5·26	2·23	0·94	0·40	0·17
Back muscles	12·30	4·62	1·73	0·65	0·24
Log of posture holding time					
Hand grip	1·91	1·01	0·11	−0·79	−1·69
Biceps brachii	2·13	1·23	0·33	−0·57	−1·47
Body pull	2·48	1·64	0·80	−0·04	−0·88
Body torque	1·66	0·80	−0·06	−0·92	−1·78
Back muscles	2·51	1·53	0·55	−0·43	−1·48

that an error of measurement was made when the maximal muscle force was determined, which seemed plausible, since assessment of the maximal force was a short-lasting event, which could have been affected by different biomechanical and situational factors. The difference in holding times amongst the five muscle groups at 100% load ranged between 0·17 and 0·41 min (about 10–25 s). It would appear that the maximal force determined was not the maximal for all the muscles, and most likely that the better assessment was made for the muscles with shorter holding times. This implies that the force (load) percentages below the maximum were not correctly established for those muscles with the longer holding times at 100% of maximum force, resulting in significant differences in intercepts on the *y*-axis.

A logical correction of this experimental error was a linear transformation of the log of holding time values, where log 0·17 (the shortest holding time at 100% of the maximum force) was taken as the reference for such a transformation (log 0·17 = −1·73). The transformed results are shown in Table 2; values for the first four columns are reduced by the amounts required to bring the fifth column values to −1·78.

Table 2.   Transformed data.

Muscles involved	Force applied(%)				
	20	40	60	80	100
Transformed log of posture holding time					
Hand grip	1·82	0·92	0·02	−0·88	−1·78
Biceps brachii	1·82	0·92	0·02	−0·88	−1·78
Body pull	1·58	0·74	0·10	−0·94	−1·78
Body torque	1·66	0·80	−0·06	−0·92	−1·78
Back muscles	2·14	1·16	0·18	−0·80	−1·78
Predicted posture holding times (min)					
Hand grip	6·17	2·50	1·02	0·41	0·17
Biceps brachii	6·17	2·50	1·02	0·41	0·17
Body pull	4·85	2·10	0·90	0·39	0·17
Body torque	5·26	2·22	0·94	0·40	0·17
Back muscles	8·50	3·19	1·20	0·45	0·17

Two-way analysis of variance of the transformed data showed significant effects of the muscle force involved, on holding time ($F = 87·21$, $df = 4/16$, $p < 0·01$), but no significant effects of the muscle groups on total variance was found ($F = 2·21$, $df = 4/16$, $p > 0·05$). The analysis confirmed, therefore, that an experimental error existed, and that it was the cause of part of the variance.

Since the new regression lines (Figure 1) obtained from the transformed data did not differ significantly, a general force–time relationship could be put forward, which may be applicable to all the muscles studied. The regression line obtained on the basis of the transformed data is

$$y = 2 \cdot 70 - 0 \cdot 0448x$$

This formula also allows the assessment of the force applied by the relatively most loaded muscle to maintain the isometric contraction.

Force applied (%) = (2·70 − log (holding time))/0·0448 where the force is expressed as the percentage of the maximal muscle force.

It is sometimes difficult, especially in real working situations, to assess the posture holding time, and therefore to find the postural load by use of the proposed formula. The aim of this additional study was to try to use several possible indicators of postural load, some of which might help in overcoming certain practical difficulties connected with the application of the above technique of assessment. Subjective estimation, which has been used successfully for assessment of discomfort, 'fatigue', pain etc., has in certain studies compared well with physiological changes. Tapping task regularity, which has been used as a secondary task in mental load assessment (Michon (1966) for example), was the second variable used as an indicator of postural load. One of the reasons for using it here was the idea that salvos of afferent impulses, produced as the result of isometric muscle contraction come to the cerebral cortex and might be a disturbing factor for the pace-making mechanism involved in maintaining tapping regularity.

3. Method

Fifteen female subjects (students), aged 19–24 years, took part in an experiment which involved seven different postures defined on the basis of work distance and work height, expressed as percentages of the maximal hand reach and the subject's shoulder height, respectively (Corlett and Manenica 1980). The tapping takes in each of the seven postures required regular tapping with the index finger on a microswitch connected to an Apple II computer. Time intervals between successive taps were registered in milliseconds, and stored by the computer.

After adopting the required posture, the subjects had to tap as regularly as possible, whilst maintaining the posture. The subjects were asked every ½ minute to assess the postural discomfort on the 20-point scale proposed by Borg (1973). During the experiment pulse rate was recorded continuously for later ½–¼ minute analysis. At the termination of each posture the subject was asked to indicate, on a body chart, the part of the body which was most affected by the discomfort and made her terminate the posture.

The order of the postures during the experiment was arranged according to a Latin square design. The recovery intervals between the successive postures ranged from several hours to a couple of days, enabling the subjects to recover fully.

4.   Results and discussion

The postures, indicated by numbers from 1 to 7, and the corresponding index means of the observed variables are given in Table 3. On the basis of holding time, by the use of the suggested formulae the postural load was calculated for each posture. The load refers to the relatively most loaded muscle which is the limiting factor of the posture, expressed as the percentage of its maximal force.

Different tapping-task parameters were calculated and analysed. This showed that the means of tapping intervals stayed the same, and were not affected by the severity of the posture ($F = 0.99$, df = 6/84, $p > 0.05$). This was also obtained for the standard deviations of the tapping time intervals (SD) and the means of the absolute differences between consecutive intervals, which were used as measures of the variability and the internal consistency of the intervals, respectively ($F = 1.04$, and $F = 1.24$). The only parameter that showed effects of the postures was Michon's (1966) PML index ($F = 2.96$, df = 6/84, $p < 0.05$), but these effects could not be attributed to the relative severity of the postures as seen through some other indicators (holding time, estimated load, pulse rate and estimated discomfort). It was expected that the PML index would increase as the severity of the posture, assessed by the holding time, increased. This did not happen. On the contrary, PML index correlated significantly with the posture holding time ($r = 0.77$), which meant that longer lasting (easier) postures were associated with higher PML indices. This correlation suggests that the PML index increased in the course of time within the posture, possibly due to some local functional changes in the finger involved in tapping.

Further analysis was carried out by dividing the holding time of each of the seven postures into four equal intervals, representing 25, 50, 75 and 100% of the time. Two-way analysis of variance for repeated measures showed that pulse rate was significantly affected by postures ($F = 3.48$, df = 6/392, $p < 0.01$) and by the four time periods ($F = 10.81$, df = 3/392, $p < 0.01$), while the effects of their inter-

Table 3.   Indices obtained for seven different postures.

Posture no.	WH/WD[a]	Holding time (min)	Estimated postural load (%)	PML index	Pulse per minute	Discomfort site
1	25/25	4·27	27·86	2·32	98·6	Lower legs
2	75/75	4·25	27·97	2·40	105·6	Lower back, arms
3	100/100	3·57	31·86	1·08	105·9	Upper arms, shoulders
4	125/50	3·26	33·89	2·38	106·6	Shoulders, arms
5	150/50	3·22	34·16	2·57	108·2	Arms, shoulders, neck
6	50/100	2·35	41·20	0·81	109·8	Lower legs
7	25/100	1·57	50·20	0·32	106·0	Lower legs

[a]WH, work height; WD, work distance.

action were not significant. However, obviously such significant effects of postures and of relative posture holding time were observed on pulse rate, the use of this variable as an index of postural stress has several shortcomings. First, it seems that pulse rate does not reflect static load reliably. Manenica (1982) showed that pulse rate was at a significantly lower level during static muscle work than during a comparative dynamic exercise. Furthermore, the length of posture holding time may have effects on pulse rate as well. As shown in this study, for example, postures 4, 5 and 6 resulted in a higher pulse rate on average than posture 7, which according to the posture holding time, was more severe. This means that a comparison of different static load situations may be inappropriate when this index is used for their assessment.

Analysis of subjective discomfort estimation was also carried out for the four periods, expressed as percentages of the total holding time. The average estimation for every subject and every posture was found for each of these periods. Two-way analysis of variance showed significant changes in the estimation in the course of time $(F = 601\cdot44, df = 3/392, p < 0\cdot001)$ which could normally be expected. There were no effects of the seven postures on the estimation of discomfort through the four periods, however $(F = 0\cdot71, df = 6/392, p > 0\cdot05)$. Interaction was not significant either $(F = 0\cdot41, df = 18/392, p > 0\cdot05)$. These results suggest that postural discomfort developed equally for the same proportions of the postural holding time, irrespective of the severity of the posture (Figure 2). This allows direct comparison of different postures regardless of their absolute holding times. Regression lines fitted between these time percentages and average discomfort estimates within the particular portion of time are shown in Table 4. The general regression line between the two sets of values seems very reliable, since the correlation coefficient was $r = 0\cdot99, p < 0\cdot001$.

From such a relationship, the relative time spent in the posture, or the relative time still to hold, can be calculated from the estimates of discomfort at any point

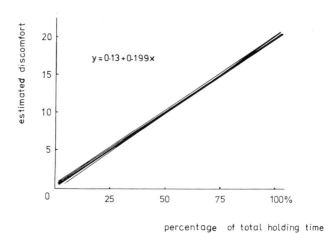

Figure 2.   The relationship between postural discomfort and the relative holding time.

Table 4.   Changes of discomfort estimation in the course of time, and regression parameters for different postures.

Posture no.	Percentage of holding time				Regression line	
	25	50	75	100	*a*	*b*
I	4·20	10·27	15·27	19·87	−0·24	0·203
2	4·33	11·33	16·00	19·87	0·03	0·205
3	4·53	10·67	15·33	19·13	0·27	0·198
4	4·81	11(13	15·53	19·40	0·29	0·198
5	5·33	10·73	15·13	19·67	0·00	0·196
6	4·67	10·07	15·20	19·20	0·04	0·196
7	5·13	10·73	16·20	19·33	0·33	0·199
Overall	4·72	10·70	15·52	19·50	0·13	0·199

in time. If, for example, we ask the subject to estimate postural discomfort on a 20-point scale, and tell him to report when discomfort increases by one point on the scale, the elapsed time between the two judgements will represent 5% of the total posture holding time. Knowledge of the total holding time then allows calculation of postural severity by the use of the previously obtained force–time regression line.

Suppose the subject estimated discomfort as 5 on a 20-point scale. After 0·20 min (12 s) he reported that the discomfort reached the value of 6. According to the proposed scheme, a value of 0·20 min represents 5% of the total posture holding time, which means that total posture holding time will be 4 min. By use of the force–time relationship it can be calculated that the force applied by the relatively most loaded muscle was 29·32% of its maximal force.

In conclusion, it may be said that this study showed once more that subjective judgements may be reliable and relatively stable parameters, especially in situations where complex psychophysiological and/or biomechanical interactions can be expected. The technique for postural load assessment proposed here seems easy to learn and easy enough to use. Its application does not require highly qualified personnel or expensive equipment. Furthermore, the severity of postures that occur for short times, but frequently during the work day, and for which the cumulative effects could cause problems within the human locomotor system, may be assessed by the technique. The technique may be applied in a real working situation as well as in the laboratory.

Chapter 24
Comparison of Some Indices of Postural Load Assessment

Vladimir Taksić

Department of Psychology, Faculty of Science and Arts, Zadar, Yugoslavia

1. Introduction

Posture is one of the most important factors of work, and may affect the physical or psychophysiological state of the operator. Inappropriate working postures can result in unnecessary static effort, overloading of the muscle groups involved, 'fatigue' and postural pains. Because of this, it is important to establish optimal working postures which will result in the well-being of the worker, eliminating short-term and long-term health consequences of bad postures. Improved postures will also indirectly contribute to working efficiency, productivity and the humanization of work.

There are, however, some methodological problems when trying to assess the load on the person of different working postures. The difficulty seems to be the choice of the indicators of postural load. One of the indicators, which can be used in experimental situations only, is the measurement of postural holding time and recording of body areas affected by pains and the intensity of such pains. In the laboratory such a procedure can include the measurement of other indices, such as pulse rate, blood pressure changes, changes in respiration and EMG of the muscles involved. The use of these variables has limitations when applied to static work and in actual field conditions.

Static and dynamic workload have a different effect on some bodily systems, especially the muscular, cardiovascular and respiratory systems (Manenica 1982). Very often, because of these differences in effects, the usefulness of the indicators is rather limited and specific to types of work, making their comparison difficult.

One frequently used index is subjective assessment, which is based on introspective transformation of the effects of postures on some established scale. The validity of such an approach was shown by Borg (1962, 1973, 1978), who found high correlations between subjective load assessment, on the one hand, and work effort on the other. It has been shown that such an approach could be generally used for different aspects of work evaluation, as well as for assessing individual reactions in the same work situations.

The aim of the study reported here was to compare different indicators of postural load. The problem was of a methodological nature and included the definition of the relevant criteria and examination of the discriminative validity of different indicators.

2. Method

A group of 15 female students, aged 19–22 years, took part in one experiment, which consisted of a tapping task performed in seven different postures. The postures were set up using combinations of working height and working distance, which were defined on the basis of percentages of individuals' shoulder height and maximum extended arm length, respectively (Corlett and Manenica 1980). Postures were defined as follows:

Table 1. *Definition of postures for a tapping task.*

	Posture						
	1	2	3	4	5	6	7
Working height (shoulder height = 100%)	25	50	150	125	100	75	75
Working distance (arm length = 100%)	100	100	50	50	100	75	25

The tapping task consisted of the subjects being asked to tap at their own speed as regularly as possible, on a microswitch, which was connected to an Apple II computer where time intervals between successive taps were recorded and stored.

The task was performed by all subjects in seven different postures in a randomized experimental design. Pulse rates were recorded continuously by the use of standard ECG apparatus. Furthermore, subjects were asked every 30 s to assess discomfort on a 20-point scale (Borg 1973).

3. Results and Discussion

The results analysed here are for maximum holding time (HT), pulse rate (absolute values, APR, and differences from the resting pulse rate values, RPR), and postural discomfort or pain (PD), each of the last three being recorded after 30 and 60 s of holding the posture. These results were included in the analysis because they were available for every posture including the shortest-held one (Table 2).

One-way analysis of variance, with one independent variable of 'posture', showed a significant effect of the seven postures on all indices. The F-ratios and their levels of significance are shown in Table 3. As can be seen all the indicators were significantly discriminative amongst the postures. To establish their discriminative validity more precisely, a Newman–Keuls multiple *post hoc* comparison was carried out. The results of this analysis are shown in Table 4.

Table 2. *Means (M) and standard deviations (SD) of seven measures for seven postures.*

Posture	HT		APR 30		APR 60		RPR 30		RPR 60		PD 30		PD 60	
	M	SD	M	SD	M	SD	M	SD	M	SD	M	SD	M	SD
1	94·2	43·0	100·4	8·8	108·0	7·1	17·33	9·30	24·93	9·58	8·86	5·52	13·73	5·58
2	141·1	93·1	101·9	15·7	110·0	17·5	18·79	10·84	26·93	11·92	6·200	4·69	10·60	5·31
3	193·3	115·7	98·4	15·1	101·6	13·6	16·39	8·32	19·60	7·60	3·00	2·77	5·53	3·41
4	195·8	91·8	90·1	10·6	95·3	13·7	8·53	7·83	13·73	7·40	4·00	3·50	6·59	4·45
5	214·0	82·6	102·8	17·4	104·3	17·0	23·06	12·75	24·53	10·29	4·33	3·08	7·80	3·91
6	255·1	103·3	100·4	12·0	103·3	14·8	20·93	7·59	23·86	8·76	3·06	1·09	6·86	2·69
7	256·1	122·8	102·4	16·7	103·9	16·3	17·0	9·06	18·66	6·74	3·79	1·89	7·26	3·36

Table 3. F-ratios for different indices of postural load.

Indices	HT	APR 30	APR 60	RPR 30	RPR 60	PD 30	PD 60
$F (6/84)$	9·03	3·72	4·70	4·54	5·22	8·23	11·72
p	0·001	0·01	0·001	0·001	0·001	0·001	0·001

Table 4. Number of discriminations amongst pairs of postures via different indicators (Newman–Keuls procedure).

Indices	HT	APR 30	APR 60	RPR 30	RPR 60	PD 30	PD 60
$p < 0·01$	7	0	2	2	4	5	6
$p > 0·05$	7	3	2	3	5	6	8

According to these results, the greatest discriminative validity was shown by posture holding time, on the basis of which seven pairs of postures can be differentiated, and by subjective assessment of postural discomfort, with discrimination amongst eight pairs of postures. The poorest discriminative validity was shown by absolute pulse rate. There could be several reasons for this. First, this index does not take into account individual differences in resting pulse rate. Secondly, pulse rate is not a very good indicator for static effort, because, as Myhre and Andersen (1971) pointed out, the continuous isometric contractions may cause occlusion of blood vessels in the muscles. Furthermore, the pulse rate was recorded after a relatively short time period which was not sufficiently long enough to activate the cardiovascular system.

An attempt to overcome this problem was made by the use, also as an index, of the differences between recorded and resting pulse rate. It was found that this index (RPR) was somewhat more discriminative than APR (Tables 4 and 5). The criterion validity of the indices shows the correlation between them and some other accepted indication of postural load. For the purpose of this analysis, maximal holding time within the posture was taken as the criterion of postural load (Table 5).

To find out the predictive validity of the indicators, step-wise backward multiple regression analysis (Gorman and Primavera 1981) was performed. This kind of analysis isolates these combinations of predictors with significant beta-ponders (or standard partial regression coefficients) which explain the variance of the criterion variable the best (Guildford and Fruchter 1976). This regression analysis (Table 6) shows that subjective estimation of discomfort after 30 and after 60 s (PD 30 and PD 60) had the best predictive validity, together with differences of pulse rate from rest taken after 60 s in the posture (RPR 60). This combination of predictors may explain 58% of the variance of the criterion variable but, as can be seen from Table 6, the subjective assessment after 60 s (PD 60) explained 50% of the variance by itself.

Table 5.   Cross-correlation coefficients amongst different indicators.

	APR 30	APR 60	RPR 30	RPR 60	PD 30	PD 60
1. HT	−0·09	−0·19	−0·16	−0·32*	−0·57*	−0·71*
2. APR 30		0·92*	0·46*	0·42*	−0·09	−0·04
3. APR 60			0·30*	0·49*	−0·02	0·02
4. RPR 30				0·83*	0·01	0·07
5. RPR 60					0·11	0·16
6. PD 30						0·92*
7. PD 60						—

*$p < 0.01$.

Table 6.   Regression analysis with holding time as the criterion variable.

Indices	R	R^2	df	F	p	Partial r
Step 4	0·76	0·58	3/101	45·86	0·001	
RPR 60						−0·28
PD 30						0·25
PD 60						−0·55
Step 5	0·74	0·55	2/102	61·54	0·001	
RPR 60						−0·30
PD 60						−0·71
Step 6	−0·71	0·50	1/103	104·7	0·001	
PD 60						

4.   Conclusion

It can be concluded that pulse rate indices are not good indicators of postural load, at least not after such short time intervals within the load as examined here. Some studies have shown that cardiovascular indices become more reliable as the holding time increases (Manenica 1982).

This present study showed, however, that subjective assessment of discomfort was the most discriminative index, after a short time in the posture, and also showed a high correlation with posture holding time. This technique is further recommended because it is easier to use than those that include physiological measurements.

Chapter 25
Posture Recording by Direct Observation, Questionnaire Assessment and Instrumentation: a Comparison Based on a Recent Field Study

D. Baty, P. W. Buckle and D. A. Stubbs

Ergonomics and Materials Handling Research Unit, Robens Institute, University of Surrey, Guildford, U.K.

1. Introduction

Many methods exist for recording posture in the work environment. These have been reviewed elsewhere (Colombini *et al.* 1985 b, Stubbs 1984). The reliability of such measures made using these methods must be considered when establishing the extent to which they provide a true assessment of the postures under investigation. In addition, the relationship between different measures should be quantified in order to aid the comparison of results from different methodologies, especially when new measures, or those borrowed from a different field of research, are used. This paper examines such quantification and comparison, using as examples three measures implemented during a recent field study of posture and back pain in nurses (Stubbs *et al.* 1984). These measures were:

1. Direct observation of posture during a shift.
2. Questionnaire assessment of posture.
3. Instrumented assessment of stooped postures using an inclinometer.

2. Methodologies

Direct observation

Time sampled observations were made of nurses undergoing the activities of a normal working shift. The data were recorded on a printed coding sheet (Figure 1). Postures for inclusion were chosen because of their association with the development of musculo-skeletal disorders at work (Buckle 1984). Consideration of the literature suggested the inclusion of standing, sitting, bending, twisting, pushing, pulling, lifting and carrying as possible back-pain risk factors.

    In addition, 'bracing' (i.e., supporting oneself), 'supporting' (i.e., supporting someone else, for example walking with a patient), walking, kneeling and squatting were included because of their biomechanical importance for this

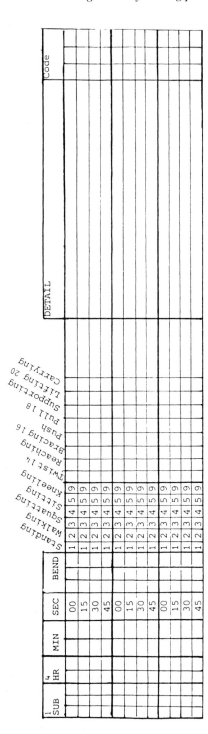

Figure 1.   Posture coding sheet.

particular investigation. Observations were made for 46 subjects for the duration of a full shift. Observers worked in pairs alternating each 20 min, and recording the postural activity of the subject every 15 s when cued by an auditory signal.

Reliability of direct observations

In order to obtain a measure of reliability between the seven observers, paired tests were run before and during the period of the field observations using a synchronized auditory cue for both observers. Twenty-one such tests were carried out with different combinations of observers. Each pair observed together for a 20 min period, under normal conditions for the field work. Percentage agreements were assessed on the basis of counting and identifying observation of a particular behaviour by both observers at the same time as 'correct', and an observation by either observer alone as 'incorrect', then,

$$\text{percentage agreement} = [\text{correct}/(\text{correct} + \text{incorrect})] \times 100\%.$$

The subjects were nurses working in the speciality of geriatric nursing in four health regions. The sample of nurses was chosen to match by grade the national distribution of nurses in geriatric hospitals.

Self-reported data and observational data

Questionnaire data were obtained by interview at the end of the shift observed. The questionnaire is described elsewhere (Stubbs *et al.* 1984). Some static and dynamic postural components of the job, as perceived by the subject concerned, were selected for comparison with the observational data. The 46 subjects were asked to estimate the amount of time spent during the day in the following activities: standing, sitting, kneeling, walking and stooping or bending. Each estimate was subsequently compared with the observed levels of activity for that subject during that shift. Postural activities were expressed by subjects in real time (i.e., hours and minutes) and subsequently converted to a percentage of the working day or shift.

Instrumented assessment of trunk inclination

An instrumented assessment of trunk inclination was made continuously throughout the shift using an inclinometer and recorded on an Oxford Systems Medilog ambulatory recording system. The calibrated inclinometer was placed on the subjects' sternums, 2 cm below the sternal notch, and subjects were asked to maintain their normal standing posture for a few minutes prior to starting the normal activities of the shift, in order to obtain a baseline for subsequent analysis.

Data recorded on tape were analysed on a Northstar Horizon microcomputer, with output data displayed as percentage times spent in each of the ranges $0-15°$, $16-30°$, $31-45°$, $46-60°$, $61-75°$ and $> 75°$ for the whole shift. Seven subjects were monitored for a full shift using the inclinometer, together with direct observation.

In addition, each subject gave a subjective estimate of the duration of stooped activity as part of the questionnaire assessment at the end of the shift.

3. Results

Results of the full direct observation study are reported elsewhere (Stubbs *et al.* 1984). The results reported here concentrate on the methodological aspects of the study: reliability of the direct observations, comparison with subjective estimates and the problems encountered with the inclinometer.

Reliability of direct observations

Whilst detailed assessment of trunk inclination in the laboratory had provided acceptable results, it was apparent from the field trials that it was not possible to obtain good reliability between observers in this more complex setting. However, when data were broken down into only two categories (less than 15° and greater than 15°), an acceptable level of reliability was found (see Table 1 a). This was therefore used in the study as a quantitative criterion for a bend having occurred. Reliability results for observations of 'stand', 'walk', and 'sit' carried coefficients of reliability of between 80 and 97% (see Table 1 a). Whilst squatting and kneeling postures occurred only infrequently in any 20 min trial period, 22 out of a total 23 observations of 'squat' and 10 out of 11 observations of kneeling activity were in agreement (Table 1 b).

Results for the other variables under investigation are shown in Table 1 b. The total number of agreements has been compared with the total number of observations and reveals large differences between categories of postural behaviour. A value of 75% was chosen to represent an acceptable standard of interobserver reliability. With this in mind, only the observations of 'bend', 'stand', 'walk', 'kneel', 'sit' and 'squat' were used for further analysis in the main study (Stubbs *et al.* 1984), although 'support', 'push' and 'pull' were reported along with their associated reliability coefficients. Data for the remaining variables were not reported.

Questionnaire results and comparison with observation results

The results of the subjective assessment of postural activity are shown in Table 2. (Note categories 'stand', 'sit', 'kneel' and 'walk' are mutually exclusive, but that 'stoop' can occur concurrently with any of them.)

A Wilcoxon signed-rank test for matched pairs was carried out to compare these data with those gathered by direct observation. Significant differences were found for the variables 'stand', 'sit', 'walk' and 'kneel'. Whilst significant differences between the two different measures of standing and walking do exist, of perhaps greater importance are the interactions between the estimates of postures. For example, subjects' estimates of walking and standing are almost reversed when

Table 1.   *Reliability results for direct observation.*
a. *Frequent behaviour categories.*

Observation	Mean agreement (%)	SD (%)
Bend[a]	81	15
Stand	90	7
Walk	80	11
Sit	97	4

[a]Bends > 15°.

b. *Infrequent behaviour categories expressed as absolute frequency over 1200 observations.*

Posture	Observations			Agreement (%)
	Correct	Incorrect	Total	
Squat	22	1	23	96
Kneel	10	1	11	91
Push	23	13	36	64
Pull	15	9	24	63
Support	16	8	24	67
Lift	9	10	19	47
Brace	59	82	141	42
Carry	23	53	76	30
Reach	20	74	94	22
Twist	19	74	93	20

Table 2.   *Observed and estimated working postures (as a percentage of the working day).*

	Observed (n = 46)	Estimated (n = 46)	Difference	Significance[b]
Stand	53·1 (9·9)[a]	19·1 (14·1)	+34·0	0·0001
Sit	23·9 (11·2)	19·2 (12·6)	+4·7	0·01
Kneel	1·3 (2·8)	8·9 (10·7)	−7·6	0·0001
Walk	19·5 (4·1)	52·9 (18·4)	−33·3	0·0001
Stoop[c]	26·7 (7·2)	32·2 (22·6)	−5·5	N.S.
Stand + walk	72·6 (11·6)	72·0 (15·4)	+0·6	N.S.

[a]Figures in parentheses are standard errors.
[b]Wilcoxon signed-rank test.
[c]Bends > 15°.

compared with the observed results. Similarly, the combined estimate of standing + walking is very close to the observed figure, approximately 72%. It is also worth noting that whilst a significant difference between the estimates for sitting was recorded the absolute difference between the means was less than 5%.

Positive correlations were shown between estimaters and observations of stooping ($r=0\cdot3$, $p>0\cdot05$, df=44), standing+walking ($r=0\cdot4$, $p<0\cdot01$, df=44) and kneeling ($r=0\cdot4$, $p<0\cdot005$, df=44) but no significant correlations were found for standing, sitting or walking.

Inclinometer results and their comparison with observation results

Inclinometer data from seven subjects are reported in Table 3. The degree of stoop has been considered in the same two categories as used for the observation data (i.e., bends less than $15°$ and bends greater than $15°$). These data are compared with the observation data for the same subjects and are expressed as proportion of the day spent stooped at an angle greater than $15°$.

Inspection of Table 3 shows a similarity both in terms of magnitude and direction of differences between the two measures of stooping activity. The correlation between the two ($r=0\cdot63$, $n=7$ pairs) is not, however, significant. A paired t-test also shows significant differences between the two ($p<0\cdot005$).

Table 3.   *Observed and inclinometer assessments of stooped postures.*[a]

Subject	Inclinometer (%)	Observed (%)	Difference
A	16·5	6·2	+10·3
B	27·2	14·6	+12·6
C	32·5	18·8	+13·7
D	36·9	11·6	+25·3
E	37·3	34·6	+2·7
F	48·2	19·0	+29·8
G	53·1	27·7	+25·4

[a]Bends $>15°$.

4. Discussion

In order to provide an overview for this discussion, the major points raised are summarized in Table 4. In this paper the use of three methods of recording postural activity at work has been outlined. It is clear that differences exist between the methods to assess posture, in terms of their output. These differences are discussed below. It would also seem that the use of several overlapping measures in one study provides a measure of quality control over each measure. For example, it may be said in retrospect that the inclinometer had obvious flaws, but it was the comparison with the observation data that prompted a more careful and critical look at the instrumented data.

The results of the reliability trials for the direct observations produced a wide range of agreement coefficients which were dependent on the posture observed. In addition, interobserver reliability varied considerably between observers. It is clear that this is a product of a number of interacting problems. Some postural behaviours are harder to define than others; for example, it is generally easier to say whether someone is sitting down, than it is to say precisely when a person is twisting. In addition, the number and type of variables to be observed will affect the ease with which each individual behaviour itself is observed; the environment and training will also have a direct effect on the ability of each observer to cope with the observation task.

Using this method of assessing reliability, some postures, for one or more of the reasons outlined above, provided less acceptable data than others. It was felt that the variables 'twist', 'brace', 'reach', 'lift' and 'carry' proved too unreliable for use in this particular work. It may be that in another environment, with better definitions of each posture category and more rigorous observer training, some of these could be used effectively. Conversely, it would seem from this work that assessment of trunk angle and of overall posture is acceptable with the direct observation method.

It may also be interesting to note that whilst interobserver reliability is one of the most common measures of reliability in the assessment of behaviour, it does not actually assess accuracy unless it is related to some predefined standard. Interobserver reliability is therefore a good measure to report, only if the reliability of the instrument has been previously established to some standard measure.

The application of direct observation techniques in the field of postural recording is important, both in terms of an economical method of recording data in the field situation and for the analysis of video material in the laboratory. Activity or work sampling has been in existence for over 50 years as a standard work-analysis tool in industry. In addition, adaptation of micromotion study techniques has been made recently, using Therblig analysis (see Barnes 1968), modified to analyse hand movements in relation to repetitive strain disorders (Armstrong *et al.* 1982). In a similar way, the assessment of gross postural activity reported here draws on techniques developed for research into animal and human behaviour. It is suggested that greater consideration is given in the future to such issues as the selection of observation categories, type of sampling and reliability. In addition, full use should be made of behavioural research literature (e.g., Altmann 1974, Sackett 1978), since the problems and techniques are the same. New technology now allows the researcher to use digital encoding to replace the old pencil and paper method of recording. Several research groups have reported the use of such an approach, in which field observations of behaviour are keyed into a portable microcomputer, with both RAM and off-line tape storage. On return to the laboratory, the tape-recorded data may be off-loaded onto larger machines for analysis (Brown *et al.* 1984, Flowers and Leger 1982).

The questionnaire study has been limited by the number of subjects and the restricted number of postural activities, but it has illustrated the importance of assessing the validity of questionnaire data in this type of study. Even allowing for

Table 4. *Summary table.*

Measure variable	Direct observation	Questionnaire assessments	Inclinometer
General	Careful consideration of the choice and definition of categories for observation should be made Observer training should be carried out until the recorded data, assessed against some objective measure, and inter-observer reliability reach the required standard	Subjective assessments are only of use if previously validated in some way	A good measure in the laboratory environment and for the measurement of static work postures or those involving ony slow changes Solid attachment of the device to the subject is essential In the field situation, the device should be damped to exclude artefacts resulting from high acceleration components
Bend	Adequate measure of stooped posture if coded simply into two categories (e.g., $\leqslant 15°$ and $> 15°$) to provide a quantitative criterion for a bend having occurred		

Stand	Relatively easy to code, but the transition between stand and walk (and vice versa) is common and leads to errors from timing of events in interobserver reliability trials	The reliability of these has been questioned, but may depend upon the job under investigation. It is suggested that some jobs will be easier to self-assess than others; for instance variable jobs with many unpredictable elements may be more difficult to assess than more routine work
Sit	Easy to code, less transitions than stand or walk	The use of subjective postural estimates for quantification of postural activity requires further research
Squat Kneel Push Pull	Infrequent occurrence, but data suggests that both are easy to observe reliably Moderately easy to observe in the dynamic sense (e.g. pushing a trolley), but it is difficult to define these categories in the static sense	
Lift Carry	Careful definition of a minimum load for these events would provide relatively easy observation categories (e.g., only only record lifts with loads over 10 kg in weight). In many work situations (e.g., packing one particular box of products) this would be relatively easy	
Reach	Difficult to code even if a very strict definition is applied, e.g., only code if 'reach distance > 0·5 arm length in an envelope between ±45° to the horizontal'. Complex definitions like this need more training	
Twist	Very difficult to define for direct observation and not easy to measure in the dynamic anthropometric sense	

some variation in the reliability of the observation data, the findings reported here tend to cast doubt on results from other studies where risk factors have been determined at an absolute level following a questionnaire assessment. In previous studies with suitable control groups it seems likely that the estimates of relative risk are in the right direction and in the correct proportion, but their interpretation in the light of this study may be biased; the quantification of risk factors can be no better than the accuracy of the subjective responses used to provide that quantification. Without some measure of concurrent validity to assess the extent of possible inaccuracies in the subjective assessments it is not possible to place full confidence in the coefficients of relative risk derived from them. It is felt that future studies of this type should include separate validations of subjective assessment of postural activity to provide a measure of any possible differences.

It was noted that subjects' estimates of time spent on their feet were closer to the observed figure for the two combined than either standing or walking alone. It is suggested that two factors influenced this result. Firstly, that the questionnaire was presented to a subject set who maintained a high level of activity for a large part of the day, both in static postures and whilst moving around, and that in such cases it may be difficult to discriminate between the two activities. Secondly, that 'standing still' may have implied 'standing still doing nothing'. It is possible that in an environment where activities are less varied, subjects' recall of the day's postural activity would be better than those reported here; it may therefore be of interest to examine variation in such subjective estimates with respect to occupation, and perhaps also with sex and age.

Results from the inclinometer study revealed a pattern of stooping activity bearing only superficial similarity to that obtained by the direct observation methodology. This is partly accounted for by difficulties in attaching the inclinometer to the subject such that it remained securely fixed throughout a full shift. More importantly differences arose from using an inclinometer based on an accelerometer, with an output proportional to the bending moment of a beam under gravity. This relationship holds for static postures and very slow changes of angle but with fast changes of angle an acceleration (positive or negative) is imposed over and above g, creating an artefact, leading to misleadingly high angles of stoop being recorded. A specific problem occurs with walking, when accelerations transmitted from heel strike may be registered by the inclinometer and confused with recordings of trunk flexion. In spite of these difficulties, it is felt that such an inclinometer is a useful tool for posture recording during laboratory work, and with electronic damping to exclude high acceleration components has since been found to be suitable for monitoring posture in the work environment.

Acknowledgement

The authors owe particular thanks to Mr. Bob Fernandes of the University of Surrey for the design of the ambulatory inclinometer system, and for the provision of electronics expertise throughout this study.

Chapter 26

Cooperation between Ergonomists and Workers in the Study of Posture in Order to Modify Work Conditions

S. Montreuil and A. Laville

Laboratoire de physiologie du travail et d'ergonomie du Conservatoire National des Arts et Métiers, Paris, France

1. Introduction

Two groups of workers, from different mass-production industries, requested a team of ergonomists to analyse the effects of their postures on their health, and the origin of such effects with respect to work conditions. The action-investigation which resulted from this worker–ergonomist association embraced certain general principles.

Owing to the development of ergonomic knowledge it is recognized that there are many factors of individual variability (Wisner 1971). The work actually performed by workers is often very different from that stipulated by the work-study and organization and methods departments in firms (Daniellou *et al.* 1983, Leplat 1978, Wisner 1981). Actual work frequently runs counter to the work organization principles applied to large-scale production. Since the worker has his own knowledge of the condition of his body, his health and the way he actually carries out his work, he is thus the essential source of information in order to transform working conditions. The posture taken by a worker results from the compromises he makes in view of his individual characteristics, his work activities and the conditions of work.

As the worker is the best qualified spokesman about his work and its requirements on himself, what should be the procedures to enable him to diagnose the postures he takes and their determinants? Resolving this would be the first important step towards transforming working conditions. The tools for investigating working conditions should be understandable and usable by the worker himself, and devised in view of the constraints the worker is subjected to when he is to act as an 'investigator' in his own work environment.

On the other hand ergonomists have a knowledge of the psychophysiological mechanisms on which postural regulation is based, of the importance of posture in work activity and of the consequences of unbalanced postures for health. They also have the methods to analyse work conditions and connect them to the psychophysiological functioning of the operator.

It is thus by the association of both types of knowledge (for the workers, empirical but concrete, for the ergonomists, theoretical and organized) that an analysis of postures, of their determinants and of their consequences can be undertaken.

2.   Basis of the study

A view of the relationship between working posture and activity

Any occupational activity can be seen as being focused on the achievement of a set production programme with defined materials and time scale. Posture, the arrangement of body segments in space, is but one of the components among many implemented to perform the activity. Study of the body's internal arrangements refers to knowledge in anthropometry, biomechanics and physiology.

Yet posture is not only a matter of standing and sitting but also of 'turning' the body segments with various objectives: searching for and locating outside information so as to identify shapes, relating them to one's body (in remote or near space), preparing the body segments with a view to an action in the environment and performing such action. In other words, it is a matter of organizing space in relation to one's own body so as to locate oneself in it, to move and to act (Paillard 1974). Thus, the regulation of posture stems from the dynamic viewpoint in that it is an active means of gaining spatial knowledge. It is a base for spatial actions within the limits set both by the body's own internal conditions and the arrangement of the environment. In industrial production the range of these limits is comparatively narrow on account of strictly imposed work conditions of space and time. This has been studied in a number of production sectors.

Examples of some posture determinants through observation results

Sewing of garments

A group of female workers have the task of sewing protective gloves: they are on piecework and have very high production speeds. A systematic observation of these workers (Teiger *et al.* 1973) showed that they keep their head and chest leaned forwards in a virtually fixed posture for all the time they are working. This off-balance increases very slightly towards the end of the workday, especially for the tallest workers, while electrical activity at the nape of the neck rises.

The posture of these female workers is determined by a set of fixed locations for a few body segments: feet on the treadles, sitting, hands near the needle, head at a fixed distance from the job and continuously turned towards the sewing area. The self-arrangement of body segments is no longer free to any degree. To this restriction of the actions in space, constraints of work accuracy and speed are added. The result is a continuous postural immobilization and a high muscular cost, explaining the local pains felt by the workers.

It should be noted that no female workers over 25 years are to be found in this dressmaking shop and the workers' heights are clustered around the average. It is possible that speed and accuracy constraints, together with a fixed arrangement in space, account for this gradual selection by age and height.

Work on a display screen

Two investigations were carried out on VDU work. One covered simple data-capture tasks (Pinsky *et al.* 1979) which consisted of transcribing figures (identification number, statement amount, etc.) from a paper document to a computer via a keyboard and display. This transcription is performed very quickly (8000–10 000 characters/hour) and output is monitored. The other study (Dessors *et al.* 1979) deals with data-capture and encoding tasks, consisting of classifying, coding, then capturing complex data on a computer's keyboard and display, but without strict output constraints.

A systematic observation of postures of female operators from both groups shows very little variety of positions taken and only slight variations in these over time among the simple data-capture operators. Among the encoding and capture operators inter- and intra-individual postural variations are more significant. Likewise, variation in the eye–display, eye–keyboard and eye–document distances are smaller for the first group.

In such situations, where there is more flexibility as to where actions are performed than in sewing, it is the high performance speed of the simple task that brings about the highest postural immobility. Postures are taken in relation to where the information needed for performing the work is presented, and where the actions take place. This is done with respect to one's own body which itself has its own individual features and whose functional condition alters along with work duration and worker age. Marcellin and Valentin (1971) showed that the age difference between operators from two car-industry workshops was explained by the greater postural requirements in one of the workshops.

Thus, posture proves to be a compromise between various factors, of which the main ones are shown in Figure 1.

Worker's operating knowledge

A worker has an operating knowledge of his work: that is, he develops a mental representation of the task to be achieved and the means to achieve it, taking into account the knowledge he has of the working of his body, his abilities and his limits (Ochanine 1978). Thus, an operator having to fasten an exhaust pipe under the bodywork of a car knows that he takes it with both hands so as to deliver the pipe horizontally, that he then slides it to insert it into the motor exhaust manifold; and that it does not always fit correctly at the first attempt. Moreover, the fitting of some exhaust pipes will compel the operator to force the pipe on through a pushing and rotating movement.

In addition to knowing the phases of the operation and how to proceed, the operator knows that in order to be efficient his handling actions are associated with a standing posture, arms up, to reach the floor of the hanging car. He must turn his

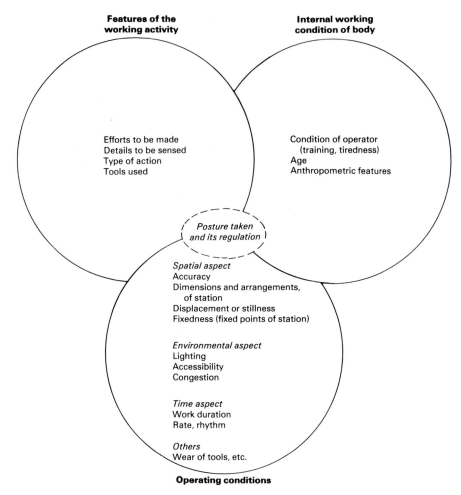

Figure 1. Factors determining the posture taken by a worker.

head round to direct his eyes to the end of the pipe, allowing for the closeness of the floor and the lighting. He also knows that he uses his feet to direct his body so as to increase his pushing strength onto the pipe when fitting it, and that this move should not last long as it could cause pains in his upper limbs and reduce the force applied and the accuracy of motions.

Hence, the worker has an operating knowledge of the way to achieve his task, together with a 'technical' knowledge and knowledge of his own functioning.

3. Approach leading to an action on working condition

We can see that many components come into play when determining postures in a working situation. If one acts to alter postures this will have a consequent effect upon the work and its performance. As it is the workers themselves who know best the pains they experience, the work to be done and its conditions for performance, they are in the best position to identify the determinants of postures and the alterations which would alleviate postural constraints or stresses. Ergonomists have a knowledge of the operating mechanisms of the body: this enables them to explain the empirical and 'experimental' knowledge of workers. Moreover, they can suggest well-founded means for work analysis, work-postures analysis and the analysis of work-posture effects on health.

The diagnosis, by workers, of postures and their determinants with a view to a transformation of work conditions, implies a certain procedure. Overall, it is a matter of training towards action, which requires passing through several stages. Through two experiments carried out with worker groups, the process appears to involve the following stages, each one implying a particular action.

Stage	Target	Action aiming at
1. De-individualize pains or symptoms experienced		Recognizing symptoms, and pains as collective phenomena
2. Being able to analyse, in detail and in concrete terms, the work sutuations and their consequences for health		Training workers in ergonomic analysis
3. Make a diagnosis on postures and their determinants, in work situations		Developing tools, measuring instruments, analysis patterns
4. Transforming work conditions		Interpreting results of analyses in the light of the work situation, and making a choice

The procedure set out here is based on long-term studies conducted with two worker groups in different contexts. The first was female workers in three factories who packaged pharmaceutical products. They performed repetitive tasks on a high-rate machine, which required an immobile standing position with a significant degree of static muscular work. The second group consisted of workers of a large car manufacturer. In this case, work was carried out on a line with work cycles of under 2 min. Operators' tasks required maintaining tiring postures, such as upraised arms—hyper-extension in the case of the overhead line work—or squatting postures—twisted in operations performed within the interior and the boot of the moving car.

A study group was set up in both cases. The groups included both workers, who were elected members of the Health and Safety and Work Conditions Committee, and ergonomists. Their general aim in the medium and long term was to identify the critical elements of work related to postures which bring about symptoms of ill-health and pains, in order to take action so as to remove or alleviate them.

First stage: survey of collective symptoms

Operators complained of postural symptoms, such as pain in the back, or of functional problems, such as the impossibility of keeping some postures at the end of a working day. The first time the groups met they realized that postural discomfort is a rather widespread phenomenon. At this stage the information often emerged from discussions the group members may have had informally with other workers. The following question then arose: to what extent are these symptoms connected with the individual features of workers, and to their state of health?

At this point it proved essential that the groups' members should gain some basic physiological knowledge of the operation of the joints and of muscular work. Viewing this as a part of training towards action, the ergonomists tried to link such knowledge with working conditions. For example, it was shown that back-pain problems can occur without lifting heavy loads; a repetitive action which uses the same muscle group in the same relatively immobile posture also entails a physiological cost. A sheer physiological approach in the early stages could serve to increase disproportionately the perception of the part of individual differences in health in the appearance of symptoms. So, the explanations relating to muscle and joint stresses should refer to overall working conditions, both in order to establish links between conditions and symptoms, and also in order that this knowledge should be more relevant to the real everyday experience of workers in general, and group members in particular. The collective aspects of the phenomenon would not be perceived as such until a first inquiry had been carried out at the working places by the group members themselves. Such investigation may consist of:

1. Taking photographs of reported painful postures at about 50 workstations.
2. Brief observations of postures and their alterations during the day, at certain workstations.
3. Drawing up and disseminating a questionnaire about the health of workers in connection with the main features of their task.

Photographic recording was carried out at the car factory, and first required the permission of the management. The workers of the group themselves chose the stations to be photographed over 2 days. The recording also required the workers' agreement. At every station the 'photographers' discussed with the worker the postures he maintained and their painfulness. Photographs were taken as slides and also with a Polaroid camera. Thus, a point could be shown at once to the worker, allowing a discussion of the difficulties connected with the work from the point of view of postures. Comments recorded ranged from "it's natural, that's my job" to "I'm 52, I have to quit the line".

Postural observations and health questionnaires were administered by the female workers who package pharmaceutical products. Further to the ergonomists' talks on the methodology of ergonomic activity analysis and various observation methods, female workers in the group decided on the type of observations and the choice of stations. Two types of observation were carried out:

1. The identification of the determinants of posture based on a reconstruction of the working environment of a female operator assigned to a machine typical of this class of industry. The posture is determined by the actions, and perception of information, necessary for correct machine operation.

2. The alteration of postures during the working day, according to point recordings of the head, trunk and leg positions at a seated workstation in front of a conveyor belt.

At the same time, group members and ergonomists drew up a questionnaire about the health condition of the female operators, bringing out also the salient features of the task. The aim was to identify to what extent operators suffer pains and other health problems related to the postures necessitated by their work. Subsequent discussion of results by the group showed that although these operators considered themselves to be in good health, the large majority of those who answered the survey reported pains in the back, a good number of them taking medicines or consulting doctors to alleviate either these pains or emotional complaints. Nobody suspected that such phenomena were so widespread.

The group's female worker members decided to construct (with the help of ergonomists) an exhibition so that the survey results would be known by their fellow workers, and also to explain certain physiological processes accounting for the pains. This exhibition was shown at the three pharmaceutical-product factories of the area at public events open to various people.

At the end of this first stage, the conclusions were alike in both groups. That is, it is not only individual fragilities or health problems which bring about pains. There are collective working conditions which speed up the process of the onset of pains, and may even be the prime cause. This fact being noted, the groups then proceeded to analyse in detail these working conditions, in order to be able to transform them.

Second stage: worker's training

For the second stage the groups applied to specialists for assistance in the identification of the causes of postural symptoms in work, and also to establish means of corrective actions. At the very start of the process, training in basic physiological knowledge was given to the groups by ergonomists, with reference to the requirements of the task and the postural determinants. Examples were taken from the actual working environment of the participants in the session. Thus, workers were enabled to come to view differently the tasks they had to perform; they acquired an ergonomic view.

For a particular working station with a cycle of below 2 min, an analysis of the way the operator performs the task takes place in about 2 hours. Shown in detail are the extent to which a search for task information (either tactile or visual), the preparation of the body segments for the action to be performed and a lack of space determine, for the large part, the posture taken. Workers were trained to analyse closely the complexity of every situation, allowing for the interaction of its various components.

An example was the work performed in assembling the steering column at one station on the car line. It had a 90 s cycle, where the operator had to fit the column axle into the intermediate axle already installed on the car. The operator maintained he could not manage to perform the task without pain, and that he could not see easily what he was doing. He said the hardest operation was to fit both ends of the axle.

The general posture was standing, feet outside the car, trunk bent far down and rotating inside the car. The right arm, slightly flexed, was directed upwards so that the right hand could catch and turn the upper part of the steering axle under the instrument panel. The left arm was flexed and directed towards the floor of the car to hold the lower part of the column. The head was tilted sideways, rotated to direct the eyes to the underside of the dashboard at the place where both components were fitted (Figure 2).

The time during which this posture could be maintained was determined by the speed at which the operator could perform the fitting, which depended upon the quality of fit of the two components. The very unbalanced posture was constantly altered as the line, which moved forward continuously, caused the upper body to move in the inside of the car while the feet remained on the ground outside. The feet followed this continuous advance in short steps (discontinuous motion).

Figure 2.   Posture taken when assembling the steering column.

Therefore it could be seen that the posture was determined by:

1. The places the hands must, and could, reach.
2. The place requiring visual inspection.
3. The spatial features of the inside of the car and its height above the ground.
4. The continuous advance of the site of the manual operation in relation to the foothold on the ground.

Following the detailed analysis of the operations carried out and the posture taken to perform them, ergonomists directed the discussion in the group towards the reasons which determine the location of the operator's hands and head (eyes) and the time for which they must be held at this location during the assembly.

Which actions could improve the situation? It was concluded that, in this situation, the parts to be fitted do not offer enough tactile guiding marks. The operator works blindly and gropes around to find the correct fitting of every column end. Thus, the time for which this posture, which the operator deems the most painful, must be held is lengthened unnecessarily.

Moreover, on the principle that it should be possible for the operator to see what he is working on, while keeping an upright and balanced posture, the solution would be of a quite different order. This station should be located elsewhere in the production process or else the car should be slanted so that this operation could be performed in a normal standing position. Whether involving a major or relatively minor transformation of the working situation, it becomes obvious that the available solutions will only be successful if based on the analysis of the interaction between posture and work.

The same bases apply to diagnostic tools for analysing postural constraints. Such tools must be devised in view of the working situations encountered in the industrial field, of the types of action which will be taken following the diagnosis and also of the constraints upon a worker who becomes an investigator.

Third stage: preparation of the diagnostic tools for analysis of work postures

For the last 10 years or so, several investigators have dedicated themselves to developing patterns of analysis for working postures. These are useful for assessing the postures taken by people and their painfulness. However, they are less useful as an aid to transforming the working conditions for they scarcely detail inter-relations between posture and work activity.

It is known that work situations are difficult to analyse, and a system of analysis prepared with reference to a particular situation would be difficult to generalize if it is based upon specific operations to be found in a given type of work. In our experiments, the tools prepared were based on the elements of the theoretical model discussed at the start of the paper. Moreover, as these experiments took place with the intention of transforming work conditions, postures were analysed with reference to the work activity so as to identify the work elements that can and should be altered in each situation. It should be noted that the groups' work was not concerned with the accurate quantification of the painfulness of one posture in relation to another.

Analysis instruments must be basic, concrete and quick to fill-in, as they will be used by investigating workers who will have little time at the working places.

According to the theoretical model, the criteria selected to prepare the tools are as follows:

1. What should the operator be able to see in order to do the work? (Locations of work; closeness of detail.)

2. What should the operator do and where should this be carried out? (Location of upper limbs; efforts to be made; accuracy required.)

3. For how long? (Time for which postures must be kept.)

These criteria are at the root of the analysis of all the operations of a cycle at a workstation, or of a suboperation of the cycle in which the operator takes up what he considers to be the most painful position.

First, the investigating worker should question the operator about his state of health and eyesight, collecting information also on age, height, length of service at the workstation, and the most painful posture taken to do the work. The items which must then be described in detail are:

1. Operation(s). This consists of describing the operation while stressing the features of the activity: type of action, efforts to be made, visual inspection required, tools used, type of part to be fastened, etc.

2. Variability. According to the information collected from the operator or from the experience of the investigating worker, this item allows one to anticipate some problems which occur when performing the operation, problems that are due to the hazards that appear during work (accidents, parts of uneven quality, uneven supply of parts, previous operations unfinished or imperfect, etc.)

3. Posture. The most efficient way to describe the posture has proved to be where the investigating worker notes the position of the eyes and hands and then reconstructs, on a three-dimensional articulated model, the position of the head, the trunk and the lower limbs.

4. Postural determinants. The reasons for adopting every posture described should be stated. This consists of identifying the conditions of work (space, time and environmental aspects) which will determine the postural compromises made.

The full procedure is shown for the example of the steering column assembly (Figure 2) in Table 1.

The tool does not predetermine the contents of each item. The ergonomists have tried to clarify the procedures by setting out several alternatives and supplying help in categorizing the operations, the postures and their determinants. They also supplied the worker with a set of photographs showing the typical postures taken by operators in production processes. Workers could use a set of pictures of postures typical for the workstations analysed. However, allowing for the degree of accuracy required in the description for the identification of postural determinants, and the complexity of each work situation, these aids did not prove

Table 1. *Example of a work posture analysis for assembling the steering column. Frequent behaviour categories.*

Operation	Variability	Posture	Factors determining the posture
To fit the column axle into the intermediate axle already pre-installed on the car 1. Matching the two extremities 2. Inserting one axle in the other by matching the grooves within each axle	Quality of the parts Correctness of the previous operation (pre-installation of the lower component) Time require to assemble the two axles will be different according to these difficulties and the posture must thus be maintained for a variable length of time	Standing Feet outside the car Chest bent far down and rotated Upper right limb slightly flexed, directed upwards Upper left limb flexed and directed towards the floor Head tilted sideways and rotated Short steps	Height of the car: 60 cm The right hand must catch and turn the upper component of the steering axle under the instrument panel The left hand holds the lower component The eyes are directed to the underside of the dash-board where both components are fitted The continuous advance of the line

very useful. Therefore, the only appropriate tool for the groups of investigating workers appears to be a very general memorandum or check list. Its use will only be fruitful provided the understanding and the application of the theoretical model have been assimilated fully during the training stage.

Fourth stage: transformation of work conditions

Transformation of work conditions is the ultimate aim of this whole procedure. In the working situations which have been analysed to date, the actions aimed at transforming work conditions to reduce postural stresses are at different levels. In our experiments, the choice of the level and even of the action to be taken does not belong to the ergonomists, who act as outside consultants. Their role is to set out the various alternatives available in the situation, and their implications.

The decisions as to which alternative courses of action to take at the working stations belong to the people first concerned, i.e., the workers and the employers. The actions chosen are often determined by socio-economic constraints on the firm, over which the ergonomists have no control. However, they will be able to demonstrate that the difficulties met by the operators in doing their work have consequences for the company's performance as well as for the workers' health.

4. Conclusion

The posture taken by a worker results from a compromise he makes between constraints of his own characteristics, those of his actual work and of the working conditions. The workers know these factors exist but need theoretical knowledge to understand their importance and their interrelations.

By their participation in the process of workplace and posture assessment, workers do not turn into ergonomics specialists. However, pooling the new knowledge they gain with their own existing experience permits them to discuss and evaluate greater or lesser modifications to workplaces. The final decision-makers in the firm should then take into account this more ergonomic view and must be more careful in considering the effects of working conditions on workers' health.

Posture is an integral part of working activities. In our experiments, the individual analysis of working conditions is necessary in order to cope with the complexity of each individual situation, in view of the interactions of the various factors determining the posture taken. In order to improve the reliability of the process described here, it seems important to expand knowledge on all the physiological functions which may regulate posture in occupational activities. The implementation of this knowledge to determine the relationship between the critical factors of the work and workplace and the physiological functions involved in the regulation of posture would allow the preparation of tools which are generally applicable in the field of diagnosis and intervention in work situations.

Acknowledgements

This study is the result of both ergonomists' and workers' action. It has been initiated and followed by IN.P.A.C.T. (INstitut Pour l'Amélioration des Conditions de Travail—Confédération Française Démocratique des Travailleurs CFDT).

SECTION 5

SEATS AND SITTING

The introduction of computer terminals as general office tools, with their consequences for changes in the content of jobs, has caused major changes in office activities and an increase in the proportion of time that many office workers spend sitting. The result has been to reveal the inadequacies of current workplace arrangements for the long-term operation of keyboards. The different work tasks of the traditional secretarial worker and the data-entry operator have thrown into sharp relief the differing seating requirements, demonstrating that their superficial resemblance (for they are both keyboard operators) is not sufficient grounds for taking decisions about the desirable arrangements for all keyboard users, regardless of what their other tasks may be.

The controlling nature of the tasks to be done on the design of the seated workplace is a point which is very evident from most of the following papers. It is a point which is always worthy of emphasis. The new interest in seating, which tends to emphasize the office and has arisen in great part from the introduction of keyboard workstations in offices, can cause us to forget that there are many millions of workers doing other jobs, all equally in need of seating and all with requirements which are to a great extent set by the work tasks, equipment, etc., with which they are engaged.

The effects of the job on the design of the seat has had less attention than the relationship between the passive seated posture and the shape of the seat itself. Early work by Åkerblom (1954), which has had a major effect on subsequent thinking, illustrated, for example, an easy chair with lumbar support and an obtuse seat–back angle in which the subject was knitting. Concern with under-thigh pressure, pressure distribution across the seat surface and the position of the lumbar support was extensive, leading to important standards for what were, effectively, dining chairs (BSI 1965). A study of a range of office chairs (Shackel *et al.* 1969) demonstrated sharply the inadequacy of one standard for office-type seating, when different chairs amongst the group tested were preferred for different purposes. The limitations of the single standard, the underlying structure of which had unduly influenced a group of experts who had been asked to judge the chairs, led to their predictions concerning the adequacy or otherwise of the various chairs being quite at variance with the users' preferences. The inadequacy of their model of sitting led to inaccurate predictions concerning performance.

The experience of the insufficiency of models is not a strange one to any ergonomist, frequently made aware of inadequacies in this respect by experimental results which do not elucidate the real world. It is certainly a factor in the differences of view held by workers in the field of seating, that the extent of the models differs between different workers. Some models are more comprehensive than others, linking posture, forces, subjective measures and performance, whilst others, for example, concentrate primarily on the biomechanic aspects.

Even the attempts at overall models have, as yet, major weaknesses. It is difficult to incorporate the effects of variations in posture and forces over the working day, as will be evident from earlier sections of this book. The proposals in Chapter 28, by Eklund and Corlett, are deficient, too, in that they do not cope with those two major influences on performance and health, motivation and time on task; probably because there are, as yet, no clear measures of these factors, only evidence for their importance.

However, there *are* good reasons for studying a limited part of the field, a major one being in the search for relationships between a selected range of the factors involved. It is arguable, for example, that until we know the effects of muscle activity in loading the spine, and the effects of these loads in changing the disposition of the various body segments, attempts to make a useful predictive model of seating are premature. Andersson's review, which immediately follows, discusses a range of studies which has provided many of the recent data for our current understanding of the postural problems of seating. Their importance can be seen from the contributions they make to the work of all the other authors in this section; these data are basic to further advances in the understanding of seating and its problems.

The two chapters following the review (Eklund *et al.* and Colombini *et al.*) provide two studies where a number of methods have been used to investigate each of the problems discussed. The former chapter brings to bear a range of measures to investigate, in particular, the effect of the angle of the head on spinal load when subjects are seated. The latter chapter is also, in part, concerned with the effect of head position and introduces as a dependent variable a measure of discomfort. The comparisons between the various measures, in both chapters, are supportive and the studies suggest that, within the limitations of the sensitivity, repeatability and accuracy of each of the methods, there can be a limited substitution between them.

The chapter by Mandal presents a logical outcome from a study of the postures adopted by people at their workplaces. He demonstrates the well-known effects of an upright sitting posture with horizontal seat surface in reducing lumbar lordosis, and the maintenance of that lordosis by means of a sloping-forward seat. He avers that a backrest is not needed if full lumbar lordosis is maintained and supports his contention by reference, *inter alia*, to the widespread use of chairs to his design. Chapter 31, by Bendix, whilst recognizing the importance of seat slope and of the maintenance of lumbar lordosis, reports studies where a backrest was preferred, and used, with beneficial reductions in back loads. The relationships between the desk and seat, as outlined by Bendix, whilst being acceptable for short periods, demonstrate that no seat is satisfactory if used continuously, and can give very different impressions of comfort if trials are only pursued over short periods of time. A true test of a chair must include a period of observation of its use extending over several days, together with discomfort records and work activities. Otherwise a test routine such as that described by Drury and Coury (1982) should be followed, in which control is maintained over some of the interactions and judgements are anchored. With some agreement on test procedures it is probable that a number of the current arguments on seating would disappear; certainly there would be fewer wild claims for 'the universal ergonomic chair' and similar nonsense.

The final chapter in this section, by Lepoutre and co-workers, presents a study where the subjects were able to provide a more than usual amount of input to the investigation. At the end of his chapter Mandal comments that perhaps "we in future consult the consumers and let them decide what type of furniture they prefer". This is what this last chapter has done, in a controlled manner and demonstrating clearly the futility of specifying a seated workspace without recognizing that individual tasks and human differences will have effects on the quality of the seating posture. It is also notable that they suggest the possibility of using the controlled assessments of subjective feelings as indicators of adequate seat height, rather than postural measures, the latter being difficult to interpret.

As must be evident from many of the chapters in this book, the problems of the interpretation of posture are amongst the most intractable, to a great extent because clear relationships between posture, effort and disease are still uncertain and require extended and expensive studies to identify. The apparently logical interpretation of biomechanical studies or of measures of spinal loading rests on the reasonable assumption that less load is better. As yet we know little about the effects of variations in load, which are the norm in any work situation, and the specification of criteria involving load fluctuations, against which postural loadings can be evaluated, is a prerequisite for further advances in the study of working postures. The emphasis on the controlled evaluation of subjective data must be right, therefore, if design errors are to be minimized, but these are not a satisfactory substitute for more extensive studies on the lines of Westgaard, in this book. There have been studies in the ergonomics literature over many years, e.g., Floyd and Ward (1966), van Wely (1970) and Giles (1981), which have linked poor work design and bodily damage arising from the resultant postures but, as yet, the major epidemiological work remains to be done.

Chapter 27
Loads on the Spine during Sitting

Gunnar B. J. Andersson

Department of Orthopaedic Surgery, Sahlgren Hospital, S-413 45 Göteborg, Sweden

1. Introduction

Sitting is a posture in which the weight of the body is transferred to a supporting area by mainly the ischial tuberosities and their surrounding soft tissues. Depending on the chair and posture, some proportion of the body weight is also transferred to the floor, the work surface, the backrest and the armrests of the chair. The advantages of sitting compared with standing postures are that sitting: (*a*) provides stability required on tasks with high visual and motor control components; (*b*) is less energy consuming; (*c*) places less stress on the joints of the lower extremities; and (*d*) lowers the hydrostatic pressure on the lower extremity circulation.

As will be apparent biomechanical considerations must be kept in mind when designing a sitting workplace to achieve these advantages, and not cause undue stresses to particularly the back and shoulders. When biomechanical aspects of sitting are considered the spine is usually emphasized, but the lower and upper extremities must also be considered. Here I will mainly consider the spine.

Although there are large individual variations among people in the shape of the spine when assuming different standing and sitting postures and when using different chairs, there are common distinguishing features (Åkerblom 1948, Andersson and Örtengren 1974 c, Andersson *et al.* 1979, Carlsöö 1963, Keegan 1953, Schoberth 1962).

When standing erect, the lumbar spine is lordotic partly because the vertebrae and discs are thicker anteriorly than posteriorly, and partly because the upper surface of the sacrum is at an angle to the horizontal plane. As the sacrum is firmly attached to the pelvis it follows that a rotational movement of the pelvis changes the sacral–horizontal angle and thus influences the shape of the lumbar spine. A forward rotation of the pelvis causes increased lordosis in order to maintain an upright trunk posture, while a backward rotation causes the lumbar spine to flatten and sometimes kyphosis can develop. Typically when sitting the pelvis is rotated backwards, and the lumbar lordosis flattens (Figure 1).

A group of muscles at the back of the thighs, the hamstring muscles, influence the configuration of the lumbar spine and pelvis because they run from the lower

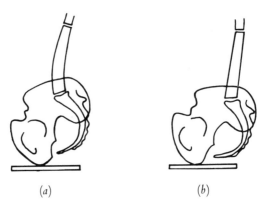

(a)   (b)

Figure 1.   When sitting down (b) from standing (a) the pelvis rotates backwards and the lumbar spine flattens.

limb to the pelvis, crossing both the hip and knee joints. Extension of the knee joint may, for example, result in backward rotation of the pelvis and flattening of the lumbar spine.

In general, the posture of a seated person depends not only on the design of the chair, but also on the task to be performed. Anterior sitting postures are adopted most often when desk work is performed, while posterior positions are preferred for resting. The height and inclination of the seat of the chair, combined with the position, shape and inclination of the backrest, and the presence of other types of support combine to influence the resulting posture, as does the location and slope of the work area. Obviously, it is important to provide not only a 'good' chair, but a chair which is also functionally adapted to the task of the occupant. This is particularly important when seated work is considered because even minor changes in the dimensions of the workspace can change the required seated posture considerably.

As it is unlikely that there is a single ideal posture, and furthermore no body posture can be maintained indefinitely, it becomes important that alterations in one's posture also are permitted by the chair. As early as 1924 this basic criterion of a good chair was stated by Vernon (1924) and has later been stressed by several investigators (Åkerblom 1948, Carlsöö 1963, Engdahl 1971, Floyd and Roberts 1958, Keegan 1953, Kroemer 1971, Kroemer and Robinette 1969). To facilitate postural changes between sitting and standing an intermediate posture, *semisitting* is sometimes suggested. In *semisitting* a higher than normal chair is used, usually with a forward-sloping seat that a person leans on, dividing the weight to be borne between the buttocks and feet.

2.   Sitting and the lumbar spine

As stated above, the configuration of the lumbar spine is altered when sitting down from standing. This creates forces within the spine motion segments which need to be considered. The cervical spine is also influenced by sitting, as the field of vision needed to perform a task may require the head to be in a fixed position. I will here concentrate on the lumbar spine and review some of the radiographic and bio-mechanical data available. For additional review, please refer to Chaffin and Andersson (1984).

Radiographic data

Radiographic studies all confirm that the pelvis rotates backwards and the lumbar spine flattens when sitting (Akerblom 1948, Andersson *et al.* 1979, Burandt 1969, Carlsöö 1972, Keegan 1953, Rosemeyer 1972, Schoberth 1962, Umezawa 1971).

Åkerblom, Keegan, Schoberth and others found that the flattening of the lumbar lordosis in sitting can be prevented by the use of a well-designed low-back support. Andersson *et al.* (1979) studied in what way different types of lumbar supports placed at different levels of the lumbar spine influenced a number of angles, including the lumbar lordosis angle, as well as the influence of changes in backrest inclination on those angles (Figure 2).

When moving from a standing to an unsupported sitting position, lumbar lordosis decreased by an average of 38°. This mainly occurred by backward rotation of the pelvis (average 28°). The remaining 10° were changes in the vertebral body angles of the two lower lumbar segments. There were also small angular changes between the vertebral bodies of L1 and L2, and between L2 and L3. Changes in the sacroiliac joint angles were found to be about 4°. The results are summarized in Table 1.

When a back support was used, there was an increase in the total lumbar angle (increased lordosis) as well as in the individual lumbar vertebral body angles. An increase in backrest inclination from 90 to 110° had a slight decreasing effect on the

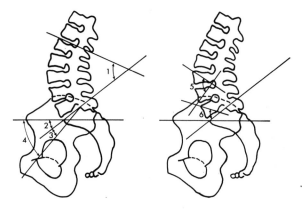

Figure 2.   Definitions of angles studied: 1, total lumbar angle; 2, sacral–horizontal angle; 3, sacral–pelvic angle; 4, pelvic–horizontal angle; 5, vertebral body angles L1/l·, L2/L3, L3/L4, L4/L5; 6, the L5/S1 angle.
Adapted from Andersson et al. (1979).

Table 1.   Mean angle and standard error of the mean in standing and unsupported sitting in eight subjects.

Angle[a]	Standing	Unsupported sitting
Total lumbar angle (1)	59·2 (2·9)[b]	22·2 (5·9)
Sacral–horizontal angle (2)	38·0 (1·9)	17·1 (4·1)
Pelvic–horizontal angle (4)	63·8 (4·9)	35·7 (2·2)
Vertebral body angles (5)		
L1/L2	80·2 (3·2)	79·4 (2·9)
L2/L3	87·6 (3·4)	82·8 (2·9)
L3/L4	89·6 (4·8)	81·6 (3·0)
L4/L5	94·2 (3·6)	84·0 (2·6)
L5/S1 angle (6)	58·4 (1·8)	53·8 (2·6)

[a] Angle numbers correspond to those used in Figure 2.
[b] Numbers in parentheses are standard errors.

total lumbar angle. The main postural change induced by an increase in the backrest–seat angle was a backward rotation of the pelvis, causing the sacral–horizontal angle to change from a mean value of 34° to a mean value of 3° and the pelvic–horizontal angle from a mean of 53 to about 22°. There was minimal motion in the sacroiliac joints and in individual lumbar vertebral body angles.

The amount of lumbar support had a marked influence on the total lumbar angle, which changed from a mean of 9·7 to 46·8° when the lumbar support was increased. The curvature of the lumbar spine, when a lumbar support was placed 4 cm in front of the plane of the backrest and the backrest inclination was 110°, closely resembled the lumbar curve of the standing position. The location of the lumbar support with respect to the level of the spine (L1/L2 or L4/L5) did not significantly influence any of the angles measured in the lumbar region.

Based on this study the following conclusions can be made. The shape of the lumbar spine during sitting seems to be a result of rotation of the pelvis. To balance the trunk in standing, lordosis is usually required as the sacral end-plate is almost always inclined forwards. When sitting, the sacral end-plate is in a more horizontal position and the normal lumbar lordosis becomes flattened. Schoberth (1962) found that a sacral–horizontal angle of 16° or more was a requirement for normal lumbar lordosis in relaxed unsupported sitting.

In order to prevent flattening of the lumbar spine in sitting, and resulting deformation forces, several alternatives have been suggested. A forward-tilted seat was proposed by Burandt (1969), Carlsöö (1963) and Mandal (1975), while Rosemeyer (1972) suggested fixation of the pelvis on the seat by a pillow support. More common, however, are suggestions of supports for either the whole back or the lumbar spine only. The influence of a lumbar support seems to be of greater importance in this respect than a particular backrest inclination, or the height of the backrest. Arguments for and against a forward-sloping seat will be given elsewhere in this volume and will not be repeated here.

The location of the lumbar support with respect to the level of the lumbar spine was of little importance to the shape of the lumbar curve in the study discussed above but may of course be critical to perceived comfort. The finding is of practical importance, particularly when a chair permits changes in backrest–seat inclination but has a fixed lumbar support. The backrest inclination is usually changed by rotation about a fixed axis, which, according to Snorrason (1968 a) is often located about 17 cm posterior to the rotational axis of the body at the ischial tuberosities. A simple geometric calculation shows that the lumbar support moves about 4·5 cm upwards with respect to the lumbar spine when the inclination of the backrest is increased from 90 to 105°, almost a whole lumbar segment. Additional backward inclination increases the upward movement of the lumbar pad even more, thus reducing the support needed in the lower lumbar area (Figure 3).

The knee flexion angle is also important in seat design as mentioned previously. The same is true for the hip flexion angle. Keegan (1953) obtained radiographs of one subject in a lateral recumbent position and found that the lumbar curve flattened when a 90° trunk–thigh angle was obtained, compared with a 135° angle. He attributed this to the combined actions of the various muscles that rotate the pelvis and thus influence the lumbar curve.

Disc-pressure data

Nachemson and Morris (1964) published data on *in vivo* disc-pressure measurements of subjects standing and sitting without support. The pressures measured when standing were found to be about 35% of those when sitting. These findings were later confirmed by Okushima (1970), Nachemson and Elfström (1970) and Tzivian *et al.* (1971).

In the early 1970s a series of disc-pressure measurements were made in the third

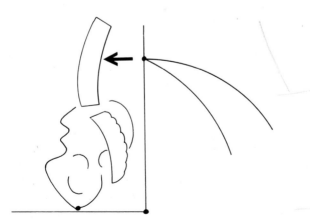

Figure 3.   When a backrest with a fixed lumbar support is inclined backwards the support moves up along the spine.
Adapted from Andersson et al. (1979).

lumbar discs of subjects standing and sitting in different chairs, and with
different chair supports (Andersson 1974, Andersson and Örtengren 1974 a, b, c,
Andersson *et al.* 1974 a, b, c). These studies confirmed that the disc pressure is
considerably lower in standing than in unsupported sitting (Figure 4). Of different
unsupported sitting postures the lowest pressure was found when sitting with the
back erect (in lordosis). The reason for the increased pressure in sitting postures can
be attributed to an increase in the trunk load moment when the pelvis is rotated
backwards and the lumbar spine and torso is rotated forwards, as well as to the
deformation of the disc itself caused by lumbar spine flattening.

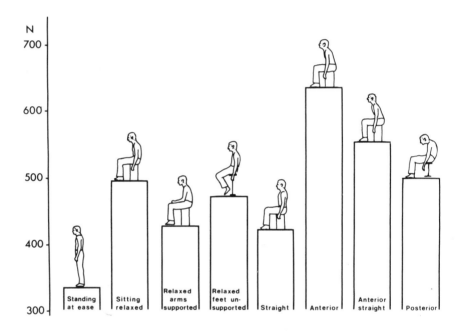

*Figure 4.   L3 loads calculated from disc pressures measured in standing and unsupported sitting.
Based on Andersson et al. (1974 b).*

When supports were added to the chair, disc pressure was found to be influenced by several of the support parameters studied. Inclination of the backrest
resulted in a decrease in disc pressure, especially when tilting the backrest from
vertical to 110° (Figure 5). An increase in lumbar support resulted also in a decrease
in disc pressure. The decrease was generally larger when the backrest–seat angle
was small. The use of armrests always resulted in a decrease in disc pressure, less
pronounced, however, when the backrest–seat angle was large (Figure 6).
    The results can be interpreted as follows:

    1. An increase in backrest inclination means an increase in load transfer to the
backrest and reduced disc pressure.

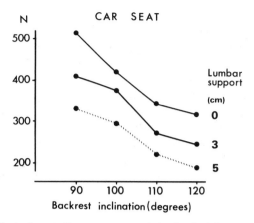

Figure 5. *When the backrest inclination is increased the disc load decreases. A change in lumbar support towards lordosis reduces the load further.*
Adapted from Andersson et al. (1974 c).

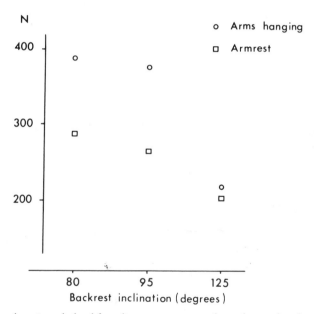

Figure 6. *Loads on L3 calculated from disc pressures measured in a chair with and without armrests. The use of armrests has its greatest unloading effect when sitting upright.*
Adapted from Andersson and Örtengren (1974 b).

2. The use of a lumbar support changes the posture of the lumbar spine towards lordosis, and hence reduces the deformation of the lumbar motion segments.

3. The use of armrests supports the weight of the arms reducing the disc pressure.

Studies were also made of typical seated office work (Figure 7). When writing at a desk a decrease in disc pressure was noted, compared with other tasks. This was expected since the arms were supported by the desk. Other office activities, such as typing and lifting a phone at arm's length increased the pressure due to the larger external load moments placed on the spine during such tasks.

Figure 7.   Loads on the L3 disc calculated from disc-pressure measurements during seated office work. Adapted from Andersson and Örtengren (1974 a).

Muscle activity

Electromyography has been used to study the activity of back muscles when sitting. Generally, similar activity levels have been recorded when standing and sitting (Andersson and Örtengren 1974 c, Andersson *et al.* 1974 a, Carlsöö 1963, Floyd and Silver 1955, Rosemeyer 1971). There is also agreement that, in sitting, the myoelectric activity decreases when (*a*) the back is slumped forward in full flexion (Åkerblom 1948, Floyd and Silver 1955, Jonsson 1970 a, Lundervold 1951 a, b, Schoberth 1962) and (*b*) the arms are supported (Andersson and Örtengren 1974 c, Andersson *et al.* 1974 a, Carlsöö 1963, Floyd and Ward 1969, Rosemeyer 1971) or a backrest is used (Åkerblom 1948, Andersson and Örtengren 1974 c, Andersson *et al.* 1974 a, Carlsöö 1963, Floyd and Ward 1969, Knutsson *et al.* 1966, Rosemeyer 1971).

   Of the different chair support parameters the backrest has been found to be particularly important, with the EMG levels decreasing as the backrest–seat angle is increased (Andersson *et al.* 1974, Hosea *et al.* 1985, Knutsson *et al.* 1966,

Rosemeyer 1971, Yamaguchi *et al.* 1972) (Figure 8). Andersson *et al.* (1974) also found that the myoelectric activity not only decreased in the lumbar region, but also in the thoracic and cervical areas of the spine when backrest inclination was increased. When the angle was greater than 110°, however, there was little further effect on the EMG levels. Hosea *et al.* (1985) found significant decreases in both

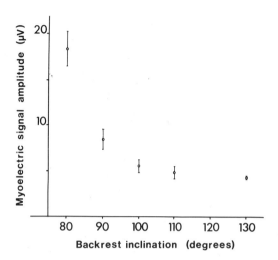

Figure 8.   Myoelectric activity was reduced when the backrest–seat angle was increased. Adapted from Andersson and Örtengren (1974 c).

thoracic and lumbar level myoelectric activities with increasing backrest inclination, with a minimum at 120°. Their tests were performed during the actual driving of cars. Knutsson *et al.* (1966) studied the effect of an additional lumbar support on EMG levels and found that muscle activity was reduced, both when the support was in front of and behind the plane of the backrest. Andersson *et al.* (1974), on the other hand, found the influence of a lumbar support to be small. Hosea *et al.* (1985) recorded significantly less activity with a 5 cm positive lumbar support than with a 3 and 7 cm support. In fact, 7 cm resulted in a large increase in activity.

Yamaguchi *et al.* (1972) found that muscle activity decreased when the seat inclination was increased backwards. This can be due to more effective use of the backrest. Hosea *et al.* (1985) could not detect any influence of seat inclination. The influence of different locations of the backrest on myoelectric activity was studied by Lundervold (1951 a, b, 1958) and Floyd and Roberts (1958). They found a comparatively lower activity when the back support was placed in the lumbar region than in the thoracic, and thus confirmed a finding by Åkerblom (1948) that a support in the lumbar region was as 'effective' as a full back support.

Carlsöö (1963) studied the importance of the angle of the knees on EMG activity. Flexion beyond 90° was found to increase myoelectric activity while there was a decrease with extension of the knees. This is probably due to the hamstring muscles pulling on the pelvis. The effect of both the height of the seat and of the table has also been investigated. Too high or too low a seat has been found to increase muscle activity (Laurig 1969, Lundervold 1951 a, b, 1958) and the vertical distance between the seat and the table appears also to be an important factor (Andersson *et al.* 1974, Carlsöö 1963, Chaffin 1973, Floyd and Ward 1969, Laurig 1969, Lundervold 1951 a, b, 1958, Schoberth 1962).

3.   Discussion

It appears from disc pressure and myoelectric data that the load on the spine increases when sitting without a back support, compared with when standing, and that this is mainly due to a change in the shape of the lumbar spine. Reduced low-back stress levels can be expected, however, by the use of proper back supports. The most important factor in reducing low-back stress is the backrest, and the most important parameter in backrest design is its inclination angle. By addition of a lumbar support the stress on the back can be further reduced, particularly when sitting upright. Such support should be placed in the lumbar region to achieve greater lordosis when sitting. In order to provide as much comfort as possible it should be adjustable in both height and size. The evidence presented also suggests that support of the arms can reduce low-back stress further as well as assuring that the seat is adjustable to the proper height.

Acknowledgements

This review is in part based upon original research supported by The Swedish Work Environment Fund and Volvo AB.

Chapter 28
Experimental and Biomechanical Analysis of Seating

J. A. E. Eklund

Division of Industrial Ergonomics, Linköping Institute of Technology,
S-581 83 Linköping, Sweden

and E. N. Corlett

Department of Production Engineering and Production Management, University of
Nottingham, University Park, Nottingham, NG7 2RD, U.K.

1. Introduction

There appears to be an increase in the number of jobs involving seated manual activities. Reports have shown that discomfort, pain, disease and other impairments arise from such seated tasks as vehicle driving, light mechanical assembly, sewing-machine operating and VDU work (Grandjean and Hünting 1977, Kelsey 1975, Kroemer 1971, Van Wely 1970, Westgaard and Aarås 1984). Many methods for measurement and criteria for evaluation of various aspects of the ergonomics of chairs have been proposed (Åkerblom 1948, G. B. J. Anderson 1974, Eklund and Corlett 1984, Pottier 1969, Shackel et al. 1969). Increased loadings are associated with increases in discomfort, pain and disease (Andersson 1981, Corlett and Bishop 1976). Disc pressures and muscle loads vary when sitting on different chairs, and these are generally higher in work chairs than when standing. The loads also vary with the task performed for any given chair (Andersson 1974, Corlett and Eklund 1984, Eklund et al. 1983). In order to decrease the effects of workloads on the body, proposals have been made for new designs of workchairs (Åkerblom 1948, Mandal 1981, Schneider and Decker 1961).

Much of the development and research on chairs has been concentrated on office chairs and vehicle seats; little has been done in the field of industrial seating. Difficulties have arisen due to the enormous variety of tasks and equipment in industry. Several projects have, however, dealt with the development of a particular seat for a particular application, for example light mechanical assembly work, control-room work, VDU work and fork-lift truck driving. Because of the enormous number of applications, to investigate every application would be very expensive and time consuming. What is needed is to recognize the important interactions between work tasks and seat requirements, so that a limited number of categories can be devised in a systematic classification.

2. Classification of seat requirements

The first major question is, therefore, the understanding of the functions of the chair features and how they relate to the task. Work tasks necessitate a number of actions to be performed by the worker in a more or less defined or constrained manner, which puts various demands on the body. Examples of these demands are the need to see processes or objects, to position the hands in order to manipulate objects and equipment, to exert forces with the hands and to sustain them for certain periods of time. Also, the workplace and workpiece design, dimensions and organization interact by causing constraints, whilst different individuals will have different capabilities and will handle their situation with different strategies; it is obvious that training and skill modify the behaviour and strategy, when performing the task. One example is the comparison of the posture of a skilled vehicle driver and of a beginner.

From this discussion it is obvious that the task and the workplace determine the postures and create a pattern of loadings on the structures of the body of the individual. The chair is one component affecting these loads. Changed chair design can therefore be used as a means to modify loads on body structures; in other words, to reduce uncomfortable, painful and potentially harmful loads. The effects which can be obtained in a given work situation by optimizing the chair are limited. In many cases, improvements should be made to the task or the workplace design as well, rather than just the chair.

Systematic analysis of seat requirements should start with the demands and constraints of the task and the workplace. This will focus attention on relevant factors of the seat. These, in turn, will aid in the choice of methods of measurement, and the criteria for evaluation will also be more evident as a result of the analysis (Table 1).

The system outlined here provides the basis for the understanding of seating in relation to work. The methods of measurement are devised not only for the evaluation of the effectiveness of work seats, but can also be used for the evaluation of working forces on the spine and other body parts.

It is clear that the two most important factors are to see the work and to actually perform the task with the hands. This means positioning and angulation of the head and the hands, which, to a large extent, then define the body posture.

The approach of grouping similar jobs into categories by using the analysis described decreases the number of chair evaluations needed in order to systemize design recommendations for 'work-seat categories'. It also gives a possibility for modular chair design, which allows a smaller number of components to cover most of the needs of various chair designs for industrial workplaces.

Better knowledge is needed concerning the interaction between different design features of chairs, and the task and the workplace. More knowledge is also needed in order to obtain more efficient design and modification changes of tasks and workplaces.

Table 1.　Factors influencing the design of a workseat and the responses arising from them.

Sources of demands and constraints	Initial responses	Subsequent responses
Task		
Positions, directions, precision, forces, repetitiveness, temporal pattern, time and space constraints	Spinal load Spinal shape Muscle load, e.g., back, neck, shoulder, abdominal, legs, arms Surface pressures Pressure on internal organs Changes in blood distribution	Discomfort Pain Disease Performance change
Workplace		
Size, height, distances, angles, visibility, job aids, body support, clothing, space constraints, e.g., leg clearance		
Individual		
Anthropometry Physical capacities		
Measures		
Workplace dimensions Work weights Work forces Work reaches Work time patterns Anthropometry Strength	Biomechanical load EMG Shrinkage Rating scales Dilation of body parts Linear measurements	Rating scales Clinical examinations Epidemiology studies

3.　Methods for seat evaluation

In order to reveal reductions in those loads which give rise to adverse effects, suitable tools for their evaluation must be devised. Three methods have been employed in this project, namely body-height shrinkage, a method for biomechanical calculations in sitting postures and discomfort ratings. The first two methods were developed for the purpose of seat evaluations and were chosen due to their field applicability, low costs for the equipment needed and the possibility of employing the methods on a larger scale in the future, without having to use highly specialized and trained personnel.

The height shrinkage and biomechanical methods evaluate the effect of, and the load on, the spine, respectively. It should be one of the functions of a good work seat that it reduces this load by keeping the forward extension of the arms to a minimum, supporting the back so that work forces may be transmitted through it whilst, at the same time, achieving low-back muscle activity (Corlett and Eklund 1984). Hence, these two methods should enable an effective back support to be identified by comparative studies.

Body-height shrinkage

This method, and the measurement procedures required for it, were recently described by Eklund and Corlett (1984).

The spinal discs have been shown in many studies to have visco-elastic properties. In addition to the elastic compression which arises immediately when discs are loaded, over longer time periods they also compress gradually, or creep. This creep or loss of disc height is similar to an exponential function, where the rate of height loss decreases over time. When the discs are unloaded, they will recover, in a manner also similar to an exponential function. Increased load will increase the rate of creep.

The daily loadings people experience on their backs cause body-height shrinkages of approximately 15 mm. This height loss is regained during the night's sleep. The rate of total disc height loss, measured as body height loss, will therefore reflect the load on the spine (Figure 1). The method also makes it possible to take the temporal pattern of loading into account, by making measurements at appropriate intervals during the day. Similar procedures and equipment have also been used by Reilly *et al.* (1984), who have demonstrated that the method permits body height to be measured with a standard deviation of less than 0·7 mm.

It has also been shown in several experiments that the method is sensitive

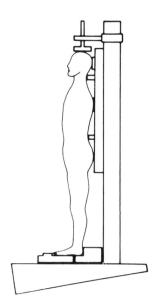

Figure 1.   Diagram of the arrangement for measuring changes in stature.

enough to discriminate between different chair designs, when evaluated under controlled circumstances for time periods as short as 30–45 min (Eklund and Corlett 1984).

A static biomechanical model for evaluation of loads in seated work tasks

The development of a biomechanical model is important both to complement other methods which may be used and also for aiding the theoretical understanding of the problems of sitting. It is also valuable in prediction of the loads likely to arise in specific body parts, under various conditions.

A model has been developed which allows the compressive and shear loads in a horizontal plane at any chosen spinal level to be calculated, for example at L3. It also permits the calculation of the momental load in the sagittal plane around any chosen disc, and the moment around the shoulder joint in the sagittal plane. The spinal compressive load may also be calculated.

Input data are obtained from photographs and force measurements. The magnitude and direction of all external forces acting on the body, reaction forces from floor, seat and backrest and the body weight of the sitter are the data required. Dempster's data (1955 b) on the weight and location of the mass centre of body segments have been used for the calculations in the model (Figure 2).

In order to make the collection of data easier, some simplifications and a two-step calculation can be introduced. If the lower legs are vertical and no active leg muscle force is used, the frictional component of the floor reaction force can be assumed to be 0 ($f = 0$). The use of the three equations of stability allows three forces to be calculated instead of measured.

The three equations of stability are:

$$F_f \sin f + F_s + F_e \cos e + F_h = 0 \qquad (1)$$

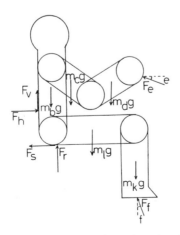

Figure 2.   A diagram showing the forces used for the biomechanical model.

$$F_f \cos f + F_r + F_v + F_e \sin e + mg = 0 \tag{2}$$

$$F_f d_f + F_r d_r + F_e d_e + F_h d_h + F_v d_v + F_s d_s$$
$$+ (m_b d_b + m_c d_c + m_d d_d + m_l d_l + m_k d_k)g = 0 \tag{3}$$

where m is the total body weight, m_b is the weight of the trunk, neck and head, m_c is the weight of the upper arms, m_d is the weight of the lower arms and hands, m_l is the weight of the thighs, m_k is the weight of the lower legs and feet, F_f is the reaction force from the floor, F_r is the vertical reaction force from the seat, F_s is the horizontal reaction force from the seat, F_v is the vertical reaction force from the backrest, F_h is the reaction force from the backrest, F_e is the external force, d_i is the lever arm between force i and point for moment calculation ($i = $ b, c, d, l, k, f, r, s, v, h, or e), and g is gravity.

One or more horizontal planes can be chosen for calculations, for example L3. A free body diagram is constructed at each chosen level or plane for the calculation (Figure 3).

The model equations are:

$$C = (m_c + m_d + m_x)g - F_e \sin \quad - F_v \tag{4}$$

$$S = F_e \cos \quad - F_h \tag{5}$$

$$M_{L_3} = -F_e d_e + (m_b d_b + m_c d_c + m_d d_d)g + F_v d_v + F_h \tag{6}$$

$$M_{shoulder} = (m_c d_{sc} + m_d d_{sd})g - F_e d_{se} \tag{7}$$

$$C_x = C + (M_x/p) \tag{8}$$

where m_x is the weight of the trunk, neck and head above the chosen plane, d_s is the lever arm between the force and the shoulder joint, p is the lever arm between the erector spinae and the centre of disc x, C_x is the compressive force in the plane, S_x is the shear force in the plane, M_x is the moment around disc x in the sagittal plane, $M_{shoulder}$ is the moment around the shoulder joint and C_x is the compressive force on disc x. The signs of force and moment are according to normal conventions. The reaction force from the backrest is distributed as pressure over the contact surface. Calculations cannot be made for planes within this area of contact, unless the pressure distribution is known or can be estimated.

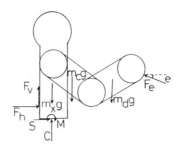

Figure 3.   Free-body diagram for calculation of forces and moments at the L3 level.

Experimental chair

In order to provide input data for the biomechanical model, an experimental base of a chair was constructed. It was instrumented with eight strain-gauged load cells, four under the seat and four behind the backrest. The load cells were designed to measure the perpendicular reaction force but with a minimum of cross sensitivity for shear forces. Separate recordings of each load cell allowed the total reaction force and its position to be calculated. Different prototypes of seats and backrests could quickly be fitted on the chair. Height adjustments of both the seat and backrest could be made with 25 mm increments.

Calculations of loads for different chairs in specific tasks enable a comparison of the effectiveness of chair design features. This method can therefore also be used as a method for seat evaluations.

Discomfort assessment

A main criterion for ergonomic design of a product is that the users accept it. User assessment is nowadays considered important and valid; if used together with other methods, it can be a confirmation that the other factors measured were relevant for the situation.

One method of assessing postural discomfort was presented by Corlett and Bishop (1976). Often, the assessment of discomfort is administered together with body mapping, a method where the perceived discomfort is referred to a part of the body. The subject is asked if he or she experiences discomfort from any part of the body. If so, the subject is asked to mention all body parts with discomfort, starting with the worst, the second worst and so on until all parts have been mentioned. The subject is also asked to assess total discomfort, and discomfort from the worst, or a particular, body part, using a five- or seven-point scale. The scales are graded from no discomfort at all to maximal discomfort (Figure 4 (*a*) and (*b*)).

4.   A seat evaluation study

This study has focused upon three work-task categories, from which musculo-skeletal impairments have been reported. The first category is of tasks demanding exertion of forward push forces, represented by metal grinding and glass cutting. The second category is of tasks demanding vision to one side, represented by sideways sitting during fork-lift truck and other work vehicle driving. The third category is of tasks with restricted knee space or with the work area well forward of the sitter. Examples are repetitive tasks on various machines and some assembly tasks. Observations of musculo-skeletal impairments from the back, neck and shoulders have been reported as a result of all such tasks.

From a theoretical point of view, it can be argued that more of a forwardly exerted force can be transmitted to a high backrest than to a low one. Trunk muscle forces and loads on the spine would be lower as there would be less need for muscular stabilization of the trunk. One can also argue that a low backrest permits

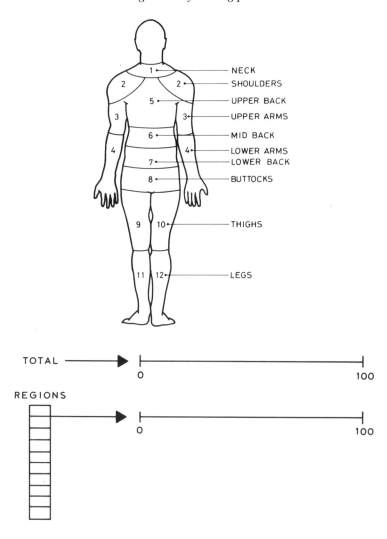

Figure 4. Body map and scales used for discomfort assessment.

a more even spread of the effects of rotation of the thoracic and cervical spine, which would result in less rotation of the neck, compared with a high backrest, in the task demanding vision sideways. Higher trunk muscle activity would also be needed in order to overcome reaction forces from rotation against the high backrest. It can also be argued that a sit-stand seat, in comparison with a conventional chair, increases lordosis and also makes it possible to come nearer to tasks with restricted knee space, two ways in which the spinal load may be lowered.

The hypothesis is that a high backrest is more beneficial in forward pushing tasks, a low backrest is more beneficial in tasks which demand vision to one side and that a sit-stand seat is beneficial where tasks with restricted knee space or with the work area far away are undertaken.

The purpose of the study was not just to evaluate the chair design factors mentioned above, but also to compare the three methods, body-height shrinkage, biomechanical analysis and discomfort assessment.

Laboratory arrangements

In the forward pushing task, the subjects pressed a metal plate by gripping a handle with both hands. The force exerted on the plate was measured and displayed on a digital display in front of the subject, in order to allow the force to be maintained at 25 N. The force was directed 15° below the horizontal plane, in order to compensate for the weight of the work object. The plate was placed approximately 50 cm above the seat surface and positioned to give a 30–35 cm viewing distance. The subjects were allowed 15 s rest every 2 min. The instrumented chair was used with a horizontal seat in combination with either an 18 cm lumbar support or a 38 cm high backrest. The thighs of the subjects were adjusted to the horizontal by the use of a footrest.

In the sideways-viewing task, the subjects watched a video film, with the television set placed 90° to the left of the subject and the chair. The subjects were instructed to look at the screen all the time. The instrumented chair with a horizontal seat was used together with a 30 cm high backrest or a 47 cm high backrest. A footrest was used in order to adjust the thighs of the subjects to the horizontal.

The task with restricted knee space was an assembly task, screwing nuts on bolts. The work area was approximately 20 cm × 20 cm and placed just above elbow level and 5 cm further away than the knee restriction. The subjects worked at their own pace and their performances were measured. The 18 cm lumbar support was provided on the instrumented chair. A horizontal seat or a sit-stand seat was used. The horizontal seat was used together with the footrest in order to get horizontal thighs. The thighs were adjusted to slope 30° forwards in the sit-stand seat. The subjects performed all three tasks with the two chair arrangements for each task. The order was balanced.

Both the sleeping time of the subjects on the night before the experiment, as well as the morning activities before the experiment were controlled. Each experimental session was performed for 45 min in the morning, 75–90 min after the subjects got up from bed. Body height was measured before and after each experimental session. Discomfort assessments and body mapping were made at the start, in the middle and just before the end of each of the sessions. Photographs were taken in the beginning, in the middle and at the end of the sessions. The forces acting on the floor, seat and backrest were measured every 4 min during the experiment. After having performed both experimental sessions for each task, the subjects were asked to choose their preferred chair design, and to compare the discomfort they experienced for the two chair designs. The sideways-viewing task was also video filmed from above for subsequent analysis of head and shoulder twist (Figure 5).

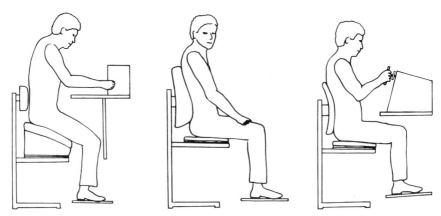

Figure 5.   Typical postures for the pushing task (right), the sideways-viewing task (middle) and the task with restricted knee space, the sit-stand seat (left).

Altogether 10 subjects participated. Their ages varied between 24 and 42 years, with an average of 32 years. Their heights varied between 152 and 191 cm, with an average of 175 cm. Statistical analyses were made using a one-tailed paired *t*-test except for discomfort data, where the Wilcoxon test was used.

5.   Results

In the forward pushing task, the low backrest caused more body-height shrinkage. On average, the subjects shrank 1·37 mm with the low backrest and 0·66 mm with the high backrest. The result is significant at the 5% level (one-tailed). The subjects reported significantly more discomfort when they used the low backrest, and seven out of eight subjects preferred the high backrest. One subject assessed the two backrests as equivalent.

During the forward force exertion the force transmitted through the feet decreased, and there was a tendency for the low backrest to cause a greater decrease of force transmission through the feet and also a more forward bent posture. The mean horizontal backrest force was 22% of body weight for the low backrest and 17% for the high one. The centre of the backrest force was higher for the high backrest. It could also be seen that the resulting shear force, in a plane just below the backrests, was higher for the low backrest.

In the sideways-viewing task, the high backrest caused an average of 1·44 mm body-height shrinkage, and the low backrest 0·88 mm, but the difference was not significant. The subjects reported significantly more discomfort when they used the high backrest. Six subjects preferred the low backrest and two preferred the high.

Analysis of head and shoulder rotation showed that the shoulders were more rotated when the low backrest was used. Consequently, neck rotation was less pronounced. The movement was not pure rotation, but there were also components of sideways moving or bending.

In the assembly task with restricted knee space, the subjects shrank 0·93 mm in the sit-stand seat and 2·41 mm in the conventional chair. The difference was significant at 5%. The subjects reported significantly more discomfort when sitting in the conventional chair. It must be noted that they also reported significantly more discomfort from the lower back in the conventional chair. Seven subjects out of eight preferred the sit-stand seat.

The lumbar support was not used by the subjects when doing the task sitting on the sit-stand seat, and it was hardly used in the conventional chair either. In upright sitting, without performing the task, significantly less horizontal backrest force was measured for the sit-stand seat. The load on the feet was comparable, for the two seats, when the task was performed. The subjects sat a little further away from the task and also had a slightly more bent posture when using the conventional chair. Therefore the lumbar spinal moment was lower for the sit-stand seat.

6. Discussion and conclusions

The conclusions that can be drawn from the experiments are that a high backrest is beneficial in forward pushing tasks, a low backrest is beneficial in sideways-viewing tasks and sit-stand seats are beneficial in assembly tasks with restricted knee space.

In the forward push task, both the more forward bent posture and the pronounced decrease in force through the feet when the push started, can be seen as compensatory mechanisms. It seems to be more likely for adverse effects to occur when the situation produced compensatory mechanisms.

Those subjects who wore glasses experienced more discomfort in the sideways-viewing task and also had to twist their necks more in order to see well. The design of glasses therefore seems to be an important factor for people working in tasks which demand sideways viewing.

Increased trunk–leg angle in the assembly task seems to have a substantial influence on the load on the spine and on the perception of discomfort from the back.

The model for chair evaluation outlined in this paper has been shown to be of value. The three methods are all effective, but have somewhat different fields of application, sensitivity and costs of use. They are particularly interesting to use together because one measures the effect of the loads on the body objectively, one theoretically evaluates the load and can be used predictively, whilst the third method relates directly to how people perceive the chairs.

Shrinkage measures the effect on the whole spinal column, and is therefore less sensitive for loadings on smaller parts of the spine. The method demands experiments of at least 30 min duration. If the differences between two situations are small, the number of subjects or experiments has to be increased.

The biomechanical method is quick and inexpensive. Also very short experiments can give much valuable information. The method has considerable potential for further development.

Discomfort assessment is a very inexpensive and sensitive method. It is also suitable for field studies. The length of the experiment sessions must, however, not be too short, because it often takes some time until the discomfort is noted.

The experiments described here have proposed seating arrangements which have been broadly matched to the physical demands placed on the sitter by the tasks. For comparison, in each of the three tasks, two forms of seat were proposed. By using the task demands to define the chair shape it has been shown that reduced body loads and increased comfort can be obtained. The load changes have been demonstrated both by measuring their effects on the spine and by calculating the resulting forces from the reaction forces on the environment.

Chapter 29

Biomechanical, Electromyographical and Radiological Study of Seated Postures

Daniela Colombini and E. Occhipinti

Unit of Occupational Preventive Medicine, Municipality of Milan, Local Health Unit 75/8, Milan, Italy

C. Frigo and A. Pedotti

Bioengineering Centre, Polytechnic of Milan, "Pro Juventute Don Carlo Gnocchi" Foundation, Milan, Italy

and A. Grieco

Occupational Preventive Medicine Service, University of Milan, Milan, Italy

1. Introduction

An increasing number of people spend their working day sitting down. Many of them complain of back and neck pain which they attribute to the fixed posture demanded by their job (Andersson 1981, Kelsey and White 1980, Magora 1972, Ohara *et al.* 1982). One way of dealing with the problem is to use more comfortable seats with ample possibility of adjustment, and equipped with back and arm supports.

In general, the attention of ergonomics research has been focused on defining 'the most correct' position, even though the most important problem at the workplace is not the posture in itself but for how long it is maintained (Laville 1980). So, even in ergonomically designed workplaces, it can be observed that operators tend to try to vary their position according to the requirements of the job instead of trying to find the so-called 'optimal' position (Cantoni *et al.* 1984). Therefore, the ideal workplace is one which allows the widest body mobility, which can be achieved by proper design and work organization.

In this respect it seemed useful to analyse a number of seated postures which are commonly encountered in office occupations, so as to evaluate the stresses imposed on the different anatomical structures of the spine for each posture examined and to obtain more detailed data for workplace and job design. A biomechanical method supported by electromyographical (EMG) and radiological (Rx) analysis was used for the study, which was aimed at:

    1. Quantifying the stresses on lumbar (L3/L4) and cervical (C6/C7) discs for various sitting postures with arms supported by a work table, with and without back support.
    2. Analysing the activity of lumbar back (near L3) and upper trapezius muscles.
    3. Analysing the position of pelvis and lumbar spine in the different postures examined.

The results will be discussed with a view to improving working posture comfort through the introduction of posture variability.

2. *Methods*

Subjects

Ten healthy volunteers (five men and five women) were studied.
Their average characteristics are reported in Table 1. Four subjects (two men and two women) also underwent radiological examination.

Table 1.   *Average characteristics of the subjects examined.*

	Number of subjects	Mean age (years)		Mean stature (cm)		Mean weight (kg)	
Male	5	32·4	(3·5)[a]	175·4	(3·5)	72·6	(10·6)
Female	5	29	(4·3)	161·8	(4·6)	54·8	(4·6)

[a]Figures in parentheses are standard deviations.

Postures studied

Six sitting postures were studied, characterized by different trunk positions. For each posture, two different head positions were considered: head upright (for instance, looking at a visual display unit (VDU)) (position A); and head bent forward (for instance, reading a sheet on the worktable) (position B). The postures examined are schematized in Figure 1. The arms were always supported by the worktable; the centre of pressure was assumed to be located on the forearms a few centimetres from the elbow. In postures 1–3 no back support was provided; the trunk was upright (posture 1), relaxed with elimination of lumbar lordosis (posture 2) and upright but with the seat plane tilted forward by about 20° (posture 3).
    In postures 4–6 a back support was provided: in posture 4 the trunk was upright and the back support was perpendicular to the seat plane; in posture 5 the trunk was 'slumping' and the buttocks were slid forward, the back support was as in posture 4; in posture 6 the back support was reclined by about 110° with respect to the seat plane. Seat, worktable and back support were adjusted to the appropriate height according to the following criteria: seat plane at knee height; worktable at elbow height; backrest above the L2 vertebra. The VDU was placed 40 cm inside the border of the table in order to allow full support of the forearms.

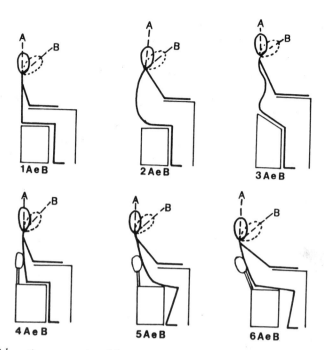

Figure 1.   Schematic representation of the seated postures examined (A and B refer to the upright and bent forward head positions, respectively).

Biomechanical study

A piezoelectric force platform (a Kistler 9261A) mounted on the floor under the chair was used to detect the resultant of the ground reactions. The subject could rest his/her feet on the platform (measuring 60 cm × 40 cm) in all the situations studied. Postural geometry was detected by a TV camera picking up the lateral view of the subject. A special purpose device (Digivec) elaborated the signals from the force platform so as to compute, on-line, the amplitude, the inclination and the point of application of the ground reaction resultant, and superimpose the image of the vector (on an appropriate scale) on the image of the sitting subject (Ambrosini *et al.* 1985). Retroreflective markers, applied to the main 'repere' points on the subject, made for a better identification of the anatomical segments and relative centres of gravity.

The resultant images were recorded on a videotape and further analyses were performed on single frames photographed from a TV monitor. An example of the images so obtained for a single subject is given in Figure 2.

The study was conducted on the sagittal plane and allowed determination of the loads on the lumbar (L3/L4) and cervical (C6/C7) intervertebral discs for each subject and posture. The method for determining lumbar loads (P_{L_3}) is illustrated in Figure 3 in the more general case in which both the arms and the back are supported. As shown in the figure, whilst the seat lay upon the force platform (its

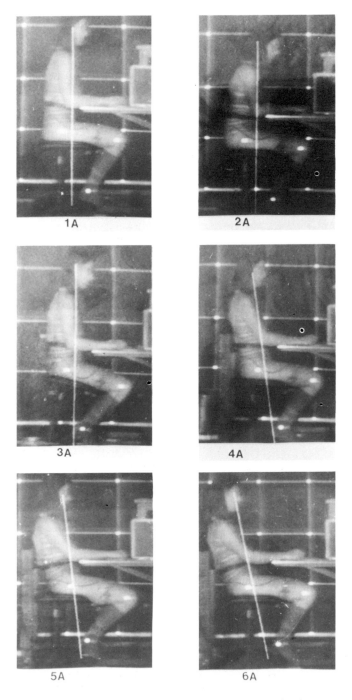

Figure 2.   Images obtained by the Digivec system in the six postures analysed. Superimposed vector represents the ground reaction force.

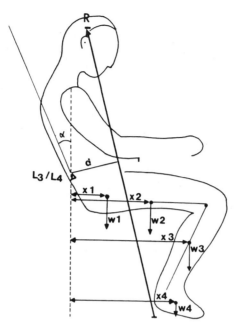

Figure 3.   Scheme for the lumbar load computation using the ground reaction vector. The moment at L3/L4 level is firstly computed (M_{L3}), the back muscle force (F_{mL3}) is obtained by supposing a lever arm of 5 cm; the compressive force on the L3/L4 disc is the sum of the back muscle force and the normal component of the above L3/L4 body weight (W cos α).

weight was previously deducted from the measured load) both the worktable and the back support lay outside. Therefore the ground reaction detected by the force platform was the resultant of the following forces: (a) the weight of the subject; (b) the force of the backrest; and (c) the force of the table supporting the arms. One method of calculating the reactive moment at the L3/L4 level is to consider the moments acting on the part of the body above the L3/L4 disc. This would require the external forces (backrest and table) to be known as well as their lever arm with respect to L3/L4 and the weight and lever arm of all the anatomical segments above the L3/L4 disc (Occhipinti et al. 1985). Alternatively, mechanical torques of the forces above L3/L4 (external forces and body segment weights) are equal to the moment of the ground reaction R, minus the moments due to the weights of the body segments below the L3/L4 level. In the images obtained from the Digivec the ground reaction is shown superimposed on the subject's image on an appropriate scale, so it is convenient to use this second method of calculating the reactive moment.

The required anthropometric data were obtained by scaling values reported in the literature (Dempster 1955 b) to the subject's weight; it was also considered that the weight of all the segments above the L3/L4 disc was 57% of body weight (Ruffs 1950). Stresses on the lumbar intervertebral discs were obtained by referring the loads to the area of the L3/L4 disc of each subject.

Estimation of disc areas was based on the following equation (Colombini *et al.* 1985 a):

$$A_{L3/L4} = 0.95 + 0.002\, W_{SK} \tag{1}$$

where $A_{L3/L4}$ is the area of the L3/L4 intervertebral disc (cm^2) and W_{SK} is the weight of the skeleton (g). Skeleton weight was then computed by the following equation (Matiegka 1960):

$$W_{SK}\,(g) = \frac{a + b + c + d}{4}\, H \times 1.1 \tag{2}$$

where a, b, c and d are the diameters of the wrist, elbow, knee and ankle (cm) and H is the stature (cm). For each position considered the following parameters were computed:

1. Mean load (P_{L3}/cm^2) value, and t-test ($p < 0.05$) to evaluate differences among them.
2. Mean values of the arm and back-support forces (vertical and horizontal components) expressed as percentages of the body weight.

Cervical loads were computed by means of the following formulae:

$$M_{C6} = W_h X_h \tag{3}$$

$$F_{mC6} = M_{C6}/2 \tag{4}$$

$$P_{C6} = F_{mC6} + W_h \cos \alpha \tag{5}$$

where M_{C6} is the moment at the C6/C7 level (kg cm), W_h is the weight of the head (kg), X_h is the distance between the head's centre of gravity and a vertical line crossing the C6/C7 disc (cm), F_{mC6} is the force of the head extensor muscles (kg), α is the forward flexion angle of the head (degrees), P_{C6} is the compression force on the C6/C7 disc (kg) and the lever arm of the head extensor muscles = 2 cm.

The relationships between anatomical parameters and disc surface area are not yet available for the cervical vertebrae, thus the loads were not referred to a surface but merely expressed in force units (kg).

Electromyographic study

Bipolar surface electrodes were placed on the right erector spinae muscle (L3 level, 3 cm from the midline of the back) and on the right trapezius muscle (2 cm from the scapular edge), parallel to the fibres' direction.

Electrodes were connected to a telemetric device and signals were fed into a 12 channel paper recorder (SAN-EI VISIGRAPH 54). Maximum voluntary contraction (MVC) against resistance of the examined muscles was evaluated by measuring the mean amplitude of the trace. Myoelectric activity was then recorded in all the examined postures and the following parameters were computed:

1. The difference (as a percentage) in the back muscle signal amplitude, for each subject, between postures 2–6 and posture 1, used, in this case, as a reference; relative mean values and *t*-test ($p < 0.05$) to evaluate differences among the mean values were also calculated.

2. The difference (as a percentage) of trapezius signal amplitude, for each subject, between head position A and corresponding head position B for the examined postures; relative mean values were also calculated.

Radiological study

This study was performed on four subjects in order to evaluate variations of the sacral angle and lumbar lordosis in four different positions. These were: sitting with the trunk upright (see postures 1 and 4); sitting with the trunk relaxed (see postures 2 and 5); straight trunk sitting with seat-plane tilted forwards 20° (see posture 3), with the addition of an extra position, standing (0).

The following parameters were measured on each radiogram (Figure 4):

1. Sacral angle (SA), in degrees, determined by connecting the tangent line to the S1 plate with a horizontal line (Caillet 1973).

2. Depth of lumbar lordosis (DLL), in millimetres, which is the perpendicular distance from the line *AB* connecting the L1 and L5 vertebral bodies and the

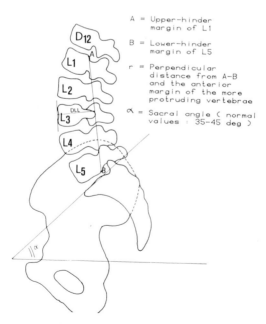

Figure 4.   Schematic reproduction of lateral roentgengram of pelvis and lumbar region in standing upright posture. Conventional definition of sacral angle (SA) and depth of lumbar lordosis (DLL) is shown.

anterior margin of the most advanced lumbar vertebral body (Suzuki and Endo 1983).

Data were analysed in order to obtain the mean differences for the SA and the parameter DLL (as a percentage), respectively, between the standing position and each one of the seated postures. Tilting of the pelvis is defined front-upward when SA and DLL are increased, back-downward when the same are decreased; the two parameters in fact change in the same direction during movement (Caillet 1973).

3. Results and discussion

Lumbo-sacral spine

Biomechanical study

Mean values and standard deviations of the force acting on the L3/L4 intervertebral disc are reported in Figure 5. The significance levels of the differences between means are coded and tabulated in the same figure.

It can be observed that:

1. All the postures with the back supported (4, 5 and 6) show a significantly

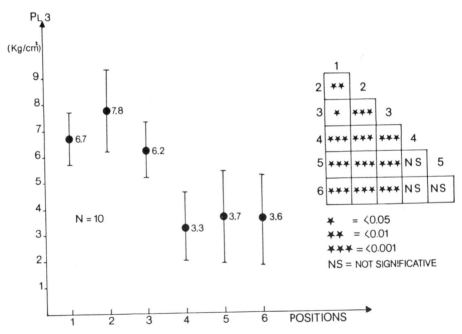

Figure 5.   Lumbar load on the L3/L4 disc for the six examined postures (kg/cm²): mean values and standard deviation. Table inset refers to the significance levels of the differences between pairs of mean values (t-test).

reduced stress ($p < 0.001$) on the lumbar disc with respect to the unsupported postures, 1, 2 and 3 (approximately half).

2. No significant differences were found among the three back-supported postures 4, 5 and 6 as regards the lumbar stress.

3. Differences between the mean lumbar stresses computed in postures 2 and 1 ($p < 0.01$), 3 and 1 ($p < 0.05$) and 3 and 2 ($p < 0.01$) were significant. The greatest lumbar stress occurred in posture 2 and the lowest in posture 3.

Data reported in the literature (Kroemer 1983) show that the inflow–outflow nourishing mechanism of the invertebral disc inverts at a critical value of 80 kg of intradiscal pressure. Considering that intradiscal pressure is about 1·35 times the hydrostatic pressure imposed on a lumbar disc (Schultz *et al.* 1982 b) and referring these data to the L3/L4 discal area of an average subject weighing 70 kg (discal area: 18·3 cm), we obtain a value of this critical hydrostatic pressure of about 3·3 kg/cm. Our data show that back-supported postures produce lumbar stresses ranging around this value, while this is not the case in postures 1, 2 and 3.

Figure 6 reports the differences between the ground reaction measured with no external supports and the ground reaction measured in each of the postures considered. Values are normalized on weight and averaged over the ten subjects examined. In postures 1, 2 and 3 the difference, due to the arm being supported on the table, is a purely vertical vector. In postures 4, 5 and 6 the horizontal and vertical components of such a difference are measured separately. On the vertical component, the effect of the arm support can be reasonably estimated as the mean of the values obtained in postures 1 and 2; the remaining force, consisting of a vertical and horizontal component, can be attributed to the back support.

These data are reported as a guideline for the analysis of seated postures with external supports in cases where the instrumentation used here is not available and biomechanical computation is required (Occhipinti *et al.* 1985).

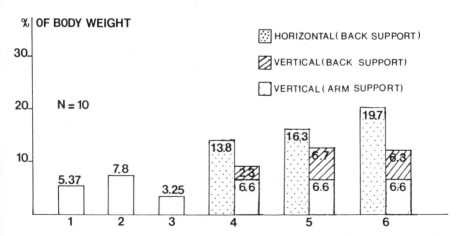

Figure 6.   External forces applied by worktable and backrest: mean values expressed as percentage of the body weight. Vertical and horizontal components are considered separately.

Electromyographic study

Myoelectric activity of the extensor back muscles, as referred to the MVC, showed values ranging from 10 to 15% in the postures with no backrest and from 8 to 11% in those with back support. Posture 1 in general showed the greatest myoelectric activity. Relative variations between the different postures examined were thus referred to posture 1. Figure 7 shows percentage variations of myoelectric activity averaged over the ten subjects. A table inset in the figure reports the significance level of the difference between the means. Lower myoelectric activities were recorded in postures 2 and 5. In all likelihood, reduction or inversion of the lumbar lordosis, increasing of dorsal kyphosis and back-downward tilting of the pelvis cause a stretching of the muscles that can thus exert, despite little or no electric activity, a prevalent elastic action. This is independent of the presence of a backrest. Similar observation can also be made for posture 6 (backward-sloped backrest). In posture 3 (forward-sloping seat) a relative reduction of myoelectric activity was observed with respect to posture 1, although the back is not supported here either. It should be noted in this case that the centre of gravity of the body is close to the lumbar fulcrum (ground reaction vector closer to the L3/L4 intervertebral disc) and thus a lower force is required on the back muscles. The smallest decrement in myoelectric activity with respect to posture 1 was recorded in posture 4 (similar to 1 except for the backrest).

The muscular contraction recorded in this posture can be explained by the need to maintain the trunk erect with a physiological lordosis. The reaction force of the backrest is prevalently horizontal and little vertical support is given to the body.

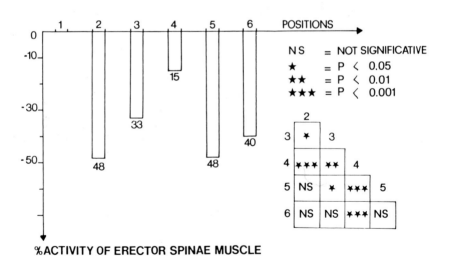

%ACTIVITY OF ERECTOR SPINAE MUSCLE

Figure 7.   Mean values of percentual reduction of myoelectric activity, in each of the postures examined relative to posture 1. The table inset reports the significance levels of the differences between pairs of means.

Radiological study

In Figure 8 an example of radiograms obtained in the four positions considered is reported. With respect to the standing erect position (0), all the seated positions considered show a back-downward tilting of the pelvis. The measured parameters were as follows:

   1. Position 1 (upright trunk): SA and DLL were reduced by 14° and 14·5%, respectively. Physiological lumbar lordosis in general completely disappeared.
   2. Position 2 (relaxed trunk) shows a mean reduction of the SA by 26° and of the DLL by 31%. The lumbar column shows an inversion of the physiological lordosis.
   3. Position 3 (forward-sloping seat) shows a mean reduction of the SA by 22° and of the DLL by 22%. In this case too the lumbar column shows an inversion of the lordosis, but relatively lower than in the previous case.

Cervical spine

The aim of this analysis was to examine the main biomechanical differences between positions A and B of the head in all the six postures examined. It should be noted (Figure 9) that the spatial position of the head is dictated by the position of the reading plane (VDU screen vertical; sheet on the table horizontal) and is quite independent of the trunk position. Angular bendings with respect to the vertical

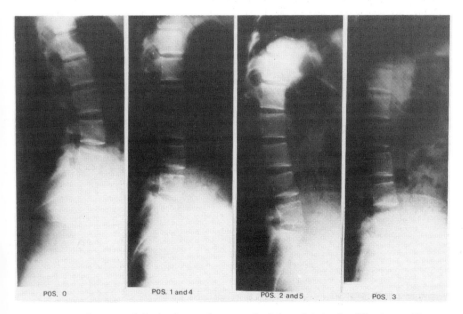

POS. 0   POS. 1 and 4   POS. 2 and 5   POS. 3

Figure 8. Radiograms of the lumbo-sacral spine and of the pelvis in the following positions: 0, standing upright; 1, seated with the trunk upright; 2, seated with the trunk relaxed; 3, seated with forward tilted seat (20°).

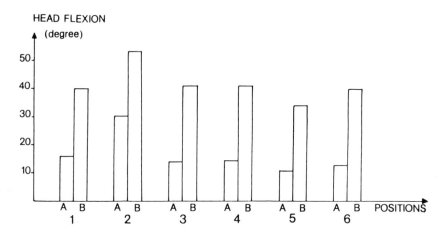

Figure 9. Mean forward flexion angles of the head in all the postures examined; A and B refer to upright head and to forward bent head positions, respectively.

are always in the range 11–16° for head position A and 34–41° for head position B with the exception of posture 2, where the values measured were 30 and 63°, respectively. Figure 10 shows the mean values of cervical load (kg) in head positions A and B in each of the six trunk postures examined. Cervical loads were constantly and significantly ($p < 0.01$) lower in head positions A with respect to B positions. Differences between head positions A in all the six postures were not significant; the same holds for positions B. The lowest cervical loads were observed in posture 5; the greatest in posture 2.

Myoelectric activity recorded on the superior trapezius was in the range

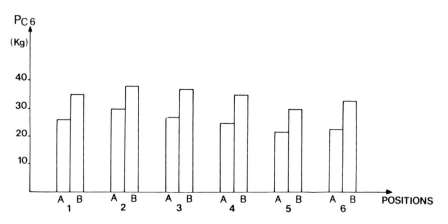

Figure 10. Mean compressive load on the C6/C7 intervertebral disc in A and B head positions in each of the postures examined. A and B refer to upright head and to foward bent head positions, respectively.

10–15% MVC, thus showing good agreement with data reported in the literature (Onishi *et al.* 1982). Head position B, with respect to A, shows an increment of myoelectric activity between 3 and 8%. The considerable variability of the recorded signals, together with the complexity of the muscular system supporting the head, suggests that different techniques of analysis should be used, and thus the data reported here should be considered as indicative only.

4. *Conclusions*

On the basis of the analyses performed, some conclusions can be drawn for the different situations considered:

1. Postures without back support are less favourable, as far as lumbar stress is concerned, than back-supported postures. Of the former, posture 3 seems to be the more advantageous because the discal load is lower, the physiological lumbar lordosis is inverted only to a small extent and the myoelectric activity of the back muscles is reduced relative to posture 1. Nevertheless the present study did not analyse the long-term effects, particularly as regards other anatomical parts such as lower limb muscles and articulations. Posture 2 was the least advantageous because of the higher values of the discal stress, both at the lumbar and cervical level, and because of the marked inversion of the lumbar lordosis.

2. Postures with the back supported are very similar to each other as far as the lumbar and cervical loads are concerned. Of these, posture 4 requires a greater muscle involvement of the lumbar back muscles even if the physiological lumbar lordosis is not inverted as in postures 5 and 6.

3. The stress on the cervical spine does not depend on the trunk position (except for posture 2); flexion of the head, as when reading a paper on the work-table, requires a greater effort of the neck muscles and produces a greater load on the cervical disc compared with the straight position of the head.

A further investigation of the above postures should define the temporal limits of acceptability, also taking into account nourishing mechanisms of the invertebral discs (Kroemer 1983). In principle, work situations which allow relatively great possibilities of alternating muscular and ligament involvement are preferable.

In conclusion the following suggestions can be made:

1. Seated postures with the back supported are more advisable, taking care that trunk support can be varied from time to time. In any case it must be considered that hydrostatic pressure, characterizing these postures, may prevent the outflow of catabolites.

2. For the above reason alternative postures should be assumed for short periods in order to create lumbar stresses above $3 \cdot 3$ kg/cm (60 kg) and facilitate catabolite outflow. In this respect, standing upright and seated postures that do not produce inversion of the lumbar lordosis while involving the back muscles are to be preferred.

3. Prolonged involvement of the cervical spine in forward bending of the head should be avoided by periodic pauses in which the neck is actively mobilized and by the use of a proper headrest.

The proposed solutions can be achieved by appropriate design of the workplace and by active involvement of the employees so that they are aware of the importance of frequent postural changes. In situations characterized by unavoidable postural fixity (data input for example) it is advisable that periodic pauses be planned specifically.

Chapter 30
Investigation of the Lumbar Flexion of Office Workers

A. C. Mandal

Finsen Institute, Copenhagen 2100, Denmark

1. Introduction

Almost half the population of the industrialized world is thought to be suffering from some form of back complaint. Apparently there is a general agreement that straining of the back is an essential factor in provoking backache. In the author's opinion nothing will give as long lasting strain as the fact that most of us spend a good deal of our lives in a seated posture with a flexed back.

2. The right-angled, upright posture

During the last 20–30 years there have been attempts to improve the seated work position for all age levels by replacing old furniture with newer types of tables and chairs. The so-called upright, right-angled position—namely with the joints of the hip, knee and ankle at right angles—has for some unknown reasons been considered the correct position.

Above all *lumbar support* has been considered to be the means to improve seated posture. But this is rather illogical as the lumbar support only carries about 5% of the body weight— and *only* in the reclined posture. In the forward bent posture,in which most precision work is done, there is hardly any effect of the lumbar support. Moreover, when sitting reclined on a 5° backward-sloping seat, you will have to bend even more in your neck to get visual contact with items lying on the table. Consequently, office workers today have far more complaints from the neck and shoulder than from the lumbar region.

No one has given any real explanation as to why this particular posture should be better than any other posture. Nevertheless, *it has quite uncritically been accepted by all experts* all over the world as the only correct one. Very little interest has been attached to the seat, which carries about 80–95% of the body weight. Its influence on the posture of the body therefore must be much more important.

3. *Concepts of correct sitting posture*

There is worldwide unanimity of opinion with respect to the 'correct' sitting posture, namely that the body should be upright and the back straight. However, nobody is able to sit in this posture while working.

Sketches representing models of sitting postures from several countries are presented in Figure 1. These sketches constitute the basis of: (*a*) anthropometry (Oxford 1969) (AUS); (*b*) international standardization (ISO 1978) (ISO and CEN); (*c*) training of designers (Dreyfuss 1955) (DREY); and (*d*) training of people to sit 'correctly' (Snorrason 1968 b) (DK).

It is surprising that no explanation appears to have been given substantiating why one should sit in this particular manner, nor is information supplied regarding the method by which the sketches were constructed. The Danish sketch (DK) may be the most interesting of them all. It clearly recommends that one should sit with a 90° flexion of the hipjoint and a concavity in the small of the back. As I will try to demonstrate later, no normal person is able to sit in this posture while working. This sketch has simply depicted a skeleton sitting on a chair. The sitting posture of a skeleton, however, is not helpful in solving the problems encountered by persons. First, a skeleton can sit staring into space for the whole day; secondly, it has no muscles or tendons to hamper flexion of the hip joint; and thirdly, an iron rod has been inserted through the vertebrae to sustain concavity in the small of the back. If preventive procedures were founded on such an insubstantial basis, the outcome would be doomed to failure.

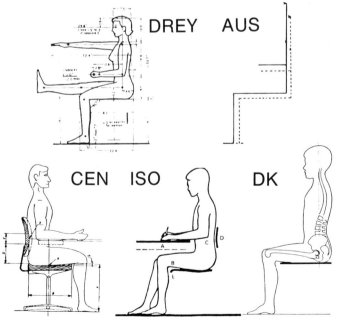

Figure 1.   Schematic models prepared in various countries representing 'correct posture'.

4. Posture training

In Scandinavia enormous efforts have been made to teach people in schools, offices and factories better sitting postures—hoping that this will prevent the increase in the number of back sufferers. In fact, we have tried to adjust people to furniture, and this is absurd. This instruction has mainly been given by physiotherapists, and it might be of interest to see how they themselves sit while working.

Figure 2 shows photographs taken with 24 min intervals by an automatic camera (ROBOT) during the final examination of physiotherapist students in Copenhagen. Most of them sit with almost maximum flexion of the backs in postures which have nothing to do with the upright ones in the instructions.

5. Anatomy of the seated person

The conformation of a seated person has been unknown to most doctors, furniture designers and physiotherapists; however, the German orthopaedic surgeon, Hanns Schoberth, has carried out some excellent research on problems of sitting posture (Schoberth 1962). The diagrams in Figure 3 were taken from his book.

When standing (*a*) there is almost a vertical axis through the thigh and the pelvis, and a concavity, or lordosis, is present in the small of the back. When a person is seated (*b*) the thigh is horizontal, the hip joint is flexed by about 60° and the pelvis has a sloping axis. The lumbar region then exhibits a convexity, or

Figure 2. Danish physiotherapists during their final examination. Photographs at 24 min intervals.

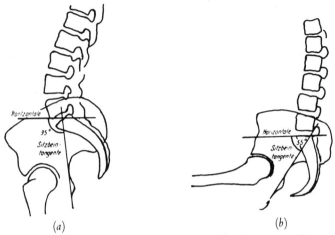

Figure 3.   Normal anatomy of the lumbar region when standing (a) and sitting (b).
From Schoberth (1962).

kyphosis. Schoberth found, in X-ray examinations of 25 persons seated upright, an average flexion of 60° in the hip joint and a 30° flexion in the lumbar region. In similar studies, Åkerblom (1948) found an average flexion of 35° in the lumbar region of 20 persons sitting upright.

The American orthopaedic surgeon, J. J. Keegan, took X-ray photographs of individuals lying on their sides (Figure 4) (Keegan 1953). He considered a posture

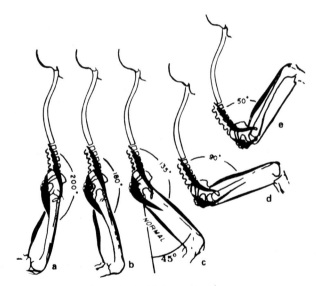

Figure 4.   X-ray photographs of an individual lying on his side.
From Keegan (1960).

of about 45° flexion of the hip (Figure 4 (c)) to be normal, because this is the position one assumes when lying relaxed on one's side. In this conformation there is complete balance between the muscles at the front and back of the pelvis. When standing (Figure 4 (a) and (b)) the muscles in front become more tense and at the back they become more relaxed. This results in increased lordosis in the lumbar region. When sitting (Figure 4 (d) and (e)) the muscles at the back are more tense and those in front are more relaxed; the small of the back is normally convex, i.e., it displays kyphosis, which is due partly to the tense hamstring muscles in the sitting position. The conformation illustrated in Figure 4 (e) is very nearly the posture adopted by seated schoolchildren for many hours each day.

When sitting on the back of a horse, one is very nearly able to assume Keegan's 'normal' posture (Figure 4 (c)), as the thighs are elevated by about 30°. When the hip joint has a flexion of 60°, one can sit upright with a vertical pelvis without bending the lumbar region.

Most children instinctively tilt forwards in their chairs when reading and writing at a table. They obviously feel more comfortable in this position, and they are able to sit with a straight back. With an elevation of 30° in the thighs and a flexion of 60° in the hip joints, they only have to bend about 10° in the lumbar region to obtain an eye distance of about 30 cm from the book. Parental training, however, has forced children to give up this practice.

6.   New European standards are harmful to the back

During the last century the average height of man has increased about 10 cm and in the same period the height of tables has for some incomprehensible reason been reduced by almost 10 cm. Since the visual distance has remained the same—about 30–40 cm for grown ups—lower tables will inevitably lead to a more constrained posture. It seems obvious that higher furniture is necessary to obtain a sitting position with sloping thighs and preserved lordosis as is found in the 'Keegan normal position' and in the horse rider.

In 1982 the European Standardization for Office Furniture (CEIN 1982) suggested that all non-adjustable tables should be of 72 cm height. No explanation for the advantage of these very low heights was given, except that it was 'to achieve good postures'. No names were given of the persons responsible for these draft standards. The influence on the posture of these low standards have never been controlled. All is based on aesthetic, moral, technical and economical considerations.

A few years ago I investigated which height of furniture consumers themselves wanted (Mandal 1982). Eighty persons were asked, and almost all preferred to sit about 15–20 cm higher than the mentioned standards, provided that the seat and the desk were sloping towards each other. In the higher position they felt less discomfort or pain in the back and they were all sitting with much straighter backs.

7. Laboratory experiment

To evaluate the effect of furniture height on the flexion of the back (hip joint and lumbar region) a person (of 171 cm height) was asked to sit and read in a chair with a 5° backward-sloping seat of 43 cm height and a table 72 cm high with a horizontal desk as advocated by CEN (Figure 5 (a)). The feet were supported by a transverse bar under the table 20 cm above ground level (to achieve the desired table height of 72 cm), and for 20 min the girl sat reading at a table height of 72 cm and a chair height of 43 cm. During this period five pictures were taken with 4 min intervals by an automatic camera (Figure 5 (a)). For the next 20 min she remained seated at a table height of 82 cm (achieved by placing the feet on a transverse bar

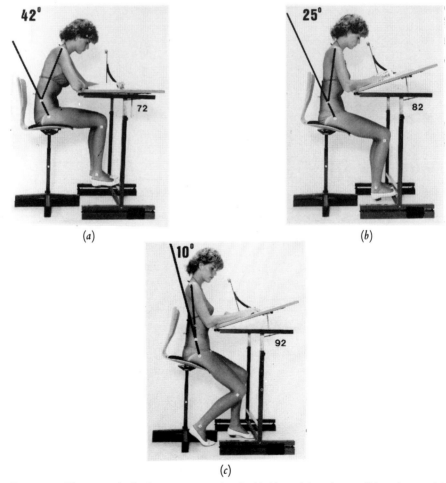

(a)   (b)

(c)

Figure 5.   Flexion in the lumbar region. Height of table/chair: (a) 72/43 cm, (b) 82/53 cm, (c) 92/63 cm.

10 cm above ground level—the chair height is now 53 cm. The desk and the seat sloped about 15° towards one another (Figure 5 (*b*)).

In this position photographs were also taken with 4 min intervals. Finally, the girl was asked to place her feet on the floor to achieve a table height of 92 cm, measured at the nearest edge (Figure (*5*)), and a chair height of 63 cm, measured at the axis of rotation. In all three situations she was asked to sit in the position she found most comfortable. No instructions were given concerning body posture or eye distance. The same experiment, using three table heights, was repeated for 10 days.

To control the flexion of various parts of the body, well-defined anatomical points were marked with spots on the skin: (*a*) knee joint (capitulum fibulae); (*b*) hip joint (trochanter major); (*e*) fourth lumbar disc (a point midway between the spina iliaca sup. anterior and posterior); and (*d*) shoulder joint (acromion).

Results

At the end of the experiment 50 photographs of each of the three situations were available. The skin marks were connected with lines on the pictures. The resulting angles between these lines (hip angle, lumbar angle) were measured. The flexion of the lumbar region was found to be an average of:

1. 42° at a table height of 72 cm.
2. 25° at a table height of 82 cm.
3. 10° at a table height of 92 cm.

The 17° reduction of flexion in the lumbar region when changing from 72 to 82 cm is highly significant ($p = 2 \times 10^{8-9}$). So is the 15° reduction from 82 to 92 cm ($p = 5 \times 10^{-8}$).

The flexion in the hip joint was found to be:

1. 57° at a table height of 72 cm.
2. 50° at a table height of 82 cm.
3. 42° at a table height of 92 cm.

The 7° reduction of flexion in the hipjoint from (1) to (2) is also highly significant ($p = 5 \times 10^{-6}$). So is the 8° reduction from (2) to (3) ($p = 7 \times 10^{-7}$). In all the reduction of flexion in the hip joint is 15°. This means that the total flexion of the back can be reduced by 47° (i.e., 32°+15°) by increasing the height of the table and chair by 20 cm, providing the seat and table top slope about 15° towards another.

8. Office experiment

Conclusions from a laboratory experiment like the above should always be considered with some reservation as the person is away from her usual surroundings and this may affect the working posture. A more realistic view of posture can of course be obtained if you can examine people in their daily work. But problems with the recording of various postures and angles have so far made this difficult.

To try to overcome this problem I have marked the same anatomical points as mentioned above by marks fixed to clothes. The persons were asked to use medium-tight jeans and one end of a 13 cm long white nylon ruler was stitched to a point just outside the hip joint. The other end of the ruler was stitched to the jeans at a point outside the fourth lumbar disc. In this way the axis of the pelvis and the hip joint was marked (Figure 6). Besides the acromion being marked by a white tape and the capitulum fibulae marked by a circular bandage (velcro), both were marked with a small black spot for measuring.

For this experiment all 10 secretaries from the Departments of Surgery and Plastic Surgery, Finsen Institute, were used. Height limits were 160–178 cm. Three different situations were examined.

1. Read/write at a table height of 72 cm with a 5° chair 49 cm high (CEN recommended).

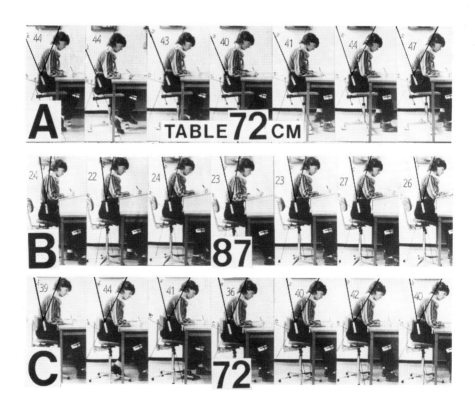

Figure 6.   Flexion in the lumbar region of an office worker. Height of table/chair: (a) 72/49 cm, (b) 87/59 cm, (c) 72/51 cm. Photographs taken at two-minute intervals.

2.   Read/write at a table height of 87 cm with a tilting chair 59 cm high (Mandal recommended).

3.   Read/write at a table height of 72 cm with a tilting chair 51 cm high (Bendix recommended).

The three different sitting postures were maintained for 15 min each. During this period the subjects were left alone in the office. They were instructed to sit the way they found most comfortable and asked to concentrate on the work. The pictures were taken with 2 min intervals resulting in seven pictures of each situation. The average lumbar flexion was reduced by 17·1° when changing from European Standard furniture (Figure 6 (*a*)) to a 15 cm higher workplace with a tilting chair and sloping desk (Figure 6 (*b*)). Also the flexion of the hip joint was reduced by 7·1°. This means that the total flexion of the back (hip joint plus lumbar region) was reduced by 24·2° (Figure 7).

During a similar experiment with typewriting an average reduction of back flexion of 9·2° was found when the CEN furniture (table height 65 cm) was substituted by a table height of 72 cm and a higher tilting chair.

In Scandinavia within the last few years there have been attempts to use the tilting chair in a new way. Bendix (1983) examined a distance of 21 cm between the tabletop and the chair and positioned the persons fully back on the rear of the seat (Figure 6 (*c*)). When the seat tilted forwards, the distance, however, was reduced to 16–17 cm. When the tilting chairs were used in this way in the present experiment, an average decrease of the lumbar flexion of 1·4° compared with the CEN chair, and an increase of 15·7° compared with the tilting chair with a tall table, were found (Figure 6 (*b*)). The comfort estimation of the position in Figure

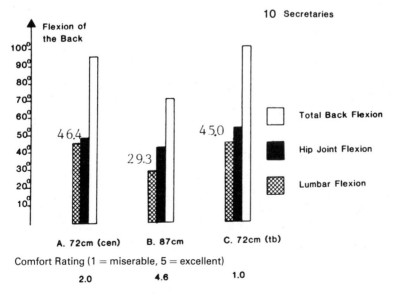

Figure 7.   *Flexion of the back in various postures.*

6(*c*) was 1·0 (1=miserable). The persons estimated the tilting chair with a tall table at a comfort rate of 4·6 (5=excellent). The comfort estimation of the CEN height was 2·0.

9.　*Conclusions*

Standardization of furniture has lead to lower and lower furniture, and this inevitably leads to more constrained sitting postures. This is probably the main reason for the rapidly growing number of back sufferers. Nobody has controlled the effect of furniture standardization and evidently it has only been to the advantage of the architects, standardization people and manufacturers. Todays miserable working postures ought to be recorded, and studies with automatic cameras give the chance to compare the postures with various furniture used during daily work. The present study indicates that our 'experts' have chosen almost the worst possible furniture. It is suggested that in the future we consult the consumers and let them decide what type of furniture they prefer.

Chapter 31
Chair and Table Adjustments for Seated Work

Tom Bendix

Laboratory for Back Research, Department of Rheumatology, Rigshospitalet, University of Copenhagen, Denmark

1. Introduction—aims in seating posture

The scope of this paper includes primarily deskwork in an office, characterized by the possibility of supporting forearms or elbows on the table. Similar workstations for certain assembly work may be included as well. Typing and visual display unit (VDU) work will be mentioned briefly, but easy-chair and conference seating are not included. The seating adjustments advocated below are based on studies of subjective preferences as well as biomechanical studies, with the aims of minimizing or preventing back pain. General aspects will be discussed first; later, advocated adjustments will be given.

First, to adjust a chair and table for seated work, it should be realized what must be attained. The following aims are considered to be important, according to various biomechanical and epidemiological investigations.

1. *The posture should be largely mobile.* Long-term sitting has been demonstrated epidemiologically to correlate with low-back pain compared with postures varying between sitting, walking and standing (Magora 1972). It has been shown biomechanically that immobility is associated with poor nutrition of the intervertebral discs (Adams and Hutton 1983 a, Holm and Nachemson 1983, Krämer 1973). Thus, no single seated posture is ideal. However, the mean of several postures taken should represent optimal loads on the various body parts.

2. The *mean posture* should be with some *intermediate lumbar curve*. The too-kyphotic (forward-flexed) curve is associated with high intradiscal pressure (Andersson 1974). Ruptures of the disc-lamella layers seem to be an important cause—directly or indirectly—of low-back pain (LBP) (Nachemson 1976). Thus, it is reasonable to believe that a posture with a relatively high intradiscal pressure over several hours a day, causes more harm to the discs than a posture with lower intradiscal pressure.

Increased lordosis, obtained by increased use of a backrest, reduces the intradiscal pressure (Andersson 1974). However, where increasing lordosis arises due to higher and forward-inclined seats, the backrest is used less (Jürgens 1969).

Moreover, a too-lordotic curve should be avoided for long-term postures. A lumbar curve similar to the standing posture is probably not to be recommended, as much LBP is aggravated during standing (Biering-Sørensen 1984). This may be explained by prolonged compression of the facet joints (Adams and Hutton 1983 b, Caillet 1981), at least in the case of degenerated discs (Miller *et al.* 1983), or by reduction of the intervertebral foramina (Panjabi *et al.* 1983).

3. The cervical spine should correspond to an *intermediate forward inclination of the head.* Subjectively, a head inclination corresponding to a viewing angle of 20–25° below the perpendicular on the trunk axis seems preferable (Grandjean *et al.* 1983). From an intermediate position, work postures with increasing forward inclinations of the head are associated with increasing neck and shoulder-girdle pains (Hünting *et al.* 1980).

4. *Active, prolonged elevation of the shoulders should be avoided* (Bendix *et al.* 1985, Hagberg 1981 c, Hünting *et al.* 1980, Ohara *et al.* 1976). However, in the case of the elbows or forearms being supported it may probably not be too harmful if the shoulders are not too elevated.

5. *The posture should be comfortable.* This is, however, difficult to interpret. People accustomed to an unhealthy posture over several years may perhaps consider it to be comfortable. Conversely, a healthy posture, if new, may be felt as uncomfortable in the first instance. The problem is the delay of several years from the onset of harmful loads on the discs to the onset of LBP.

The interaction between comfort and constrained maintenance of the posture should also be emphasised; the greater the comfort, the greater the ability to keep sitting for longer periods. However, this tendency should, according to the author's conception, be approached through suitable work organization, and not by constructing unpleasant workstations merely to encourage people to change posture.

2. How to solve these aims

Largely mobile postures

To estimate spontaneous body movements during sitting, the tilting movements of a freely tiltable seat were analysed (Bendix *et al.* 1986). The tilt movements were recorded by means of an angular transducer as indicated in Figure 1. Such movements are only induced by the user. Although some trunk movements can be performed without tilting the seat, the recording device is considered to reflect the movements of the body by and large.

The method gives two measures for movements; one indicated frequency, reflecting restlessness versus maintenance or constraint of the posture. The other measure indicates the magnitudes of the movements, reflecting the dynamics of the posture.

Comparing a seat-height adjustment of 5–7 cm above popliteal level with that of 1 cm below popliteal level, the posture during 1 hour of seated deskwork was

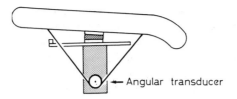

Figure 1. Tilting movements at the seat were recorded by an angular transducer under the seat, activated through a string.

estimated to be most mobile with the higher adjustment. The tilting movements tended to be greater and were performed more often. Simultaneously, the acceptability rating favoured the higher adjustment.

Furthermore, postures were demonstrated to be more mobile when the backrest was adjusted fully forwards or backwards, than when it was in a position such that the ischial tuberosities were 4 cm behind the axis for tilting—an intermediate position. All three positions were employed only at the time of initial backrest adjustment; during the sitting period, subjects were free to adopt other positions. However, the intermediate backrest position was clearly preferred on the acceptability rating to the two extreme positions.

Concluding this discussion, the seat should be positioned higher than the generally used recommendation of about 1 cm below popliteal level. The sagittal position of the backrest should be as individually preferred; any alternative to the 4 cm position should not be deprecated because such other positions seem to increase posture mobility.

Also a movable, slanting desk (see later) may facilitate a mobile posture. The papers or books most often used should be placed on the slanting desk, and papers used for writing should be placed on a horizontal table in front of the movable desk. In that case, shifts between varying postures of both head and trunk are facilitated.

Appropriate lumbar curve

The lumbar curve changes towards lordosis (extension) with increasing height and forward inclination of the seat (Bendix, 1984, Bendix and Biering-Sørensen, 1983). This gives an additional indication for a higher seat adjustment than generally used. Increasing seat height should be associated with a forward-inclining seat, because heights exceeding the popliteal level lead to the thighs inclining forwards.

Greater heights, that is with thigh inclinations of approximately 45°, as used in a saddle-chair, move the spine even further towards lumbar lordosis, although not to the lordosis arising in the standing posture (Bendix *et al.* 1985).

Another way to extend the lumbar spine is to use a slanting desk (Bendix and Hagberg 1984). The effect on the lumbar spine was demonstrated to be greater with a 45° slanting desk than with an increased seat height of 5–7 cm and concomitant seat inclination change from −5 to +5° (Figures 2 and 3).

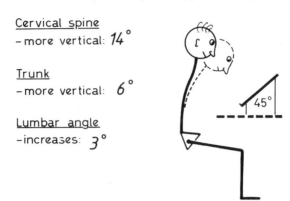

Cervical spine
- more vertical: *14°*

Trunk
- more vertical: *6°*

Lumbar angle
- increases: *3°*

Figure 2. The effects of a movable, slanting desk (solid line) on posture, compared with a horizontal tabletop (dashed line).
From Bendix and Hagberg (1984).

Seat incination

−5°⁻+5° ⇒

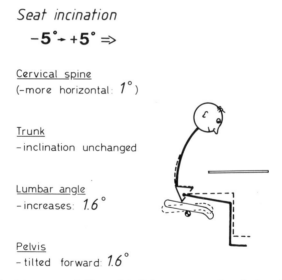

Cervical spine
(-more horizontal: *1°*)

Trunk
- inclination unchanged

Lumbar angle
- increases: *1.6°*

Pelvis
- tilted forward: *1.6°*

Figure 3. While sitting on a +5° (forward) inclining seat, approximately 6 cm above popliteal level (solid line), the lumbar curve changes towards increased lordosis compared with a −5° seat 1 cm below popliteal level (dashed line), but no effects on the inclinations of trunk or head were obtained.
From Bendix (1984).

A third way to reduce lumbar kyphosis is to increase the seat-to-table height difference (see below).

A combination of all three changes from the generally used adjustment of the seated workstation gives the greatest extension of the lumbar spine. How far this movement towards lumbar lordosis should be is debatable, and should be seen in relation to other aspects, especially subjective preferences (see below).

Cervical spine

During VDU work, the preferred viewing angle has been reported as 20–25° below the perpendicular on the trunk axis (Grandjean *et al.* 1983). Using a slanting desk of 45° an almost corresponding head position can be obtained (Figure 2).

Change of the head and cervical spine may be obtained by increasing the seat-to-table height difference. In that case—in contrast to a movable, slanting desk on a horizontal tabletop—the shoulders are elevated as well. Although this is carried out passively, a too-high elevation may increase shoulder-girdle tension as the arms may be unsupported for some time.

Varying the seat inclinations and heights, at identical seat-to-table height differences, do not influence the position of the head and cervical spine (Figure 3).

Avoid actively elevated shoulders

With touch-typing for example, with little visual requirement and without the facility of elbow or forearm support, the typewriter's interspace bar should be at the level of, or below, the elbows (Bendix and Jessen 1986, Bendix *et al.* 1985, Grandjean 1980). With greater visual requirements—i.e., sight-typing, the majority of VDU work, small-component assembly work, etc.—the bar/tabletop should be above elbow level (Grandjean 1980, Grandjean *et al.* 1983). For such work, forearm/wrist support should be used for at least short periods. With typing, a 'shelf' might be positioned parallel to and in front of the keys, and approximately at the level of the interspace bar (Bendix and Jessen 1986).

If elbow/forearm support is available, table heights above elbow level are preferable, because the spine gets more upright and less kyphotic. Passive, moderate elevation of the shoulders does not induce fatigue in the upper trapezius muscle. However, a suitable compromise between an upright posture, but not too-elevated shoulders, should be made. For this problem, it is especially important to consider subjective preferences.

Subjective preference

Whether subjective preferences reflect comfort, lack of discomfort, pleasantness or adaptation is debatable (Lueder 1983). According to Wachsler and Learner (1960), however, they should especially reflect back and neck/shoulder comfort. In the present studies the term 'overall acceptability' has been selected.

Using an easily height-adjustable and tiltable chair and table, subjectively preferred heights have been investigated with 108 subjects (Bendix and Bloch 1986). On the basis of this study it is advocated that the seat be adjusted 3–5 cm above popliteal height, and the table 4–6 cm above elbow level, while seated upright (on the chair above), with the shoulders relaxed and the elbows at 90°.

A greater discrepancy exists between various studies of subjective height preferences. The present author's interpretation of Ward and Mabey (1977) is of reported preferences of seat heights slightly below popliteal level and table heights about 3 cm above elbow level. Mandal (1982) reported seat-height preferences about 15 cm higher than that. Our observations are somewhere in between. The most likely explanation may be the varying experimental workstation designs in relation to the following biomechanical features.

As illustrated in Figure 4 the use of the backrest decreases with increasing seat heights. To optimize backrest utilization it has been recommended since 1913 that the seat should incline slightly backwards (Figure 4 (a)) (Strasser 1913). With such a seat it is, however, preferable to choose a seat height below popliteal level, as higher seat positions would cause pressure against the hamstrings.

According to the conception of Mandal (1982) a backrest should not be used while sitting on a forward-inclining seat. Consequently, he has investigated subjective preferences with a tiltable chair without a backrest. In that case it is natural that the subjects preferred the seat to be as high as possible without sliding off the seat, to obtain a balance giving minimal requirement for backrest use (Figure 4 (c)).

In the case of a backrest available on a tiltable chair, the majority of subjects preferred a compromise between the two extremes above (Figure 4 (b)) (Bendix and Bloch 1986). Due to the tiltable arrangement they might adjust the seat above popliteal level without compressing the hamstrings. They even had the opportunity to avoid contact with the backrest and adjust the seat as high as in Mandal's study, but almost nobody preferred that.

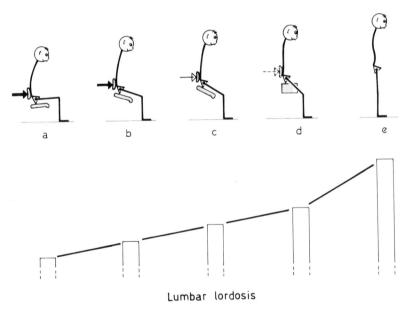

Lumbar lordosis

Figure 4. With increasing seat height the lumbar curve changes towards increased lordosis although the curve during standing is not obtained. Simultaneously, the utilization of the backrest decreased (as indicated by arrow thickness).
From Bendix et al. (1985).

In addition to the subjective preference for using a backrest, objectively the load on the back is decreased (Andersson 1974, Eklund and Corlett 1984), and the stability of the posture improves. A further argument against sitting on a too-high seat is an observed tendency for increase foot swelling during prolonged sitting with increasing seat height (Winkel and Bendix 1986).

It is advocated that the seat should tilt over a transverse axis to follow varying thigh inclinations and to facilitate posture changes. Subjectively, such a seat was preferred to fixed forward- or backward-inclining seats (Bendix 1984). From an anatomical point of view, the axis should be positioned through the knee joints. In that case, however, the seat height would vary with the tilting movements, which is not appropriate for seated deskwork. An axis under the middle of the seat (Mandal 1976) also enables seat movements (Bendix *et al.* 1986), and keeps the seat-to-table height difference almost constant. In our study above (Bendix and Bloch 1986), the subjects preferred to place their buttocks in a position with the ischial tuberosites about 6 cm behind the axis for tilting the seat.

3. Advocated adjustment of the seated workstation

Combining the subjective preferences and biomechanical aspects described above, the following criteria are advocated for seated deskwork (Figure 5):

1. Highest priority should be given to a mobile, slanting desk, inclining 35–45°, and positioned on the horizontal tabletop.

2. The seat should be tiltable and upholstered to avoid sliding off when inclined forwards.

3. Seat height should be about 3–5 cm above popliteal height (including shoeheels).

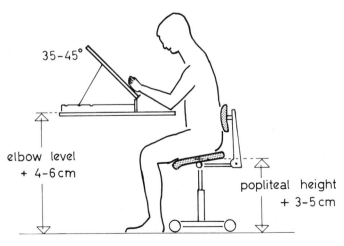

Figure 5.   Advocated arrangement of the seated workstation for deskwork.

4. The backrest should be positioned against the back while the ischial tuberosities are placed about 4–6 cm behind the axis for tilting the seat.

5. The height of the horizontal tabletop should be 4–6 cm above elbow level, measured with the subjects seated as above, the shoulders relaxed and the elbows at 90°.

6. The work should be organized to avoid any seated period exceeding 1 hour.

Acknowledgements

The studies have been supported by grants from The National Danish Rheumatism Association, The Danish Medical Research Council, The Swedish Work Environment Fund, and The Technical Faculty of the University of Luleå, Sweden. Chairs were supplied by Sedeo-Matic, Copenhagen.

Chapter 32

Experimental Analysis of a Visuo-Postural System in an Office Workstation

F. X. Lepoutre, D. Roger and P. Loslever

Laboratoire d'Automatique Industrielle et Humaine,
Université de Valenciennes et du Hainaut-Cambrésis, Le Mont Houy,
59326 Valenciennes Cédex, France

1. Introduction

The development of technical tools leads an increasing number of people to perform tasks in sitting postures, especially in the handling of information, and necessitates a keyboard and information display (video or paper). The postures imposed by these tasks are often static and may cause strains or pains in the arms, back or neck. Research into the reasons for, and consequences of, these phenomena are numerous (Dainoff 1982). However it appears that, as far as posture is concerned, although adjustable furniture is a step forward, the adjustment criteria in relation to anthropometric dimensions and the task to be done are not clearly established. Moreover, current research shows the complexity of the problems due to the interactions within the 'task–vision–posture–furniture' system and the difficulty of measuring and interpreting physiological, biomechanical and psychological phenomena in an actual environment (Grandjean *et al.* 1983) or in the laboratory (Happ and Beaver 1981).

It seemed interesting, therefore, within the research that we are doing on the vision–posture system, to undertake with French furniture manufacturers a laboratory study to model relations between furniture adjustments and 'postural comfort'. After explaining how this study, which first began in 1980, has been carried out, we present and justify the different parts of the experimental protocol (Section 2).

At first, the adjustments which we studied were the heights of the seat and the table and the distances between the backrest and the keyboard (the study of three parameters made us use a Latin square experimental plan). The experimental task was typing during half a day. Subsequently, the difficulty in identifying the phenomena led us to measure simultaneously the performance and subjective feelings of the operators as well as certain postural parameters. These were, apart from the pressure on the backrest, the curvatures and the inclination of the back, measured with an original device called a 'curvometer'. These postural measurements ($\sim 4 \times 10^6$ per subject) caused us to use specific treatments to quantify 'preferential postures' and 'postural activity'.

Analysis of the sensitivity of the variables obtained according to the different adjustments, the relative influence of the three adjustments and particular ergonomic considerations are presented in Section 3.

2. *Experimental protocol*

As a first step, we made an inquiry into a secretarial task, took many anthropometric and furniture-dimension measurements and made a bibliographical study and some preliminary experiments. These enabled us to construct, with the help of the manufacturers, an experimental protocol whose different elements (parameters, task, environment, measurements, treatments) were as follows.

Adjustments studied

Seat height

The French norm NF D 61.201 (Afnor 1973) points out "the necessity to adapt the workstation to [a worker's] particular use taking into account his dimensions and personal preferences". This adaptation is obtained by optimum seat-height adjustment. The adjustment criteria are not precise and the "seat height is adjusted in relation to the distance between the popliteal point and the floor" (Åkerblom 1948). So it seemed useful to study more precisely the influence of this adjustment on the operator's comfort, at three levels described in Figure 1.

Table height and seat–table distance

Usually, worktable height is not adjustable and is fixed, for example, at 650 mm for a typist (Afnor 1972). However, it appears that it is not possible to investigate the influence of seat height for subjects of different morphology with that fixed parameter. Indeed, it appears, for example, that a large subject cannot sit at this fixed level with his legs placed normally under the table. It seemed necessary to be able to adjust the table height and therefore interesting to investigate the influence of the seat–table distance. Three table heights were studied; the lowest just allowed room for the legs; the highest just allowed viewing of the text in the typewriter without changing posture; the third is the means of the other two.

Backrest–keyboard distance

The importance of the backrest in sitting comfort is not contested. Its effect takes several forms: mechanical, by supporting a part of the weight of the back and by maintaining dorsal curvature (Andersson *et al.* 1980, Corlett and Eklund 1984); physiological, by contributing to postural regulation, as an exteroceptive reference; and psychological, by contributing to the subject's safety/security. Moreover, backrest form and position are often talked about without being able to

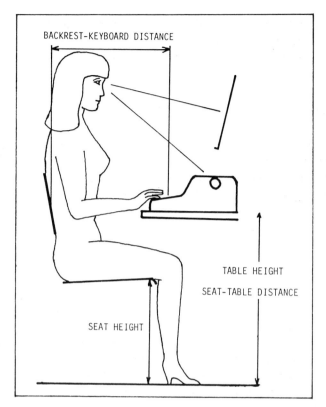

BACKREST-KEYBOARD DISTANCE

TABLE HEIGHT

SEAT-TABLE DISTANCE

SEAT HEIGHT

Figure 1.   The experimental variable adjustments.

demonstrate objectively the qualities which have been subjectively perceived. Finally, the great majority of seats for typists are provided with adjustable backrests.

It therefore seemed useful to contribute to a study of backrests by investigating the influence of the backrest–keyboard distance on the operators. The near distance used was the limit after which the subject's buttocks would slide to the front of the seat to avoid a too erect posture. The other two distances are 1 and 3 cm greater than the near one (Figure 1).

Experimental task

The experiment utilized a typing task, recopying text in a similar fashion to data entry. This is a constraining task from a postural point of view, imposing a fixed posture by reason of continuous keyboard utilization and visual fixation on paper or display. Each experiment, corresponding to one combination of the levels of the three adjustments, was performed for four 35 min periods separated by 5 min for tests and questionnaires. Each subject performed nine experiments, organized in a Latin square plan (Dugue and Girault 1969), which enabled an investigation using analysis of variance of the significance of the influences of the three adjustments on the measurements.

Invariant parameters

To ensure experimental reproducibility, precautions were taken to keep other parameters steady. Lighting was held at about 200 lx, the seat inclination at about 3° and backrest inclination at about 100°. The text support was placed above the typewriter at a constant height, a solution which was well appreciated by all the subjects (Life and Pheasant 1984).

Measures

In order to quantify, as objectively as possible, the reactions of the subjects to different furniture adjustments the measurements which were taken were of three types: performance, feelings of the subject and postural behaviour.

Performance qualities

Although it is well known that performance measures are not a sufficient indicator to quantify fatigue or comfort (Happ and Beaver 1981), it seemed necessary to take them into account to confirm results obtained by other measurements. For that reason, the number of characters typed and the number of errors made during each 35 min period were counted. Moreover, tests of reaction time to a visual stimulus were given to subjects before and after each of the four experimental periods.

Subjective feelings and judgements

It was important to obtain from subjects their feelings and judgements about the adjustments studied. A great deal of research is limited to these sorts of measurements (Stammerjohn *et al.* 1981), but this limitation can lead to distortions due to sociocultural habits and routines, and must therefore be compared with more objective measurements. To measure these subjective parameters, a form with 17 questions on it was presented after each experimental period, on which the subject indicated the strain and the feeling of pain (level and localization), and judgements about adaptation to the adjustments studied.

Postural behaviour

Besides performance and subjective feelings, it was necessary to support the results with objective postural measures of comfort or fatigue. The conventional methods are, by way of example, myoelectric activity and lumbar disc pressure (Andersson *et al.* 1980, Hünting *et al.* 1981), or articular angle measurement (Grandjean *et al.* 1984). The method used was to record continuously dorsal curvatures and inclinations in the sagittal plane, as well as the pressures on the backrest.

The measuring device for the dorsal curve, named 'curve-meter' (Lepoutre 1979), consisted of a flexible 'stick' fixed upon the individual's back with elastic belts, whose distortions were measured, in the individual's sagittal plane, with four pairs of strain gauges located approximately at the levels of vertebrae D1, D6, D10 and L3 (Figure 2). This sensor is very easy to implement, does not disturb the individual and can be worn during a whole working day. The four measurements obtained were curvatures (per metre), inverted to local curvature radii. These measurements were directly related to kyphosis angle measurements achieved with radiographs (Lecain 1979). A fifth measurement, the inclination (in degrees) of the lower part of the stick, made it possible for the dorsal curvature to be related to the vertical. It was then possible with these five measurements to represent, at any time, the shape of the individual's spine (Figure 2). Lastly, a sixth measurement, the

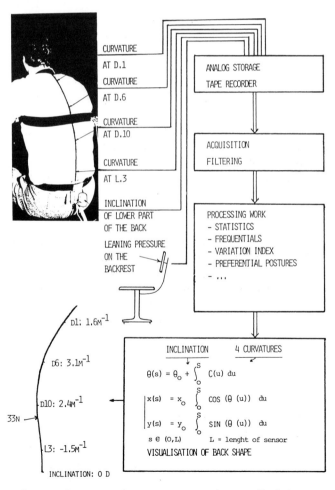

Figure 2.   Collection and processing of measurements. Visualization of back shape.

leaning pressure on the back of the seat (in Newtons), made it possible to evaluate forward and backward forces applied to the individual's biokinetic system. All six measurements were stored on a tape recorder during experiments in order to proces then off-line later.

Measurement analysis — notation

Each of the six measures defined in the preceding paragraph gave 16 800 values for each of the four 35 min experimental periods. All the measurements were central-ized and underwent filtering operations, error detection, correction, normal-ization and coding. This coding consisted of transforming raw values to a comfort score. This fundamental stage is obvious for questionnaires and performance but depends on previously established models for postural measurements. It was decided to give lower comfort scores to larger dorsal curvatures (Andersson *et al.* 1980) and to increasing posture variation indices.

At the end of the preliminary treatments, three performance indices, 17 questionnaire answers and 18 postural indices for each of the nine experiments were obtained for each subject. It was then possible to compute the statistical significance of the variance analysis and to interpret the results in terms of comfort.

3. Results

The results presented are for 10 female subjects (19–38 years old, 1·56–1·70 m height), and are concerned with mean values across the sample.

Sensitivity of measurements

In general, it appears that the adjustments studied have little effect on the measure-ments taken. This is probably because a relatively narrow range around the average adjustment was investigated and the average was, *a priori*, favourable to comfort, and adjusted to each subject's anthropometric dimensions. On the other hand, an adjustment effect was statistically verified for about 20% of the postural indices and around 12% of the performance and questionnaire indices. Although these indices were different from one subject to another, it appeared that the reaction times, the number of characters typed and the lumbar curvature at L3 were very rarely significant indices. The dorsal curvature at D6, the back inclination, and the subject's judgement about seat height were more frequently found to be significant indices.

General results

Table 1 shows the percentages of the measures across the sample which gave significant effects for the three adjustments. Generally, postural measures were the most sensitive and performance measures the least sensitive to the adjustments

Table 1. Influence of each type of adjustment on the three groups of measures.

| Measures | Adjustments | | | |
	Seat height (%)	Seat–table distance (%)	Backrest–keyboard distance (%)	Mean (%)
Postural	18	14	25	19
Performance	17	8	8	11
Questionnaire	15	14	12	14
Mean	17	12	15	15

studied. It appears that the backrest adjustment influences the postural measures most and the subjective feelings least. This may be explained by the complexity of the backrest's influence on posture (see below). On the contrary, since they all have about the same sensitivity, it will be easier to settle on good seat height in relation to performance and subjective feelings than in relation to postural measures which are more difficult to interpret. Seat–table distance, for the range studied (approximately ±4 cm), has a relatively smaller influence on all the measures.

In general, the measures at the lumbar level show less significance (11% of measures) than all the other postural measures (19%). This results from the large dispersion, between and within subjects, of lumbar postural behaviour.

Seat height

For this sample it appears that the medium and high seat heights (i.e., equal to or 25 mm higher than the popliteal point to floor + heel) produce: (*a*) better performance; (*b*) decreased fatigue and pains at the neck, arms and legs level; (*c*) decreased posture variation indices at the D1 and D6 level; (*d*) favourable ratings from subjects; but, on the contrary, (*e*) an increased D1 level curvature.

Except for the last point, it seems that a relatively high seat height is favourable. One possible reason is that the subject's weight is better distributed on the seat.

Seat–table distance

Within the limits of seat–table distance adjustments in this study it appears that a higher table (larger distance) produces: (*a*) decreased dorsal curvature (Figure 3); (*b*) decreased typing errors, which possibly results from a better visual condition (decreased eye–text distance); but (*c*) increased arm pain.

A 'good' adjustment of the seat–table distance must be found between a high table height, which gives a better dorsal posture, and a lower table, which decreased arm fatigue.

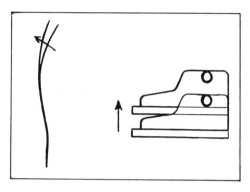

Figure 3. Seat–table distance influence on dorsal curvature.

Backrest–keyboard distance

The backrest–keyboard distance has a complex influence on the subjects (Table 2). It seems, overall, that a nearer backrest is the better adjustment, except for dorsal curvature problems. However, analysis subject by subject shows different behaviours characterized by trunk inclination and leaning pressure on the backrest. Therefore, the results obtained for the mean of the sample must be used with discretion.

4. Conclusions

If we want to improve working conditions in an office, we need a better understanding of the 'man–task–furniture' system. That is why this study had as its aim the investigation of the effects of certain fundamental parameters, such as seat height, seat–table distance and backrest–keyboard distance, on the 'comfort' or 'fatigue' of subjects.

The experimental method used employed varied and complementary measurements, such as performance characteristics, dorsal curvatures and subjective feelings and judgements of the subjects, to diminish the risk of errors or false interpretations. Analysis shows that, for this sample, it is advantageous to adjust the seat

Table 2. Backrest–keyboard distance influences.

Backrest	Far	Near
Advantages	Smaller dorsal curvature and variation index (D6) Decreasing reaction time	Smaller lumbar curvature Well appreciated by subjects
Disadvantages	Larger lumbar curvature Larger arm strain Larger number of errors (visual strain)	Larger dorsal curvature and variation index (D6)

to a height equal to or higher than the popliteal joint to floor + heel distance. A larger seat–table distance seems to improve dorsal posture but fatigues the arms. A smaller backrest–keyboard distance has many advantages, especially for lumbar curvature and task vision, but increases dorsal curvature ($\sim$D6).

However, analysis must be continued in order to observe behaviour differences between subjects, especially for backrest utilization. All these results show the complexity of the system and the risk involved in giving ergonomic recommendations independently of the task and subjects, or founded on too few measures.

Acknowledgements

This study was supported by CETIM (Centre Technique des Industries Mécaniques), 52 Avenue Félix Louat, 60300 Senlis, France.

SECTION 6

TWO CASE STUDIES

This short, final, section of just two chapters has an importance beyond its size. The case study of sewing-machine use in the shoe industry brings together, in a practical problem, many of the points in earlier chapters and illustrates the use of some of the previously discussed criteria for an immediate before-and-after assessment of the changes.

Chapter 34, working on from a similar though larger-scale case, attempts to test the value of its results in money terms. Putting aside for the moment the question as to whether the relief of human distress should be dependent on a positive financial result, it is undeniable that if such a relationship could be demonstrated it would be a strong argument in the ergonomist's armoury. Whether the financial result of ergonomics interventions was positive or negative, it must be beneficial to have opportunities to evaluate the costs of particular proposed procedures and to assess the benefits of various likely outcomes.

The evaluations by Spilling *et al.* cover estimates—and measures—of direct gains from reductions in labour turnover and sick leave, as well as training costs, against the costs of the ergonomics changes. Although it might have been legitimate to estimate productivity gains from, e.g., learning–curve assumptions or data, they have confined themselves to the most direct and evident areas of cost reduction. They do note, however, that other information leads them to believe that an important factor in labour retention was the improved work environment arising from the changes to the workplaces and factory generally.

Even though much of the text of this book has been confined to particular relationships between certain factors associated with the physical aspects of posture—the disposition of body parts, the forces present, the physiological responses and the fatigue and other components of the subjective experiences of posture—questions concerning the effects on a person of a more widespread range of influences have surfaced at times. It is clear that the understanding and betterment of working postures must include such wider considerations if the results are to be acceptable to and effective for those involved. These wider considerations, less precisely definable than many of those factors dealt with previously, present major problems in their measurement. Measurement, indeed, with its associated requirement for the definition of criteria, dimensions and the implications these have for the concept of the system's behaviour (the model), is a key to the eventual practical use of much of this research. Measurement in ergonomics must extend beyond providing the information relevant to forces, metabolic processes and the like, into the responses of people to the conditions imposed. It is performance which is the purpose of the postures after all, and to achieve performance with benefit to health must surely be the ergonomist's objective, distant though it may appear on occasion.

To think, then, in cost–benefit terms is most appropriate, but perhaps on an even wider front than the study by Spilling *et al.* A series of studies culminating with research by Schiro (1985) has set out a possible way forward by bringing together a range of measures for performance, work attitudes, labour turnover costs and environmental factors. These provide profiles of current state, repeatable over time for comparison and evidence of change. Also provided is a number of cost models for the financial evaluation of the changes. Development of methods and techniques to achieve the more comprehensive understanding of the impacts of work on people still have far to go, but will enable studies such as those in this book to be integrated within a wider frame and their full importance recognized. They will also give a better perspective for the pursuit of the specialized studies themselves, allowing us to recognize which of the many effects that physiology, anatomy, engineering, etc., allow us to measure are relevant just for those disciplines themselves, and which are truly of importance for the ergonomist in the design of human work.

Chapter 33
Postural Change Due to Adaptations of a Sewing Workstation

John Wick

G. H. Bass & Co., Wilton, ME 04294, U.S.A.

and Colin G. Drury

State University of New York at Buffalo, Amherst, NY 14260, U.S.A.

1. Introduction

Ergonomics has been applied previously to sewing-machine workstations and work methods in the apparel and shoe industries. Singleton (1960) studied the motor controls of sewing machines in the shoe industry. Singleton (1959) also studied techniques for training sewing-machine operators in the shoe industry. The Research Association of the German Apparel Industry studied the postural and work-envelope problems of apparel sewing workstations. Sillanpaa (1984) also studied the postural and work-envelope problems of apparel sewing. M. A. Ayoub did considerable (unpublished) work in 1983 for Hanes, Inc., a prominent apparel manufacturer, solving problems of workstation adjustment and sewing work methods.

Much of this work addressed the problem of cumulative trauma diseases (CTD); however, CTD remains a major problem of sewing jobs in both the shoe and apparel industries. Typical diagnoses are tenosynovitis, tendonitis, carpal tunnel syndrome, thoracic outlet syndrome, low-back pain and cervical spine injuries. All of these symptoms can result from poor posture of the trunk and upper extremities.

In this study the postures of the trunk, neck/head, upper extremities and lower extremities were analysed. Workstation design changes were developed in order to improve the working posture of the operator. The design was then tested and analysed again to determine the affect on the operator's posture.

The methodology employed was that described by Drury and Wick (1984). It is an eight-step methodology as follows:

1. *Task analysis of the original job.* The operator was videotaped performing the work from the front, back and each side. The sewing cycle was separated into task steps, each step representing a change in a body-part position. The videotape was stopped at each task step and the positions of each body part were recorded. The angles of the head/neck, trunk, upper extremities and lower extremities were

determined by stopping the videotape during a cycle of sewing showing a profile of the operator. Tracing paper was placed on the television screen and the body-part angles and a vertical or horizontal reference were traced. These body-part angles were then measured relative to the reference.

2. *Measurement of postural stress.* The Postural Discomfort/Pain scale developed by Corlett and Bishop (1976) and the General Comfort Rating scale developed by Shackel *et al.* (1969) were administered at approximately $\frac{1}{2}$ hour intervals for 5 days.

3. *Analysis of data collected, and design changes.* The body-part angles were compared with the literature recommendations for seated work. The body angles were also used to calculate the compressive forces on the spinal discs, using a static model reported by Drury *et al.* (1983). Damaging wrist motions were identified, defined as a pinch or grip with deviation, flexion or extension of the wrist. The psychophysical data from the two postural-stress scales indicated the body parts in which the operators perceived stress. Workplace design was based on anthropo-metric data (Dreyfuss 1974), the goal of the design being to maintain neutral posture in all body joints.

4. *Laboratory trial of revised workplace, with further design changes if needed.* A prototype workplace incorporating the recommended changes was tested in the laboratory by an operator, to verify that the new design provided ergonomic solutions and could be used to perform the job effectively.

5. *Production trial of the revised workplace.* The new design was tested under production conditions and data were collected to test its effectiveness.

6. *Remeasurement of postural stress.*

7. *Task analysis of revised job.*

8. *Comparison of data from the original job with data from the revised job.* The two sets of postural-stress data from before the change were compared with the two sets of data after the change, to determine that the new design had the desired positive effect and that no unwanted effects resulted.

The writing of final recommendations for implementation followed.

2. *The study*

Four operators (all female, aged 29, 29, 30 and 33 years and experience, respect-ively, 7·5, 3·5, 2·5 and 4·5 years) were studied performing the job of sewing the first row of stitching of the collar of a shoe. The collar is a piece of leather at the top of some styles of shoes from the instep on one side, around the heel and to the other side of the instep. It is sewn on one edge, folded outward and sewn again. This study is of the sewing of the first row of stitches.

The workstation consisted of a chair, a bench and a single-needle post sewing machine. The material was delivered to the operator on a transporter in tote boxes. The chair was not upholstered and did not swivel. It was adjustable only by using a screwdriver and wrench. The bench was 122 cm wide × 51 cm deep. The top was

3·8 cm thick. The base of the bench was a metal framework, adjustable for height using two wrenches.

The sewing machine was a single-needle post machine driven by a mechanical clutch motor. The machine was mounted horizontally on the bench top with the needle at the lateral centre of the bench and 16·5 cm from the bench edge nearest the operator. The point of operation (needle piercing the material) was 19 cm above the bench surface. The machine was controlled by a treadle bolted to the underbench framework and linked to the motor clutch lever by a metal rod not adjustable by the user. A pedal to the right of the treadle was used to lift the presser foot and was linked to the lifting mechanism by a chain. The foot controls were 5 cm above the floor. Needle positioning was accomplished either by skilled operation of the treadle or by using the right hand to manipulate a hand wheel located on the upper right of the machine.

After analysis of the questionnaires and the videotapes, the ergonomic problems of the job were identified and design of an improved workstation was begun. A subject operator was identified. In industry it is not usually practical to test many prototypes at one time because of disruptions to the workflow. The subject operator was identified as one who used the methods prescribed by the Industrial Engineers and who seemed to exhibit no extremes in reporting problems on the postural-stress scales.

Analysis of the videotapes indicated that the various postures assumed by the operators studied ranged from poor to acceptable. An acceptable posture for seated work was taken as being with thighs horizontal, back erect and head/neck flexed forwards 30° or less. The posture depended upon the size of the person and how the workstation was originally assembled. There was no attempt to match chair or bench height to the size of the operator. All operators assumed a trunk forward flexion posture with head/neck flexed forwards in excess of 30° (Grandjean 1982). The lumbar supports of the existing chairs were not used because of the forward flexion of the trunk. In all cases the thighs sloped downwards from hips to knees. In order to present the shoe parts to the machine, all operators deviated their wrists to the ulnar side.

Analysis of the psychophysical data collected from the questionnaires indicated a significant number of complaints of discomfort in the neck, shoulders, middle back, lower back and, in some operators, the wrists.

A prototype workstation was designed to address the biomechanical problems identified in the analyses. This prototype included an improved chair, redesigned foot controls, a fixed height for the bench, bench-mounted armrests and a base which slanted the machine towards the operator. The goal of the prototype was to improve posture and wrist position, and this was achieved by employing the improved workstation design in a properly adjusted mode.

The chair was an upholstered swivel chair, adjustable by the user for height and for backrest location. It was adjusted according to anthropometric data (Dreyfuss 1974) to place the operator at the correct height for the point of operation.

The foot control was a smaller, lightweight remote treadle, linked to an electronic motor control by an electric cable. This linkage allowed the foot control to be placed on an adjustable footrest and to be located by the operator so as to provide both maximum comfort and control. The electronic motor control allowed for needle positioning and presser foot control by 'heeling' the treadle.

The bench height was established at 76 cm. By utilizing an adjustable chair and an adjustable footrest, an operator would be expected to be at the best biomechanical advantage to handle tote boxes of material to and from the transporter system. Bench-mounted armrests were added. These armrests were adjustable for height. They were locked in the normal working position but could swivel outwards for materials handling and to enter and exit the workstation.

The sewing machine was tilted 11° towards the operator by adding a hardwood base. The height of the point of operation above the table top (19 cm) and its distance from the bench edge (16·5 cm) were such as to put it in the range of vision for an operator in an adequate posture (i.e., thighs horizontal, back erect and head/neck flexed forwards 30° or less).

The adjustable chair and footrest allowed the operator to properly position the lower extremities. For a task such as operating a sewing machine the operator's thighs should be horizontal (Murrell 1965) and the angle between the upper and lower leg near 105° (Grandjean 1980). Chairs and footrests adjustable by the user encourage the operator to assume good lower-extremity posture.

Positioning the point of operation within the vertical range of vision at the recommended distance from the eye and at the preferred angle to the line of sight (Grandjean 1982), should cause the operator to not lean forwards, preventing forward flexion of the trunk away from the lumbar support of the chair and forward flexion of the head/neck beyond the recommended range. The tilted machine allowed the work to be done with less ulnar deviation of the wrists. With the point of operation positioned to meet the above criteria, the operator would have to extend the upper arms. The bench-mounted armrests supported the upper arms and shoulders minimizing static muscular stress.

The prototype workstation was tested in the laboratory by the subject operator. After a satisfactory trial the prototype was used by the subject operator on the factory floor under production conditions for 3 weeks. During that time psychophysical data were again collected and the work was videotaped. Both the analysis of the psychophysical data collected and the task analysis indicated significant improvement in posture.

Trunk forward flexion was reduced from 17 to 1°. The improved angle allowed the operator to use the lumbar support of the chair and reduced the compressive force on the L5/S1 disc. Forward flexion of the head/neck was reduced from 46 to 37° (it was later reduced further by the operator getting a new prescription for corrective eyeglasses). Lessening of the forward flexion of the head/neck reduced static stress in the involved muscles. The angle of the thighs was changed from 9° below horizontal to horizontal. This also allowed the operator to sit more firmly into the lumbar support of the chair and relieved pressure on the back of the thighs.

The angle between the upper and lower legs was increased from 99 to 114°.

This was preferred by the operator for greater comfort and foot control. The foot to lower leg angle was increased from 71 to 86°. The foot angle changed from 86 to 104° during operation of the foot controls which is consistent with the recommendations of researchers (Murrell 1965).

The operator was noticeably less fatigued at the end of the day. Perceived postural stress improved by 20% (Wilcoxon test: $n = 5$, $d = 0$, $a = 0.031$). The compressive force on the L5/S1 disc was reduced by 82.5%. Perceived body-part discomfort was reduced by 18.5% ($\chi^2(1) = 14.799$, $p < 0.001$). Specifically, discomfort was reduced in the shoulders and back. Damaging wrist motions per day, defined as above in the methodology, were reduced for the left arm from 3947 to 3158 and for the right arm from 7105 to 5526.

3. Discussion

The postural problems of the sewing workplace have been studied by others. The results of the work to date are improved seating, adjustable sewing-machine stands and work aids to minimize reaches. These changes have caused some improvement in working posture of sewing operators. However, sewing operators still lean forwards away from the lumbar support of the chair and incline the head/neck forwards beyond 30° from the vertical. These conditions are stressful to the lower and middle back, shoulders and neck. Also, the sewing-machine stands frequently are not easily and quickly adjustable because some tools are required and/or the adjustment controls are not conveniently located. This study addressed these continuing problems.

The careful location of the point of operation, combined with bench-mounted armrests where appropriate, helped solve the head/neck posture problem. Also, with the point of operation high enough and at an improved angle, and close enough to the operator, the lumbar support of the chair could be utilized to relieve static stress in the lower and middle back.

An adequately adjustable sewing workstation needs to be adjustable by the user. If the point of operation is set at the proper height for the 95th percentile male, then all others can be adjusted to it easily by a chair and footrest adjustable by the user. In order for a footrest to be used, the foot control for the sewing machine must be redesigned to be easily moved about by attaching it to the motor with a wire or cable instead of a fixed rod.

This basic design has been tested on many other shoe-stitching jobs successfully. The design may be useful in apparel stitching as well and should be tested in that industry.

Chapter 34
Cost–Benefit Analysis of Work Environment Investment at STK's Telephone Plant at Kongsvinger

Svein Spilling and Jakob Eitrheim

Department of Environment, Norwegian Iron and Metalworkers Association, Lilletorget 1, Oslo 1, Norway

and Arne Aarås

Standard Telefon og Kabelfabrik A/S, Postboks 60, Økern, Oslo 5, Norway

1. Introduction

Complaints originating in the musculo-skeletal system are one of the most common categories of illness in Norway today. A survey of 9000 patients, treated by general practitioners, showed that 20% of all patients suffered from a musculo-skeletal complaint, accounting for 30% of occupational time lost through sick leave (Borchgreving *et al.* 1980). In 1976 it was estimated at STK that the direct daily cost due to sick leave, for the individual, the company and the community at large, was NKr 445 (Hjort 1976).

The cost of lost production amounted to at least the same value. Thus, both the economic and medical implications of musculo-skeletal illness are serious, and apparently of increasing importance relative to other illnesses.

A small STK manufacturing plant in Norway has provided us with an opportunity to carry out a quantitative study of the occurrence of musculo-skeletal illness (i.e., sick leave with a relevant medical diagnosis) among workers in work situations demanding static muscle work. In particular we could investigate the relationship between such illness and muscle strain due to working conditions (Westgaard and Aarås 1984, 1985).

This paper represents an attempt to carry out an estimation of the cost and the benefit to STK-K with regard to the improved work environment of the plant.

2. The arguments for company investment

Many companies still doubt the positive effect of applying ergonomic principles. Companies often present arguments, such as redesign, which have turned out to be positive in one company but cannot be transferred to another setting with expectation of the same positive results. Local conditions such as human resources and economic factors are more important. Each company must do its own research and redesign. Lack of knowledge and training in ergonomic principles for key groups in the company may explain why the results of investment in the work environment are not always as beneficial as expected.

Many workers even suffer musculo-skeletal illness in their working life without focusing on the workplace as the cause, or a major contributor to, development of this disease. It is, therefore, important to educate users in the consequences of poor design and in the positive features of good design, and to train them in the possible adjustments and variations in a workplace for different people and tasks. Giving operators information on ergonomic principles will increase motivation to utilize all the practical possibilities of workplace design to reduce strain and static muscle load.

Two other important groups who need education in applying ergonomic principles include industrial engineers, responsible for planning and organizing work, and engineering designers, who are able directly to influence the ways in which work is set out and equipment designed and produced. Motivation of managers to accept the initial cost of following ergonomic principles in the development of products and workplaces can be gained by giving significant information about employees' health problems, and also quantitative and qualitative data to show the economic advantages to be gained from reduced absence, reduced labour turnover, training and recruitment costs and increased production.

Such savings will be beneficial both for the company, the employees and society at large. Sickness absence is costly for the company, which has to give sick payments to the employees, the workers themselves, who cannot fully enjoy their leisure time, and society, which has to pay for the medical treatment, rehabilitation and, finally, often an early pension.

3. *Kongsvinger project*

Kongsvinger is a rural community which is located 100 km north-east of Oslo. It supports an industrial estate having a number of manufacturing plants, one of which is a telephone switching plant belonging to Standard Telefon and Kabelfabrik A/S (STK-K). The assembly and wiring of electromechanical components and racks are done at this plant. Employees are mostly women, who work at individual workstations. In the period 1967–1974 all workstands had a fixed height, which created excessive muscular load due to the need to adopt awkward postures at work. Figures 1–3 illustrate such work situations.

In 1975 an extensive redesign of all workstands was carried out, which gave each operator greater flexibility to vary working posture. The range of adjustment allowed work to be carried out both in a sitting and standing position. Figures 4–7 show such workplaces.

Our experience shows that a significant number of workers prefer to vary their work position during the day, and in particular to work in a standing position towards the end of the day. The individual ergonomic designs and allowance for

Figure 1.   *The shoulders are lifted; using heavy hand tools.*

Figure 2.   *The upper part of the body is bent over.*

Figure 3. Fine detail demands a short viewing distance and the head must be bent forwards.

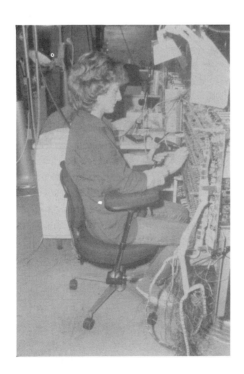

Figure 4. Adjustable workplace. Armrest supports elevated arms.

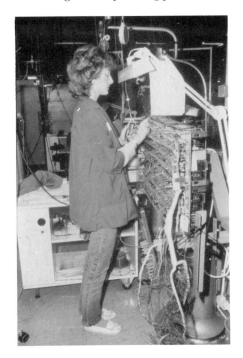

Figure 5.   Standing work position.

Figure 6.   The workstand gives possibilities for adjusting height and angle.

Figure 7.   By tilting the table top, a short viewing distance can be combined with relaxed shoulder and upright body position.

variation of the work posture has been shown to reduce static muscle load and load-related diseases (Table 1).

Long-term sick leave was relatively stable in the years 1975–1982, averaging 9·4% of possible working time for the whole period. This is similar to average sick leave in the period 1967–1974, but in this first period long-term sick leave showed a steep rise, increasing to 13·4 and 16·9% in 1973 and 1974 (Figure 8). The sharp interruption in the upward trend in 1975 coincided with the implementation of the ergonomic redesign.

Table 1.   Sick leave and labour turnover at STK, Kongsvinger (in percentage of possible working time for the average number of people employed in the time period).

	1967–1974	1975–1982
Short-term sick leave	1·3	1·5
Long-term sick leave	9·9	9·4
Musculo-skeletal sick leave	5·3	3·1
Labour turnover	30·1	7·6

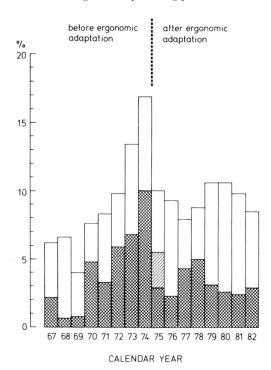

Figure 8.   Long-term sick leave (of more than 3 days' duration, as percentage of possible working time each year) at STK, Kongsvinger, in the years 1967–1982. The hatched parts of the columns indicate long-term sick leave due to musculo-skeletal illness. Single hatching in 1975 indicates musculoskeletal sick leave beginning in 1974.

Average musculo-skeletal sick leave in the years 1975–1982 was 3·1% of total production time, while it was 5·3% in the period 1967–1974 (Table 1). This reduction is statistically significant ($p < 0.01$), and even more dramatic when compared with the two years immediately preceding the ergonomic redesign (6·8 and 10·0%) (Westgaard and Aarås 1984, 1985).

Simultaneous with the reduction in long-term sick leave there was a statistically significant ($p < 0.001$) reduction in labour turnover. Turnover in the years 1975–1982 was 7·6% of total man-labour years, while it was 30·1% in the period 1967–1974 (see also Figure 9).

In contrast to long-term sick leave and labour turnover, short-term sick leave has remained stable (Table 1). Taken overall, the results indicate a clear positive effect of the ergonomic redesign in terms of improved health among the workers.

4.   The costs and benefits of the project

The relevant question is: How has the company benefitted as a result of these work environment improvements?

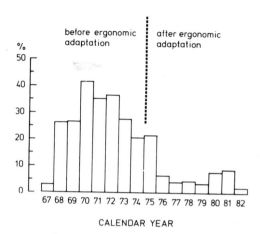

CALENDAR YEAR

Figure 9.   Labour turnover (percentage of average number of workers each year) at STK, Kongsvinger, in the years 1967–1982.

An attempt has been made to break down and analyse the investment costs and benefits. This analysis has been carried out mainly by the Norwegian Iron and Metalworkers Association. It does not consider the costs of treatment and other health services for the employees and for society at large. The investments to improve the work environment are relatively simple to assess. Such investments include purchasing and installation of worktables, ventilation and lighting as well as maintenance and running costs. The savings which the company have achieved so far, and will get in the future, due to such work environmental improvements are much more difficult to calculate. They are assessed by using statistical data from the company for the period before and after the ergonomic redesign. For this we have used the *technique of theory of investment* (Mossin 1972):

Investment analysis consists of the calculation and comparison of the business alternatives that must take place in connection with an investment decision. The underlying basis for every investment analysis is that the alternative investments are fully described with a complete income and expenditure (cash flow) analysis that relates to each. The calculation includes all income and expenditure falling within a certain selected time period, usually 1 year, such that income or expenditure groups may be identified to the end of the year.

Cash flow can be represented graphically along a time axis where negative totals represent expenditure and positive sums represent income. The cash-flow graph for investment may consist of an expenditure K at time 0 and thereafter net income of $b_1, b_2, \ldots, b_n$, at times $1, 2, \ldots, n$.

$-K$	b_1	b_2	$\ldots$	b_{n-1}	b_n
0	1	2	$\ldots$	b_{n-1}	n

An investment decision formula generally allows a choice between alternative cash-flow calculations and the decision depends on a ranking of these. Such ranking can be set up by defining a profitability forecast as a function:

$$f(K_1, b_1, b_2, \ldots, b_n),$$

in such a way that cash flow $(-K, b_1, b_2, \ldots, b_n)$ ranks before cash flow $(-K'_1, b'_1, b'_2, \ldots, b'_n)$ if and only if

$$f(K_1, b_1, b_2, \ldots, b_n) > f(K'_1, b'_1, b'_2, \ldots, b'_n)$$

A profitability forecast for cash flow is based on discount of future income. Present value, V, in such a profitability forecast, is defined as

$$V = b_1/(1 + r) + b_2/(1 + r)^2 + \ldots + b_n/(1 + r)^n - K$$

where r specifies an interest rate.

Thus present value is found as the sum of the discounted future incomes with deduction of the expenditure K, and will depend on the selected value of the discount rate. Whether or not the investment project is to be implemented is dependent upon the present value, which gives an indication of the difference between the initial investment and what would be gained from an alternative investment at a certain rate of interest.

Profitability forecasts will depend on the interest rate for the alternative investments and the eventual interest rate of the loan borrowed for the investment. Taking the present value V as the basis, which has been calculated from the capital cost, the project will be acceptable when the present value is positive in the formula above. Therefore, after time period 1, the necessary condition for the present value to be positive is that,

$$V = b/(1 + r) - K > 0; \quad \text{or} \quad b/K > 1 + r$$

Generally the condition can be expressed as follows:

$$\sum_{t=1}^{n} \frac{b_t}{(1 + r)^6}$$

We will now discuss an analysis of the Konsvinger project according to this theory. We assume that the investment has a lifetime of 12 years (1976–1988), which means a yearly amortization of 8·5%. Amortization after 1987 is not included in the calculations.

To compare the investments and savings at different times, it is first necessary to de-inflate all expenditure and benefits relative to the same base year (1976 = 100%) (Table 2). Similarly, in order to compare business alternatives with differing interest rates, it is necessary to discount future income and expenditure to obtain a profitability forecast or present value. Then we can rank the business alternatives for present value. For the Kongsvinger project a discount rate of 7% was used to calculate the present value, as recommended by the Norwegian Ministry of Finance.

No alternative investments have been calculated for this project; instead, the

Table 2. *Social cost and price index 1976–1987.*

	1976	1977	1978	1979	1980	1981	1982	1983	1984	1985	1986	1987
Social cost (%)[a]	41·6	41·5	45·0	48·1	48·1	48·1	48·1	48·1	48·1	48·1	48·1	48·1
Price index	100·0	109·0	117·9	123·6	137·1	155·8	173·4	188·0	188·0	188·0	188·0	188·0
Discounting factor 7%	1·0000	1·07	1·145	1·3110	1·4028	1·5010	1·6061	1·7185	1·8388	1·9675	2·1052	

[a] Norwegian Employer Federation statistic.

Table 3. *Investment at STK-K.*

	1976	1977	1978	1979	1980	1981	1982	1983	1984	1985	1986	1987
Ergonomic workstations	240 000	0	0	0	0	0	0	0	0	0	0	
Ventilation	0	0	0	0	0	300 000	0	0	0	0	0	0
Lighting	0	0	0	0	0	27 000	27 000	0	0	0	0	
Running costs for lighting	0	0	0	0	0	1 700	3 400	3 400	3 400	3 400	3 400	3 400

actual figures for expenditure and savings were used. When the analysis was performed in 1983, no future price index could be estimated. Therefore expenditure and savings after 1983 are calculated on the basis of the 1983 increases. After de-inflating back to the 1976 value, all expenditures and savings are calculated at the present value.

Investment and cost

Changes to the ergonomic design of the operators' workstations were carried out in 1975–1976, to ventilation in 1981 and to lighting in 1981 and 1982 (Table 3). Total investment in ergonomic workstations during 1975–1976 was NKr 240 000 in 1976 value and present value. Installation of a new ventilation system in 1981 required NKr 300 000. Lifetime for this installation is 12 years, which means that in 1988 the amortization value remains:

$$\text{NKr } 300\ 000 \times \frac{5}{12} = \text{NKr } 125\ 000$$

Investment cost until 1988 in 1981 value is thus NKr 175 000 (300 000 − 125 000). The 1976 value is:

$$\text{NKr } 175\ 000/155 \cdot 8 \times 100 = \text{NKr } 112\ 323$$

where 155·8 is the index for 1981 (Table 2). Investment cost in present value is:

$$\text{NKr } 112\ 323/1 \cdot 07^5 = \text{NKr } 112\ 323/1 \cdot 4028 = \text{NKr } 80\ 070$$

In 1981 and 1982 low-luminance light fittings were installed. This investment was NKr 27 000 in 1981 and NKr 27 000 in 1982. The amortization value in 1988 remains:

$$\text{NKr } 27\ 000 \times \frac{5}{12} + \text{NKr } 27\ 000 \times \frac{6}{12} = \text{NKr } (11\ 250 + 13\ 500) = \text{NKr } 24\ 750$$

The investment cost until 1988 for lighting installed in 1981 is thus NKr 15 750, and in 1982 is NKr 13 500. The 1976 value is:

$$\text{NKr } (15\ 750/155 \cdot 8 \times 100 + 13\ 500/173 \cdot 4 \times 100) = \text{NKr } (10\ 109 + 7785)$$
$$= \text{NKr } 17\ 894$$

The investment cost for lighting, at the present value, is therefore:

$$\text{NKr } (10\ 109/1 \cdot 07^5 + 7785/1 \cdot 07^6) = \text{NKr } 12\ 393$$

Running costs for lighting led to an increased consumption of electricity in comparison with the old lighting installation, and in 1983 this amounted to NKr 2700. Annual cost of replacement of extra lighting tubes is NKr 700. The 1976 value therefore is:

$$\text{NKr } 3400/188 \cdot 0 \times 100 = \text{NKr } 1809$$

Running costs for lighting at the present value is:

$$\text{NKr } 1809/1 \cdot 07^7 = \text{NKr } 1126$$

Similar costs can be calculated for the period 1981–1988, and taking the first year (1981) running costs to be 50% of the costs in 1983, total running costs for lighting in the period 1981–1988, at the present value, are NKr 6529. Therefore, total expenditure, K, for ergonomics, ventilation, lighting installation and running costs at the present value is NKr 338 992 (Table 4).

Table 4. Investment and savings at the present value.

Investment (NKr)		Savings (NKr)	
Ergonomics	240 000	Reduction in recruitment cost	108 812
Ventilation	80 070	Reduction in training cost	1 645 720
Lighting installation	12 393	Reduction in instructors' salary cost	812 019
Lighting, running costs	6 529	Reduction in sick payments	659 643
	NKr 338 992		NKr 3 226 194

Savings and benefits

The major contribution to savings, as already pointed out, has been a significant improvement in worker health reflected in a dramatic reduction in labour turnover and in sickness absence. Reduction of turnover gave a consequent reduction of direct costs of recruitment and training. In the period 1967–1974 the average labour turnover was 30·1%. In the next 8 years the turnover was reduced dramatically to 7·6% on average.

Several factors may influence the turnover rate. Good wages in this company compared with other companies, or a shortage of alternative work in the same district may be the most important factors. Improved workplaces may reduce the level of pain during work to a more acceptable level. One of the authors has developed and implemented the redesign of the workplaces in cooperation with production engineers. Education and training in ergonomic principles was also given. As discussed in Chapter 1 (Westgaard *et al.*) this work may have influenced attitudes and awareness of the relationships between workload and the development of musculo-skeletal illness. Such acceptance may increase willingness to take necessary corrective measures, such as short breaks or a change in work duties, for workers with symptoms of postural pain, and thus be a factor in improving the work situation.

However, interviews with workers themselves in 1978 seemed to point out clearly that improved work environment was the most important factor in reduction of turnover (Table 5), 57% assessing it as such. As a hypothesis, therefore, we will suggest that 50% of the reduction of turnover was due to improved work environment. There has been little variation in turnover after 1976, so we use

Table 5. *Workers' views on the reasons for the reduction in labour turnover. The significance of different factors is indicated as: 1, important; 2, less important; 3, may be significant.*

	Significance (number of indications)		
Reason	1	2	3
Difficult to obtain alternative employment	11	11	1
The pay rate is good	25	12	1
Improved work environment	50	8	2
Other	1	2	1
Total	87	33	5

the average turnover of 7·6% for the whole period 1976–1988. Reduction in labour turnover due to improved work environment is, therefore:

$$(30 \cdot 1 - 7 \cdot 6) \times 50\% = 11 \cdot 25\%$$

From statistical data on numbers of employees (Table 6) we can calculate how many persons would have terminated work and thus how many would have had to be recruited.

From 1983 we have assumed the number of employees to be constant at 110. We therefore suppose that 144 workers would have terminated work during the period 1976–1988. These 144 employees would have been additional to those who actually terminated their work. That means that STK-K have had 144 recruitments less in the period 1976–1988, due to improved work environment. Recruitments are costly for the company due to interviews, recruitment procedure and training.

Table 6. *Labour turnover statistics.*

Year	Number of employees	Reduced turnover (%)	Reduced turnover number
1976	94	11·25	10·6
1977	98	11·25	11·0
1978	92	11·25	10·4
1979	95	11·25	10·7
1980	117	11·25	13·2
1981	115	11·25	12·9
1982	114	11·25	12·8
1983	110	11·25	12·4
1984	110	11·25	12·4
1985	110	11·25	12·4
1986	110	11·25	12·4
1987	110	11·25	12·4
Total			144·0

Interviews and recruitment procedure

The management at STK-K has assessed that an average of 42 hours is required to recruit three persons. That means 14 hours for interviews, filling in forms, registration and so on, per person.

The Norwegian Employers Federation calculate the relevant salary, including social costs, to be NKr 120 per hour at 1982 values. Salary cost for 1982 recruitment was NKr 21 504 ($12 \cdot 8 \times 14 \times 120$). In addition, we have costs for telephones and advertisements, assessed to be NKr 200 per person in 1982. Total recruitment cost in 1982 was therefore NKr 24 064. The 1976 value was NKr 13 878, and the present value is

$$\text{NKr } 13\ 878/1 \cdot 07^6 = \text{NKr } 9246$$

Similar calculations were carried out for the other years 1976–1988, taking into account the appropriate inflation rates and a discount rate of 7%. Total saving in recruiting cost, at the present value, is NKr 108 812 (Table 7).

Training cost

According to the management at STK-K, training time is assessed to average 16 weeks per person. Training cost includes outlays for the company on wages, education and classroom space for the newly recruited operators. During training it is assessed that 40% production is achieved for each operator, meaning that the company has a production loss of 60% during these 16 weeks, when comparing a newly recruited operator with a fully trained employee. Training time (16 weeks) is 640 hours when working 40 hours per week, and loss of production during education time is thus 384 hours. The cost will therefore be (wages + social cost) $\times$ 384 hours.

We set wages to be NKr 50 per hour in 1982, and must add $48 \cdot 1\%$ of this as the social cost. Total training cost per recruited operators will, at the 1982 value, be

$$\text{NKr } (50 + 50 \times 48 \cdot 1/100) \times 384 = \text{NKr } 28\ 435$$

Corrected for inflation, the 1976 value will be NKr 16 398 and the present value is

$$\text{NKr } 16\ 398/1 \cdot 07^6 = \text{NKr } 10\ 925$$

Total saving, at the present value, for training cost is NKr 1 645 720 (Table 7).

Instructors' cost

In addition the company has to pay the instructors. If we assess that one instructor can train five operators, that means that during 16 weeks the instructor spends $3 \cdot 2$ weeks per operator. However, salary and social cost of NKr 120 at 1982 values means NKr 14 000 which correspond at 1976 values to NKr 8091, and the present value is

$$\text{NKr } 8091/1 \cdot 07^6 = \text{NKr } 5390$$

Table 7. Savings in recruiting, training and instructors' salary cost.

	1976	1977	1978	1979	1980	1981	1982	1983	1984	1985	1986	1987
Recruiting cost ex 1976 values	11 493	11 926	11 276	11 601	14 311	13 986	13 878	13 444	13 444	13 444	13 444	13 444
7% discounting factor	1·0	1·07	1·145	1·2252	1·3110	1·4028	1·5010	1·6061	1·7185	1·8388	1·9675	2·1052
Present value	11 493	11 146	9 848	9 469	10 916	9 970	9 246	8 371	7 823	7 311	6 833	6 386
Σ present value	108 812											
Training cost	173 819	180 378	170 539	175 459	216 454	211 534	209 894	203 334	203 335	203 335	203 335	203 335
Present value	173 819	168 578	148 942	143 208	165 106	150 794	139 836	126 602	118 321	110 580	103 347	96 587
Σ present value	1 645 720											
Instructors' salary	85 765	89 001	84 146	86 573	106 801	104 373	103 564	100 328	100 328	100 328	100 328	100 328
Present value	85 765	83 179	73 490	70 660	81 465	74 403	68 997	62 467	58 381	54 562	50 993	47 657
Σ present value	812 019											

Total saving for the years 1976–1988, at the present value, is NKr 812 019
The sum of training cost and instructors' salary saved amounts to NKr 2 457 739, which alone would justify the total work environmental investment.

Sickness reduction for musculo-skeletal illness

Certificated sickness absence is a great part of total social costs for a company. There are two ways to calculate the cost of sick leave. One is to calculate what the company has to pay in wages during sick leave, and the other is to estimate the company loss in production over the same period due to sick leave. It is difficult to estimate what is needed in extra hours to achieve the same production volume. The calculations are very complex and they lead us to the figures for wages payable during sickness. The wages and social cost for the first 14 days of sickness are payable by the company.

We find it reasonable to assess the difference in sick leave due to musculo-skeletal illness as a result of the ergonomic redesign, but a problem arises in that we do not know exactly the frequency and duration of the sick leave.

We have assessed time lost in production due to each musculo-skeletal sick leave to be 20 days on average. From Figure 10 we can see that very little sick leave due to musculo-skeletal illness is of less than 10 days. Assuming 20 days' average sick leave we can calculate how many such periods have occurred and it is then easy to multiply by the cost of each sick leave period to arrive at the total savings. We take the hourly wages for 1982 of NKr 50 plus a social cost at 48·1% as a base. This corresponds to NKr 42·70 at 1976 values. Seventy hours is taken as the average paid sick leave. Due to reduced sick leave in 1976 total savings are NKr 96 843. The corresponding calculations are made for the following years. Total savings for reduced sick leave due to musculo-skeletal illness are therefore, at the present value, NKr 659 643 (Table 8).

Therefore, the total saving in recruitment costs, training costs, instructor salary costs and sick payments, at the present value, is NKr 3 226 194.

5.   Conclusion

Calculated at the present value STK-K has invested a total of NKr 338 992 in improving the work environment. Due to this investment the company has obtained savings in terms of reduced cost, at the present value, of NKr 3 226 194.

For the Kongsvinger project, therefore, the present value of the investment in work environment is

$$V = \text{savings} - \text{investments} = \text{NKr} \ (3\ 226\ 194 - 338\ 992) = \text{NKr} \ 2\ 887\ 202$$

Only the cost of the project work itself is not included. This cost is difficult to estimate, but it is clear that it has been small, relative to the apparent savings for the company.

In conclusion then, a total investment in a work environment programme, at a

Table 8. *Savings due to reduced sick leave.*

	(1) Musculo-skeletal sick leave (%) for 1967–1974 on average	(2) Musculo-skeletal sick leave each year during 1975–1982	(3)=(1)−(2) Reduction in sick leave (%)	(4) No. of employees × working days (230)	(5)=(3)×(4) Total days absence	(6)=(5)×20 No. of sick leaves (average sick leave = 20 days)	(7)=(6)×70 Average paid sick leave due in musculo-skeletal illness (70 hours)	(8)=(7)×42·70 Salary including social cost (NK 42·70 at 1976 values)	(9) Discounting factor 7% per annum	(10+=(8)×(9) Savings each year due to reduction in musculo-skeletal illness
1976	5·3	2·3	3·0	21 620	648·6	32·4	2 268	96 843	1·0	96 843
1977	5·3	4·3	1·0	22 540	225·4	11·3	791	33 776	1·07	31 566
1978	5·3	5·0	0·3	21 160	63·5	3·2	224	9 565	1·145	8 354
1979	5·3	3·1	2·2	21 850	480·7	24·0	1 680	71 736	1·2252	58 550
1980	5·3	2·6	2·7	26 910	726·6	36·6	2 541	108 501	1·3110	82 762
1981	5·3	2·4	2·9	26 450	767·1	38·4	2 688	114 778	1·4028	81 242
1982	5·3	2·9	2·4	26 220	629·3	31·5	2 205	94 154	1·5010	62 728
1983	5·3	3·0	2·3	25 300	581·9	29·1	2 037	86 980	1·6061	54 156
1984	5·3	3·0	2·3	25 300	581·9	29·1	2 037	86 980	1·7185	50 614
1985	5·3	3·0	2·3	25 300	581·9	29·1	2 037	86 980	1·8388	47 303
1986	5·3	3·0	2·3	25 300	581·9	29·1	2 037	86 980	1·9675	44 208
1987	5·3	3·0	2·3	25 300	581·9	29·1	2 037	86 980	2·1052	41 317
										659 643

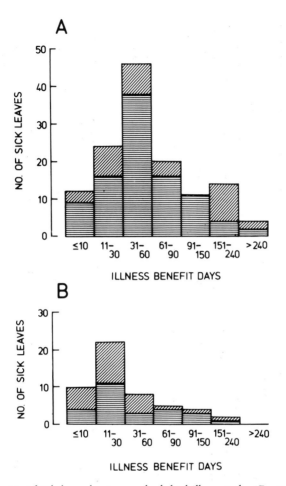

Figure 10. Duration of sick leave due to musculo-skeletal illness at the 8B system (a) and cable making (b) for the period 1967–1974. Sick leave due to a complaint located in the neck, shoulder or arms indicated by horizontal hatching, low-back complaints by diagonal hatching.

present value of less than NKr 350 000, produced significant improvements in worker health and a reduction in direct costs to STK, at the present value, of almost 3 million kroner, in the period from 1976 to 1988.

References

Aarås, A., 1982, Neck and shoulder pain, in *Low Back Pain and Industrial and Social Disablement*, Nelson, M. (ed.) (London: Back Pain Association), pp. 40–44.

Adams, M. and Hutton, W., 1983 a, The effect of posture on the fluid content of lumbar intervertebral discs. *Spine*, **8**, 665–71.

Adams, M. and Hutton, W., 1983 b, The mechanical function of the lumbar apophyseal joints. *Spine*, **8**, 327–330.

Adrian, M., Tipton, C. and Karpovich, P., 1965, *Electrogoniometry Manual* (Massachusetts: Springfield College).

AFNOR, 1972, Bureaux et tables pour dactylographie. Norme française NF D 67-605, AFNOR, Paris.

AFNOR, 1973, Sièges du bureau—Repose pieds. Norme française NF D 61-201, AFNOR, Paris.

Åkerblom, B., 1948, Standing and sitting posture. With special reference to the construction of chairs. Doctoral dissertation, Nordiska Bokhandeln, Stockholm.

Åkerblom, B., 1954, *Chairs and Sitting in Human Factors in Equipment Design* (London: Lewis).

Altmann, J., 1974, Observational study of behaviour. Sampling methods. *Behaviour*, **49**, 227–267.

Ambrosini, A., Cometti, A., Pedotti, A. and Santambrogio, G. C., 1985, A new device for real-time analysis of posture and gait. *Proceedings of the IFAC-INSERM "Human Gait Analysis and Applications"*, November, 1983, Montpellier, Rabischong, P., Liegois, A. and Peruchon, E. (eds.), pp. 77–87.

Anderson, J. A. D., 1974, Occupation as a modifying factor in the diagnosis and treatment of rheumatic diseases. *Current Medical Research Opinion*, **2**, 521–528,.

Anderson, J. A. D., 1980, Back pain and occupation, in *The Lumbar Spine and Back Pain*, Jayson, M. I. V. (ed.) (London: Pitman), pp. 57–82.

Andersson, G. B. J., 1974, On myoelectric back muscle activity and lumbar disc pressure in sitting postures. Doctoral dissertation, Gotab, University of Göteborg, Göteborg.

Andersson, G. B. J., 1981, Epidemiologic aspects on low-back pain in industry. *Spine*, **6**, 53–60.

Andersson, G. B. J., 1982, Occupational aspects of low-back pain, in *Low Back Pain and Industrial and Social Disablement* (London: Back Pain Association), pp. 45–51.

Andersson, G. B. J. and Örtengren, R., 1974 a, Lumbar disc pressure and myoelectric back muscle activity during sitting. II. Studies on an office chair. *Scandinavian Journal of Rehabilitation Medicine*, **3**, 115–121.

Andersson, G. B. J. and Örtengren, R., 1974 b, Lumbar disc pressure and myoelectric back muscle activity during sitting. III. Studies on a wheelchair. *Scandinavian Journal of Rehabilitation Medicine*, **3**, 122–127.

Andersson, G. B. J. and Örtengren, R., 1974 c, Myoelectric back muscle activity during sitting. *Scandinavian Journal of Rehabilitation Medicine*, Suppl. 3, 73–90.

Andersson, G. B. J., Jonsson, B. and Örtengren, R., 1974 a, Myoelectric activity in individual lumbar erector spinae muscles in sitting. A study with surface and wire electrodes. *Scandinavian Journal of Rehabilitation Medicine*, Suppl. 3, 91–108.

Andersson, G. B. J., Örtengren, R., Nachemson, A. and Elfström, G., 1974 b, Lumbar disc pressure and myoelectric back muscle activity during sitting. I. Studies on an experimental chair. *Scandinavian Journal of Rehabilitation Medicine*, **3**, 104–114.

Andersson, G. B. J., Örtengren, R., Nachemson, A. and Elfström, G., 1974 c, Lumbar disc pressure and myoelectric back muscle activity during sitting. IV. Studies on a driver's seat. *Scandinavian Journal of Rehabilitation Medicine*, **3**, 129–133.

Andersson, G. B. J., Örtengren, R. and Nachemson, A., 1976, Quantitative studies of back loads in lifting. *Spine*, **3**, 178–185.

Andersson, G. B. J., Murphy, R. W., Örtengren, R. and Nachemson, A., 1979, The influence of backrest inclination and lumbar support on the lumbar lordosis in sitting. *Spine*, **4**, 52–58.

Andersson, G. B. J., Örtengren, R. and Schultz, A., 1980, Analysis and measurement of the loads on the lumbar spine during work at a table. *Journal of Biomechanics*, **13**, 513–520.

Ariano, M., Armstrong, R. B. and Edgerton, V. R., 1973, Hindlimb muscle fibre population in five animals. *Journal of Histochemistry and Cytochemistry*, **21**, 51–55.

Armstrong, J. R., 1958, *Lumbar Disc Lesions* (London: Livingstone).

Armstrong, T., 1983, *An Ergonomics Guide to Carpal Tunnel Syndrome.* (Akron, Ohio: American Industrial Hygiene Association).

Armstrong, T. and Chaffin, D., 1979, Carpal tunnel syndrome and selected personal attributes. *Journal of Occupational Medicine*, **21**, 481–486.

Armstrong, T. J., Foulke, J. A., Joseph, B. S. and Goldstein, S. A., 1982, Investigation of cumulative trauma disorders in a poultry processing plant. *American Industrial Hygiene Association Journal*, **43**, 103–116.

Armstrong, T., Castelli, W., Gaynor Evans, F. and Diaz-Perez, R., 1984, Some histological changes in carpal tunnel contents and their biomechanical implications. *Journal of Occupational Medicine*, **26**, 197–201.

Asmussen, E., Hansen, O. and Lammert, O., 1965, The relation between isometric and dynamic muscle strength in man. Communication No. 20, The Testing and Observation Institute of the Danish National Association for Infantile Paralysis, Hellerup.

Åstrand, P.-O. and Rodahl, K., 1970, *Textbook of Work Physiology* (New York: McGraw-Hill).

Ayoub, M. M. and McDaniel, J. W., 1971, The biomechanics of pushing and pulling tasks. Technical Report, Texas Tech. University, Lubbock, Texas.

Barbenel, J. C., 1972, The biomechanics of the temporomandibular joint: a theoretical study. *Journal of Biomechanics*, **5**, 251–256.

Barbonis, P. A., 1979, Measurement and modelling of postural load. Ph.D. thesis, Birmingham University.

Barnes, R. M., 1968, Motion and time study, in *Design and Measurement of Work* (6th edn.) (New York: Wiley), Chapter 11.

Basmajain, J. V., 1974, *Muscles Alive: Their Functions Revealed by Electromyography* (Baltimore: Williams & Wilkins).

Baty, D., Buckle, P. W. and Stubbs, D. A., 1986, Posture recording by direct observation, questionnaire assessment and instrumentation: a comparison based on a recent field study. This volume, Chapter 25.

Becher, G., Mucke, R., Kramer, H. and Drews, I., 1983, Studies on local muscular endurance capacity of isometric working muscles using electromyographic methods. *Electromyography and Clinical Neurophysiology*, **23**, 415–423.

Begg, A. C. and Falconer, M. A., 1949, Plain radiography in intraspinal protrusion of lumbar intervertebral discs. *British Journal of Surgery*, **36**, 225–239.

Bencke, A., 1897, Zur Lehre von der spondylitis deformans—Beitrag zur wissenschaftlichen Medizin. *Festschrift an der 59. Versammlung deutscher Naturforscher und Ärzte*, Braunschweig.

Bendix, T., 1983, Posture of the trunk when sitting on forward-inclining seats. *Scandinavian Journal of Rehabilitation Medicine*, **15**, 197–203.

Bendix, T., 1984, Seated trunk posture at various seat inclinations, seat heights and table heights. *Human Factors*, **26**, 695–703.

Bendix, T. and Biering-Sørensen, F., 1983, Posture of the trunk when sitting on forward-inclining seats. *Scandinavian Journal of Rehabilitation Medicine*, **15**, 197–203.

Bendix, T. and Bloch, I., 1986, How should a seated workplace with a tiltable chair be adjusted? *Applied Ergonomics* (in press).

Bendix, T. and Hagberg, M., 1984, Trunk posture and load on the trapezius muscle whilst sitting at sloping desks. *Ergonomics*, **27**, 873–882.

Bendix, T. and Jessen, F., 1984, Håndledsstötte og tizanidin ved maskinskrivning; en kontrolleret og elektromyografisk undersögelse. Abstract Dansk Reumatologisk Selskap, 26.10.84, Copenhagen.

Bendix, T. and Jessen, F., 1986, Wrist support during typing. A controlled, electromyographic study. *Applied Ergonomics* (in press).

Bendix, T., Krohn, L., Jessen, F. and Aarås, A., 1985, Trunk posture and trapezius muscle load while working in standing, supported-standing and sitting positions. *Spine*, **10**, 433–439.

Bendix, T., Winkel, J. and Jessen, F., 1985b, Comparison of office chairs with fixed forwards or backwards inclining, or tiltable seats. *European Journal of Applied Physiology*, **54**, 378–385.

Bendix, T., Jessen, B. and Winkel, J., 1986, An evaluation of a tiltable office chair with respect to seat height, backrest position, and task. *European Journal of Applied Physiology* (in press).

Berns, T. A. R. and Milner, N. P., 1980, TRAM—a technique for the recording and analysis of moving work posture, in *Methods to Study Moving Work Posture*, Milner, N. P. (ed.) Ergolab Report 80: 23, pp. 22–26.

Berns, T. A. R. *et al.*, 1979, *Ergonomi och Underhåll* (Ergonomics and Maintenance) (Stockholm: Arbetarskyddsfonden), Dnr 78/377.

Bevegard, S., Freyschuss, U. and Strandell, T., 1966, Circulatory adaptation to arm and leg exercise in supine and sitting position, *Journal of Applied Physiology*, **21**, 37–46.

Beyer, J. and Wright, I., 1951, The hyperabduction syndrome. *Circulation* (journal of the American Heart Association), **4**, 161–172.

Biering-Sørensen, F., 1984, A one-year prospective study of low-back trouble in a general population. *Danish Medical Bulletin*, **31**, 362–375.

Bigland-Ritchie, B. and Woods, J., 1974, Integrated EMG and oxygen uptake during dynamic contractions of human muscles. *Journal of Applied Physiology*, **36**, 475–479.

Bjelle, A., Hagberg, M. and Michaelson, G., 1979, Clinical and ergonomic factors in prolonged shoulder pain among industrial workers. *Scandinavian Journal of Work, Environment and Health*, **5**, 205–210.

Bjelle, A., Hagberg, M. and Michaelson, G., 1981, Occupational and individual factors in acute shoulder–neck disorders among industrial workers. *British Journal of Industrial Medicine*, **38**, 356–363.

Björksten, H. and Jonsson, B., 1977, Endurance limit of force in long-term intermittent static contractions. *Scandinavian Journal of Work, Environment and Health*, **3**, 23–27.

BMDP, 1984, *P4V BMDP Statistical Software* (Los Angeles: UCLA Press).

Bora, F. and Osterman, A., 1982, Compression neuropathy. *Clinical Orthopaedics*, **163**, 20–31.

Borchgreving, C. F., Brekke, T. H. and Øgar, B., 1980, Musculo-skeletal illness in general practice. *Tidskrift for den Norske Lægeforening*, **100**, 439–445 (in Norwegian).

Borg, G. A. V., 1962, A simple rating scale for use in physical work tests. *Kungliga Fysiografiska Sällskapets i Lund Förhandlingar*, **32**, 7–15.

Borg, G. A. V., 1973, A note on a category scale with 'ratio properties' for estimating perceived exertion. Report No. 36, The Institute of Applied Psychology, University of Stockholm.

Borg, G. A. V., 1978, Subjective aspects of physical and mental load. *Ergonomics*, **21**, 215–220.

Borg, G. A. V., 1982, A category scale with ratio properties for intermodal and inter-individual comparisons, in *Psychophysical Judgment and the Process of Perception*, Geissler, H.-G. and Petzold, P. (eds.) (Berlin: VEB Deutscher Verlag der Wissenschaften).

Bouisset, S., 1973, Processed EMG and muscular force in normal motor activities, in *New Developments in Electromyography and Clinical Neurophysiology*, Vol. 1, Desmedt, J. E. (ed.) (Basel: Karger), pp. 547–583.

Brain, W., Wright, A. and Wilkinson, M., 1947, Spontaneous compression of both median nerves in the carpal tunnel. *Lancet*, **i**, 277–282.

Brown, K., Baker, P. and Madely, R., 1984, The use of a portable microcomputer to study behaviour. *Annals of Human Biology*, **11**, 473.

BSI, 1965, *Specification for Office Desks, Tables and Seating*. BSI 3893: 1965, British Standards Institution, London.

Buckle, P. W., 1983, A multidisciplinary investigation of factors associated with low-back pain. Ph.D. thesis, Cranfield Institute of Technology, Bedfordshire.

Buckle, P. W., 1984, Ergonomic needs in epidemiological studies of low-back pain, in *Proceedings of the Society of Occupational Medicine Symposium on Occupational Aspects of Back Disorders*, Royal Army Medical College, Millbank, London, 30 May, Brothwood, J. (ed.), pp. 24–33.

Burandt, U., 1969, Röntgenuntersuchung über die Stellung von Becken und Wirbelsäule beim Sitzen auf vorgeneigten Flächen, in *Sitting Posture*, Grandjean, E. (ed.) (London: Taylor & Francis), pp. 242–250.

Burandt, U. and Grandjean, E., 1963, Sitting habits of office employees. *Ergonomics*, **6**, 217–228.

Caillet, R., 1973, *Il dolore lombosacrale* (Rome: Leonardo Ed. Scientifiche).

Caillet, R., 1980, *Soft Tissue Pain and Disability* (Philadelphia: Davis).

Caillet, R., 1981, *Low Back Pain* (3rd ed.) (Philadelphia: Davis).

Cameron, C., 1973, A theory of fatigue. *Ergonomics*, **16**, 633–648.

Campbell, K. M., Biggs, N. L., Blanton, P. L. and Lehr, R. P., 1973, Electromyographic investigation of the relative activity among four components of the triceps surae. *American Journal of Physical Medicine*, **52**, 30–41.

Cantoni, S., Colombini, D., Occhipinti, E., Grieco, A., Frigo, C. and Pedotti, A., 1984, Posture analysis and evaluation at the old and new workplace of a telephone company, in *Ergonomics and Health in Modern Offices*, Grandjean, E. (ed.) (London: Taylor & Francis), pp. 456–464.

Cardell, H. and Melin, E., 1981, Sambandsundersökning av nack- and skulder-besvär och muskulära belastningsfaktorer hos personal inom LM-koncernen. Report BKH 9/81, Industrial Health Care Center, Ericsson, Stockholm.

Carlsöö, S., 1963, Writing desk, chair and posture of work. Department of Anatomy, Stockholm (in Swedish).

Carlsöö, S., 1972, *How Man Moves* (London: Heinemann).

Carlsöö, S., 1980, A back and lift test. *Applied Ergonomics*, **11**, 66–72.

Carlsöö, S. and Molbech, S., 1966, The function of certain two-joint muscles in a closed muscular chain. *Acta Morph. Neer.-Scand.*, **4**, 377–386.

Carlsöö, S. and Hammarskjöld, E., 1985, Guide till belastningsfaktor kroppsställning (A guide for the postural load on the body). Report No. BHF 1985:1. The Research Foundation for Occupational Safety and Health in the Swedish Construction Industry, Box 94, S-182 11 Danderyd.

CEN (Comité Européen de Normalisation), 1982, Pr EN, La Defence, AFNOR, Tour Europe, Cédex 7, 92080 Paris.

Cerretelli, P., 1973, *Fisiologia del lavoro e dello sport* (Rome: Societa Ed. Univero).

Chaffin, D. B., 1969, A computerized biomechanical model—development of and use in studying gross body actions. *Journal of Biomechanics*, **2**, 429–441.

Chaffin, D. B., 1973, Localized muscle fatigue—definition and measurement. *Journal of Occupational Medicine*, **15**, 346–354.

Chaffin, D. B., 1980, Manual materials handling: the cause of overexertion injury and illness in industry. *Journal of Environmental Pathology and Toxicology*, **2**, 31–66.

Chaffin, D. B. and Andersson, G. B. J., 1984, *Occupational Biomechanics* (New York: Wiley).

Chao, E., 1980, Justification of triaxial goniometer for the measurement of joint rotation. *Journal of Biomechanics*, **13**, 989–1006.

Chao, E., Opgrande, J. and Axmear, F., 1976, Three-dimensional force analysis of finger joints in selected isometric hand functions. *Journal of Biomechanics*, **9**, 387–396.

Chao, E., An, K., Askew, L. and Morrey, B., 1980, Electrogoniometer for the measurement of human elbow joint rotation. *Journal of Biomechanical Engineering*, **102**, 301–310.

Close, R. I., 1972, Dynamic properties of mammalian skeletal muscles. *Physiological Reviews*, **52**, 129–197.

Colombini, D., Occhipinti, E., Grieco, A. and Faccini, M., 1985 a, Estimation of lumbar disc areas by means of anthropometric parameters. *Spine* (in press).

Colombini, D., Occhipinti, E., Molteni, G., Grieco, A., Pedotti, A., Boccardi, S., Frigo, C. and Menoni, O., 1985 b, Posture analysis. *Ergonomics*, **28**, 275–284.

Corlett, E. N. and Bishop, R. B., 1976, A technique for assessing postural discomfort. *Ergonomics*, **19**, 175–182.

Corlett, E. N. and Bishop, R. B., 1978, The ergonomics of spot welders. *Applied Ergonomics*, **9**, 23–32.

Corlett, E. N. and Eklund, J. A. E., 1984, How does a backrest work? *Applied Ergonomics*, **15**, 111–114.

Corlett, E. N. and Manenica, I., 1980, The effects and measurement of working postures. *Applied Ergonomics*, **11**, 7–16.

Corlett, E. N., Madeley, S. J. and Manenica, I., 1979, Posture targetting: a technique for recording working postures. *Ergonomics*, **22**, 357–366.

Crowninshield, R. D., 1978, use of optimization techniques to predict muscle forces. *Journal of Biomechanical Engineering*, **100**, 88–92.

Crowninshield, R. D. and Brand, R. A., 1981 a, A physiologically based criterion of muscle force prediction in locomotion. *Journal of Biomechanics*, **14**, 793–801.

Crowninshield, R. D. and Brand, R. A., 1981 b, The prediction of forces in joint structures: distribution of intersegmental resultants. *Exercise and Sports Sciences Reviews*, **9**, 159–182.

Crowninshield, R. D., Johnston, R. C., Andrews, J. G. and Brand, R. A., 1978, A biomechanical investigation of the human hip. *Journal of Biomechanics*, **11**, 75–85.

Cyraix, J., 1979, *Textbook of Orthopaedic Medicine* (7th edn.) (London: Baillière Tindall).

Dainoff, M. J., 1982, Occupational stress factors in visual display terminal (VDT) operation: a review of empirical research. *Behaviour & Information Technology*, **1**, 141–176.

Daniellou, F., Laville, A. and Teiger, C., 1983, Fiction et réalité du travail ouvrier. *Les Cahiers Français*, **205**, 39–45.

Dantzig, G. B., 1968, *Linear Programming* (New York: McGraw-Hill).

Davis, P. R., 1977, Man and manual labour. *Ergonomics*, **20**, 601–609.

Davis, P. R., 1981, The use of intra-abdominal pressure in evaluating stresses on the lumbar spine. *Spine*, **6**, 90–92.

Davis, P. R. and Troup, J. D. G., 1964, Effects on the trunk of handling heavy loads in different postures. *Ergonomics: Proceedings of the second IEA Congress*, Dortmund (London: Taylor & Francis), pp. 323–327.

Debevec, F., 1973, Bolezni lumbosakralne hrbtenice s posebnim poudarkom na biomehanicnih razmerah. Disertacija, Univerza v Ljubljani.

Decoulx, P. and Rieunau, G., 1958, Les fractures du rachis dorsolombaire sans troubles nerveux. *Revue de chirurgie orthopédique et réparatrice de l'appareil moteur.*, **44**, 245–322.

Dempster, W. T., 1955 a, The range of motion of cadaver joints: the lower limb. *University of Michigan Medical Bulletin*, **12**, 364–379.

Dempster, W. T., 1955 b, Space requirements of the seated operator—geometrical, kinematic and mechanical aspect of the body with special reference to the limbs. WACD Technical Report No. 55–159, Wright-Patterson Air Force Base, Ohio.

Dempster, W. T., 1961, Free-body diagrams as an approach to the mechanics of human posture and motion, in *Biomechanical Studies of the Musculo-skeletal System*, Evans, F. G. (ed.) (Springfield, Illinois: Charles C. Thomas), pp. 81–85.

De Puky, 1935, The physiological oscillation of the length of the body. *Acta Orthopaedica Scandinavica*, **6**, 338–347.

Dessors, D., Gerard, D. and Kerguelen, A., 1979, Etude de postes de saisie sur écran de visualisation. Laboratoire de physiologie du travail et d'ergonomie du C.N.A.M., Paris.

DHSS (Department of Health and Social Security) 1982, *Social Security Statistics* (London: HMSO).

Diffrient, N., Tilley, A. R. and Bardagjy, J. C., 1974, *Human Scale* (Cambridge, Massachusetts: MIT Press).

Donskoi, D. D., 1975, *Grundlagen der Biomechanik* (Berlin: Bartels und Wernitz), pp. 176–180.

Dreyfuss, H., 1955, *Designing for People* (New York: Simon and Schuster).

Dreyfuss, H., 1974, *Humanscale 1/2/3* (Cambridge, Massachusetts: MIT Press).

Drillis, R. and Contini, R., 1966, Body segment parameters. Report No. 1166-03, Office of Vocational Rehabilitation, Dept. of Health Education and Welfare, School of Engineering and Science, N.Y. University, New York.

Drury, C. G. and Coury, B. G., 1982, A methodology for chair evaluation. *Applied Ergonomics*, **13**, 195–202.

Drury, C. G. and Wick, J., 1984, Ergonomic applications in the shoe industry, in *Proceedings of the International Conference on Occupational Ergonomics*, Attwood, D. and McCann, C. (eds.), pp. 489–493.

Drury, C. G., Roberts, C. P., Hansgen, R. and Bayman, J. R., 1983, Evaluation of a palletizing aid. *Applied Ergonomics*, **14**, 242–246.

Dudley, N. A., 1968, *Work Measurement: Some Research Studies* (London: Pitman).

Dugue, D. and Girault, M., 1969, *Analyse de la variance et plans d'expérience* (Paris: Dunod).

Dul, J., 1979, Op zoek naar een betere zithouding tijdens bureauwerkzaamheden. Thesis, No. OC-D-25, Twente University of Technology, Enschede.

Dul, J., 1983, Development of a minimum-fatigue optimization technique for predicting individual muscle forces during human posture and movement with application to the ankle musculature during standing and walking. Ph.D. dissertation, Biomedical Engineering, Vanderbilt University, Nashville, Tennessee. (Abstract in *Dissertation Abstracts International*, **44**, 3140-B).

Dul, J. and Townsend, M. A., 1985, Biomechanical analysis of two-joint muscular actions.

Dul, J., Snijders, C. J. and Timmerman, P., 1982, Bewegungen and Kräfte im oberen Kopfgelenk beim Vorbeugen der Halswirbelsäule. *Manuelle Medizin*, **20**, 51–58.

Dul, J., Johnson, G. E., Shiavi, R. and Townsend, M. A., 1984 a, Muscular synergism. II. A Minimum-fatigue criterion for load sharing between synergistic muscles. *Journal of Biomechanics*, **17**, 675–684.

Dul, J., Townsend, M. A., Shiavi, R. and Johnson, G. E., 1984 b, Muscular synergism. I. On criteria for load sharing between synergistic muscles. *Journal of Biomechanics*, **17**, 663–673.

Dul, J., Pieters, J. M. and Dijkstra, S., 1985 a, Instructional feedback in motor skill learning (paper prepared for publication).

Dul, J., Shiavi, R. and Green, N. E., 1985 b, Simulation of tendon transfer surgery. *Engineering in Medicine*, **14**, 31–38.

Eastman, M. C. and Kamon, E., 1976, Posture and subjective evaluation at flat and slanted desks. *Human Factors*, **18**, 15–26.

Edwards, R. H. T., 1983, Biomechanical bases of fatigue in exercise performance: catastrophe theory of muscular fatigue, in *Biochemistry of Exercise*, Knuttgen, H. G., Vogel, J. A. and Poortmans, J. (eds.), International Series on Sport Sciences, Vol. 13 (Campaign: Human Kinetics), pp. 3–28.

Edwards, R. H. T., Hill, D. K., Jones, D. A. and Merton, P. A., 1977, Fatigue of long duration in human skeletal muscle after exercise. *Journal of Physiology*, **272**, 769–778.

Ehnström, G., 1981, Ett försök med mikropauser bland anställda vid postgirots bokföringsavdelningar. Examensarbete, National Board of Occupational Safety and Health, Umeå.

Eie, N. and When, P., 1962, Measurements of the intra-abdominal pressure to weight bearing of the lumbo-sacral spine. *Journal of the Oslo City Hospital*, **12**, 205–217.

Ekholm, J., Schüldt, K., Harms-Ringdahl, K., Aborelius, U. P. and Németh, G., 1983, Effects of ergonomic aids upon the level of static neck muscular activity during sedentary work. *Abstracts of the First European Conference on Research in Rehabilitation*, Edinburgh, 6–8 April 1983.

Eklund, J. A. E. and Corlett, E. N., 1984, Shrinkage as a measure of the effect of load on the spine. *Spine*, **9**, 189–194.

Eklund, J. A. E. and Corlett, E. N., 1986, Experimental and biomechanical analysis of seating. This volume, Chapter 28.

Eklund, J. A. E., Corlett, E. N. and Johnson, F., 1983, A method for measuring the load imposed on the back of a sitting person. *Ergonomics*, **26**, 1063–1076.

Engdahl, S., 1971, School chairs. Report No. 24, Swedish Furniture Research Institute, Stockholm (in Swedish).

Falconer, M. and Weddel, G., 1943, Costoclavicular compression of the subclavian artery and vein. *Lancet*, **ii**, 539–543.

Farfan, H. F., 1973, *Mechanical Disorders of the Low Back* (Philadelphia: Lea & Febiger), pp. 62–92.

Feldman, R., Goldman, R. and Keyserling, W., 1983, Peripheral nerve entrapment syndromes and ergonomic factors. *American Journal of Industrial Medicine*, **4**, 661–681.

Fiske, D. W. and Maddi, S. R., 1967, *Functions of Varied Experiences* (Homewood: Dorsey Press).

Fitzgerald, J. G., 1972, Changes in spinal stature following brief periods of static shoulder loading. Report No. 514, Institute of Aviation Medicine, Farnborough.

Flowers, J. H. and Leger, D., 1982, Personal computers and behavioural observation: an introduction. *Behaviour Research Methods and Instrumentation*, **14**, 227–230.

Floyd, W. F. and Roberts, D. F., 1958, Anatomical and physiological principles in chair and table design. *Ergonomics*, **2**, 1–16.

Floyd, W. F. and Silver, P. H. S., 1950, Electromyographic study of patterns of activity of the anterior abdominal wall muscles in man. *Journal of Anatomy*, **84**, 132–145.

Floyd, W. F. and Silver, P. H. S., 1955, The function of the erectores spinae muscles in certain movements and postures in man. *Journal of Physiology*, **129**, 184–203.

Floyd, W. F. and Ward, J. S., 1966, Posture in industry. *International Journal of Production Research*, **5**, 213–224.

Floyd, W. F. and Ward, J. S., 1969, Anthropometric and physiological considerations in school, office and factory seating, in *Sitting Posture*, Grandjean, E. (ed.) (London: Taylor & Francis), pp. 18–25.

Flux, R., 1977, A laboratory evaluation of the posture recording cart (PRC). Internal Report, Ergonomics Unit, Institute of Occupational Medicine, Stanhope Bretby, Burton.

Forssberg, E., 1899, Om vexlingar i kroppslängden hos kavallerirekryter. *Militär Hälsovård*, **24**, 19–28.

Foulke, J., Goldstein, S. and Armstrong, T. J., 1981, An EMG pre-amplifier system for biomechanical studies. *Journal of Biomechanics*, **14**, 437–439.

Friden, J., Sjostrom, M. and Ekblom, B., 1983, Myofibrillar damage following intense eccentric exercise in man. *International Journal of Sports Medicine*, **4**, 170–176.

Friman, G., 1976, Serum creatine phosphokinase in epidemic influenza. *Scandinavian Journal of Infectious Diseases*, **8**, 11–30.

Friman, G., 1978, Effect of acute infectious disease on human isometric muscle endurance. *Uppsala Journal of Medical Science*, **83**, 105–108.

Fucique, J. T., 1967, The ergonomics of offices. *Ergonomics*, **10**, 589–604.

Garg, A., Mital, A. and Asfour, S. S., 1980, A comparison of isometric strength and dynamic lifting capability. *Ergonomics*, **23**, 13–27.

Giles, R., 1981, An operator's tale—frozen shoulder and how to cure it. *Applied Ergonomics*, **12**, 3–6.

Goldie, I., 1964, Epicondylitis lateralis humeri: A pathogentical study. *Acta Chirurgica Scandinavica Suppl.*, **339**, 119.

Gorman, B. S. and Primavera, L. H., 1981, Basic backstep: a simple backward-selection multiple regression program for minicomputer and microcomputers. *Behaviour Research Methods and Instrumentation*, **13**, 703.

Grandjean, E., 1967, *Physiologische Arbeitsgestaltung: Leitfaden der Ergonomie* (Munich: Ott Verlag).

Grandjean, E., 1977, Ergonomics of posture. *Applied Ergonomics*, **8**, 135–140.

Grandjean, E., 1980, *Fitting the Task to the Man: An Ergonomic Approach* (3rd edn.) (London: Taylor & Francis), pp. 31, 39, 46, 51.

Grandjean, E. and Hünting, W., 1977, Ergonomics of posture—review of various problems of standing and sitting posture. *Applied Ergonomics*, **8**, 135–140.

Grandjean, E., Hünting, W. and Pidermann, M., 1983, VDT workstation design: preferred settings and their effects. *Human Factors*, **25**, 161–175.

Grandjean, E., Hünting, W. and Nishiyama, K., 1984, Preferred VDT workstation sittings, body posture and physical impairments. *Applied Ergonomics*, **15**, 90–104.

Gregor, R. J., Hager, C. L. and Roy, R. R., 1981, *In vivo* muscle force during unrestrained locomotion. *Journal of Biomechanics*, **14**, 489.

Grieve, D. W., 1979, The postural stability diagram (PSD): personal constraints on the static exertion of force. *Ergonomics*, **22**, 115–1164.

Grieve, D. W. and Pheasant, S. T., 1981, Naturally preferred directions for the exertion of maximal manual forces. *Ergonomics*, **24**, 685–693.

Guilford, J. P. and Fruchter, B., 1978, *Fundamental Statistics in Psychology and Education*, 6th edn. (New York: McGraw-Hill).

Gutmann, G., 1968, Schulkopfschmerz und Kopfhaltung. *Zeitschrift für Orthopedie*, **105**, 497–515.

Gutmann, G., 1973, Haltungsfehler und Kopfschmerz—Die pathogenetische Bedeutung der Schulmöbel. *Manuelle Medizin*, **11**, 76–86.

Gutmann, G., 1977, Kopfgelenke und Kopfschmerz. *Manuelle Medizin*, **15**, 1–15.

Hadler, N. M., 1977, Industrial rheumatology. *Arthritis and Rheumatology*, **20**, 1019–1025.

Hagberg, M., 1981 a, Electromyographic signs of shoulder muscular fatigue in two elevated arm positions. *American Journal of Physical Medicine*, **60**, 111–121.

Hagberg, M., 1981 b, Muscular endurance and surface electromyogram in isometric and dynamic exercise. *Journal of Applied Physiology: Respiratory, Environmental and Exercise Physiology*, **51**, 1–17.

Hagberg, M., 1981 c, Workload and fatigue in repetitive arm elevations. *Ergonomics*, **24**, 543–555.

Hagberg, M., 1982, Local shoulder muscular strain—symptoms and disorders. *Journal of Human Ergology*, **11**, 99–108.

Hagberg, M., 1984 a, Delayed effects on muscular endurance, electromyography, creatine kinase and shoulder discomfort in lifting tasks. *Abstracts of the XXI International Congress on Occupational Health*, Dublin, 9–14 September.

Hagberg, M., 1984 b, Occupational musculo-skeletal stress and disorders of the neck and shoulder: a review of possible pathophysiology. *International Archives of Occupational and Environmental Health*, **53**, 269–278.

Hagberg, M. and Kvarnström, S., 1984, Muscular endurance and electromyographic fatigue in myofascial shoulder pain. *Archives of Physical Medicine and Rehabilitation*, **65**, 522–525.

Hagberg, M., Hagner, I.-M. and Bjelle, A., 1986, Shoulder muscular strength, endurance and electromyographic fatigue in ankylosing spondylitis (paper prepared for publication).

Haider, E., Luczak, H. and Rohmert, W., 1982, Ergonomics investigations of workplaces in a police command-control centre equipped with TV displays. *Applied Ergonomics*, **13**, 163–170.

Halbertsma, J., 1983, The stride cycle of the cat: modelling of locomotion by computerized analysis of automatic recordings. *Acta Physiologica Scandinavica*, Suppl. 521, 75 pp.

Hamill, P. V. V., Johnston, F. E. and Lemeshaw, S., 1973, *Height and Weight of Youths 12–18 years*. Vital and Health Statistics, Series II, No. 124, DHEW Publication No. HSM 73-1606 (Washington, D.C.: Government Printing Office).

Happ, A. and Beaver, C. W., 1981, Effects of work at a VDT-intensive laboratory task on performance, mood and fatigue symptoms. *Proceedings of the Human Factors Society—25th Annual Meeting* (Santa Monica: Human Factors Society).

Hardt, D. E., 1978, Determining muscle forces in the leg during normal human walking—an application and evaluation of optimization methods. *Journal of Biomechanical Engineering*, **100**, 72–78.

Hardt, D. E. and Mann, R. W., 1979, Technical note: a five body–three-dimensional dynamic analysis of walking. *Journal of Biomechanics*, **13**, 455–457.

Hebb, D. O., 1965, Drives and the C.N.S., in *Curiosity and Exploratory Behaviour*, Fowler, H. (ed.) (New York: Macmillan).

Herberts, P., Kadefors, R. and Broman, H., 1980, Arm positioning in manual tasks. An electromyographic study of localized muscle fatigue. *Ergonomics*, **23**, 655–665.

Herberts, P., Kadefors, R., Andersson, G. and Petersen, I., 1981, Shoulder pain in industry: an epidemiological study on welders. *Acta Orthopaedica Scandinavica*, **52**, 299–306.

Herberts, P., Kadefors, R., Högfors, C. and Sigholm, G., 1984, Shoulder pain and heavy manual labour. *Clinical Orthopaedics*, **191**, 166–178.

Herman, R. and Bragin, S. J., 1967, Function of the gastrocnemius and soleus muscle. *Journal of the American Physical Therapists Association*, **47**, 105–113.

Hirsch, C., 1955, The reaction of intervertebral discs to compression forces. *Journal of Bone and Joint Surgery*, A, **37**, 1188–1196.

Hjort, L., 1976, Yrkesskader og sykefravær—hva koster det? Rapport fra Norsk Produktivitefsinstitutt (in Norwegian).

Hnik, P., Hudlicka, O., Kucera, J. and Payne, R., 1969, Activation of muscle afferents by non-proprioceptive stimuli. *American Journal of Physiology*, **217**, 1451–1458.

Hof, A. L., 1984, EMG and muscle force: an introduction. *Human Movement Science*, **3**, 119–153.

Hoffman, G., 1981, Tendinitis and bursitis. *American Family Practice*, **23**, 103–110.

Höfling, G., 1972, *Schlechte Haltung beim Schreiben; Ursache und ihre Beseitigung* (Stuttgart: Hippokrates Verlag).

Holm, S. and Nachemson, A., 1982, Nutritional changes in the canine intervertebral disc after spinal fusion. *Clinical Orthopaedics and Related Research*, **169**, 243–258.

Holm, S. and Nachemson, A., 1983, Variation in the nutrition of the canine intervertebral disc induced by motion. *Spine*, **8**, 866–874.

Holzmann, P., 1982, ARBAN—a new method for analysis of ergonomic effort. *Applied Ergonomics*, **13**, 82–86.

Holtzmann, P. and Wangenheim, M., 1983, ARBAN—En metod för ergonomiska arbetsanalyser (ARBAN—a method for ergonomic work analyses). Report No. BHF 1983:3. The Research Foundation for Occupational Safety and Health in the Swedish Construction Industry, Box 94, S-184 11 Danderyd.

Hooykamp, P. C., 1976, Modelvorming van de relaties tussen lichaamshouding en werkvlak. Int. rapport, Eindhoven WLV, University of Technology.

Hopkins, G. O., McDoughall, J., Mills, K. R., Isenberg, D. A. and Ebringer, A., 1983, Muscle changes in ankylosing spondylitis. *British Journal of Rheumatology*, **22**, 151–157.

Hosea, H. R. *et al.*, 1985, Myoelectric analysis of the paraspinal musculature while driving. *Spine* (in press).

Houtz, S. J. and Fischer, F. J., 1959, An analysis of muscle action and joint excursion during exercise on a stationary bicycle. *Journal of Bone and Joint Surgery*, A, **41**, 123–131.

Hulten, B., Thorstensson, A., Sjödin, B. and Kärlsson, J., 1975, Relationship between isometric endurance and fibre types in human leg muscles. *Acta Physiologica Scandinavica*, **93**, 135–138.

Hultman, G., Nordin, M. and Örtengren, R., 1984, The influence of a preventive educational programme on trunk flexion in janitors. *Applied Ergonomics*, **15**, 127–133.

Humphreys, P. W. and Lind, A. R., 1963, Blood flow through active and inactive muscles of the forearm during sustained handgrip contractions. *Journal of Physiology*, **166**, 120–135.

Hünting, W., Grandjean, E. and Maeda, K., 1980, Constrained postures in accounting-machine operators. *Applied Ergonomics*, **11**, 145–149.

Hünting, W., Laubli, T. and Grandjean, E., 1981, Postural and visual loads at VDT workplaces. I. Constrained postures. *Ergonomics*, **24**, 917–931.

Hutchinson, A., 1970, *Labanotation* (Oxford: Oxford University Press).

Hymovich, L. and Lyndholm, M., 1966, Hand, wrist and forearm injuries, the result of repetitive motions. *Journal of Occupational Medicine*, **11**, 573–577.

ISO (International Organisation for Standardisation), 1978, *Chairs and Tables for Educational Institutes—Functional Sizes*. ISO/TC 136, ISO Central Secretariat, Geneva.

Jacobson, C. and Sperling, L., 1976, Classification of the hand-grip. *Journal of Occupational Medicine*, **18**, 395–398.

Jäger, M. and Luttmann, A., 1985, Biomechanische Analyse ausgewählter Tätigkeiten im Baugewerbe, in *Untersuchungen zum Gesundheitsrisiko beim Heben und Umsetzen schwerer Lasten im Baugewerbe*, Laurig, W., Gerhard, L., Luttman, A., Jäger, M. and Nau, H.-E. (eds.). Schriftenreihe Bundesanstalt für Arbeitsschutz, Forschung-Fb 409 (Bremerhaven: Wirtschaftsverlag NW), pp. 42–122.

Jäger, M., Luttmann, A. and Laurig, W., 1983, Biomechanisches Modell des Transports von Müllgroßbehältern über Bordsteinkanten. *Zentralblatt für Arbeitsmedizin, Arbeitsschutz, Prophylaxe und Ergonomie*, **33**, 251–259.

Jäger, M., Luttmann, A. and Laurig, W., 1984, The load of the spine during the transport of dustbins. *Applied Ergonomics*, **15**, 91–98.

Jenik, P., 1972, *Biomechanische Analyse ausgewählter Arbeitsbewegungen des Armes* (Berlin: Beuth-Vertrieb GmbH).

Jensen, R., Klein, B. and Sanderson, L., 1983, Motion-related wrist disorders traced to industries, occupational groups. *Monthly Labor Review*, September, pp. 13–16.

Johnson, C. E., Basmajain, J. V. and Dasher, W., 1972, Electromyograph on the sartorius muscle *Anatomical Record*, **173**, 127–130.

Johnson, M. A., Polger, J., Weightman, D. and Appleton, D., 1973, Data on the distribution of fibre types in thirty-six human muscles. An autopsy study. *Journal of Neurological Science*, **18**, 111–129.

Jonsson, B., 1970 a, The functions of individual muscles in the lumbar part of the erector spinae muscle. *Electromyography*, **10**, 5–21.

Jonsson, B., 1970 b. The lumbar part of the erector spinae muscle: a technique for electromyographic studies of the function of its individual muscles. Thesis (Göteborg: Elanders).

Jonsson, B., 1978, Kinesiology—with special reference to electromyographic kinesiology, in *Contemporary Clinical Neurophysiology*, Vol. 34, Cobb, W. A. and van Dui, H. (eds.) (Amsterdam: Elsevier), pp. 417–428.

Jonsson, B. and Rundgren, A., 1971, The peroneus longus and brevis muscles: a roentgenologic and electromyographic study. *Electromyography*, **11**, 93–103.

Jonsson, B., Hagberg, M. and Sima, S., 1981, Vocational electromyography in shoulder muscles in an electronic plant, in *Biomechanics*, Vol. VII-B, Morecki, A., Fidelus, K., Kedzior, K. and Wit, A. (eds.) (Baltimore: University Park Press), pp. 10–15.

Junghanns, H., 1979, Die Wirbelsäule in der Arbeitsmedizin, Teil I: Biomechanische und biochemische Probleme der Wirbelsäulenbelastung, in *Die Wirbelsäule in Forschung und Praxis*, Vol. 78, Junghanns, H. (ed.) (Stuttgart: Hippokrates-Verlag).

Jürgens, H. W., 1969, Die Verteilung des Körperdrücks aufs Sitzfläche und Rückenlehne als Problem der Industrieanthropologie. *Ergonomics*, **12**, 198–205.

Kamon, E., 1966, Electromyography of static and dynamic postures of the body supported on the arms. *Journal of Applied Physiology*, **21**, 1611–1618.

Karhu, O., Kansi, P. and Kuorinka, I., 1977, Correcting working postures in industry: a practical method for analysis. *Applied Ergonomics*, **8**, 199–201.

Keegan, J. J., 1953, Alterations of the lumbar curve related to posture and seating. *Journal of Bone and Joint Surgery*, A, **35**, 589–603.

Keller, A., Taylor, C. and Zahm, V., 1947, Studies to determine the functional requirements for hand and arm prosthesis. Department of Engineering, University of California, Los Angeles.

Kelly, A. and Jacobson, H., 1964, Hand disability due to tenosynovitis. *Industrial Medicine and Surgery*, **33**, 570–574.

Kelsey, J. L. and Hardy, R. J., 1975, Driving of motor vehicles as a risk factor for acute herniated lumbar intervertebral disc. *American Journal of Epidemiology*, **102**, 63–73.

Kelsey, J. L. and White, A. A., 1980, Epidemiology and impact of low-back pain. *Spine*, **5**, 133–142.

Keyserling, W. M., Herring, G. D. and Chaffin, D. B., 1980, Isometric strength testing as a means of controlling medical incidents on strenuous jobs. *Journal of Occupational Medicine*, **5**, 332–336.

Kilbom, A. and Brundin, T., 1976, Circulatory effects of isometric muscle contractions performed separately and in combination with dynamic exercise. *European Journal of Applied Physiology*, **36**, 7–17.

Kilbom, A., Gamberale, F., Persson, J. and Annwall, G., 1983, Physiological and psychological indices of fatigue during static contractions. *European Journal of Applied Physiology*, **50**, 179–193.

Kleinbaum, D., Kupper, L. L. and Morgenstern, H., 1982, *Epidemiologic Research: Principles and Quantitative Methods* (Belmont, California: Lifelong Learning), pp. 476–490.

Knutsson, B., Lindh, K. and Telhag, H., 1966, Sitting—an electromyographic and mechanical study. *Acta Orthopaedica Scandinavica*, **37**, 415–428.

Komi, P. V. and Tesch, P., 1979, EMG frequency spectrum, muscle structure and fatigue during dynamic contractions in man. *European Journal of Applied Physiology*, **42**, 41–50.

Komi, P. V. and Viitasalo, J. T., 1977, Changes in motor unit activity and metabolism in human skeletal muscle during and after repeated eccentric and concentric contractions. *Acta Physiologica Scandinavica*, **100**, 246–254.

Körner, L., Parker, P., Almström, C., Herberts, P. and Kadefors, R., 1984, The relation between spectral changes of the myoelectric signal and the intramuscular pressure of human skeletal muscle. *European Journal of Applied Physiology and Occupational Physiology*, **52**, 202–206.

Kralj, A., 1969, Optimum coordination and selection of muscles for functional stimulation. *Proceedings of the Eighth International Conference on Medical and Biological Engineering*, Chicago.

Krämer, J., 1973, *Biomechanische Veränderungen im lumbalen Bewegungssegment* (Stuttgart: Hippokrates Verlag). (Cited in Grandjean and Hünting (1977).)

Krämer, J., 1985, Dynamic characteristics of the vertebral column: effects of prolonged loading. *Ergonomics*, **28**, 95–97.

Krämer, J. and Gritz, A., 1980, Changes in body length by pressure-dependent fluid shifts in the intervertebral discs. *Z. Orthop.*, **118**, 161–164.

Kroemer, K. H. E., 1971, Seating in plant and office. *American Industrial Hygiene Association Journal*, **2**, 633–652.

Kroemer, K. H. E., 1977, Die Messung der Muskelstärke des Menschen. Forschungsbericht No. 161, Bundesanstalt für Arbeitsschutz und Unfallforschung, Dortmund.

Kroemer, K. H. E. and Robinette, J. C., 1969, Ergonomics in the design of office furniture. *Industrial Medicine and Surgery*, **38**, 115–125.

Kuipers, H., Drukker, J., Frederik, P. M., Geurten, P. and von Kranenburg, G., 1983, Muscle degeneration after exercise in rats. *International Journal of Sports Medicine*, **4**, 45–51.

Kukkonen, R., Luopajärvi, T. and Riihimäki, V., 1983, Prevention of fatigue amongst data-entry operators, in *Ergonomics of Workstation Design*, Kvålseth, T. O. (ed.) (London: Butterworths), pp. 28–34.

Kuo, K. H. M. and Clamann, H. P., 1981, Co-activation of synergistic muscles of different fibre types in fast and slow contractions. *American Journal of Physical Medicine*, **60**, 219–238.

Kuorinka, I. and Koskinen, P., 1979, Occupational rheumatic diseases and upper limb strain in manual jobs in light mechanical industry. *Scandinavian Journal of Work, Environment and Health*, **5**, Suppl. 3, 39–47, 1979.

Kurppa, K., Waris, P. and Rokkanen, P., 1979, Tennis elbow, lateral elbow pain syndrome. *Scandinavian Journal of Work, Environment and Health*, **5**, Suppl. 3, 15–19.

Kvarnström, S., 1983, Occurrence of musculo-skeletal disorders in a manufacturing industry, with special attention to occupational shoulder disorders. *Scandinavian Journal of Rehabilitation Medicine*, Suppl. 8, 114 pp.

Laban, R. and Lawrence, F. C., 1947, *Effort* (London: MacDonald and Evans).

Lamphier, T., Crooker, C. and Crooker, J., 1965, DeQuervain's Disease. *Industrial Medicine and Surgery*, **34**, 847–856.

Landau, K. and Reus, J., 1979, Körperhaltungen bei Tätigkeiten in Industrie, Verwaltung,

Landwirtschaft und Bergbau. *International Archives of Occupational and Environmental and Health*, **44**, 213–231.

Landau, K., Luczak, H. and Rohmert, W., 1975, Arbeitswissenschaftlicher Erhebungsbogen zur Tätigkeitsanalyse—AET, in *Arbeitswissenschaftliche Beurteilung der Belastung und Beanspruchung an unterschiedlichen industriellen Arbeitsplätzen*, Rohmert, W. and Rutenfranz, J. (eds.) (Bonn: Forschungsbericht für den Bundesminister für Arbeit und Sozialordnung).

Landsmeer, J., 1962, Power grip and precision handling. *American Journal of Rheumatic Diseases*, **21**, 164–169.

Laurig, W., 1969, Der Stehsitz als physiologisch-günstige Alternative zum Reinen Steharbeitsplatz. *Arbeitsmedizin, Sozialmedizin und Arbeitshygein*, **4**, 219–221.

Lecain, J. P., 1979, Autre méthode d'évaluation et de mesure de la courbure dorsale dans le plan sagittal: le courbomètre. Mémoire pour le CES de Rééducation et de réadaptation fonctionnelle, Lille.

Leebeek, H. J., 1983, Teleac cursus ergonomie. [TV course on ergonomics] (Utrecht: Stichling Teleac).

Lehmann, G., 1962, *Praktische Arbeitsphysiologie* (Stuttgart: Georg Thieme Verlag).

Leplat, J., 1978, Le diagnostic ergonomique des contraintes de travail, in *L'ergonomie au service de l'homme au travail*, S.F.P. (ed.) (E.M.E.), pp. 21–28.

Lepoutre, F. X., 1979, Analyse et traitement des mesures des courbures du dos d'un sujet dans le plan sagittal. Modélisation et application biomécanique. Thèse de Docteur-Ingénieur, Université de Valenciennes.

Leskinen, T. P. J., Stahlhammar, H. R. and Kuorinka, I. A. A., 1983, A dynamic analysis of spinal compression with different lifting techniques. *Ergonomics*, **26**, 595–604.

Leuba, C., 1965, Toward some integration of learning theory: the concept of optimal stimulation. In *Curiosity and Exploratory Behaviour*, Fowler, H. (ed.) (New York: Macmillan).

Lieb, F. J. and Perry, J., 1971, Quadriceps function: an electromyographic study under isometric conditions. *Journal of Bone and Joint Surgery*, A, **53**, 749.

Life, M. A. and Pheasant, S. T., 1984, An integrated approach to the study of posture in keyboard operation. *Applied Ergonomics*, **15**, 83–90.

Lin, H. S., Liu, Y. K. and Adans, K. H., 1978, Mechanical response of the lumbar intervertebral joint under physiological (complex) loading. *Journal of Bone and Joint Surgery*, A, **60**, 41–45.

Lind, A. R. and McNicol, G. W., 1967, Local and central circulatory responses to sustained contractions and the effect of free or unrestricted arterial inflow on post-exercise hyperaemia. *Journal of Physiology*, **192**, 575–594.

Lind, A. R., Rochelle, R. H., Petrofsky, J. S., Rinehart, J. S. and Burse, R. OL., 1975, *The Influence of Dynamic Exercise on Fatiguing Isometric Exercise and the Assessment of Changing Levels of Isometric Component*. Publication No. NIOSH 75-117 (Cincinnati: Center for Disease Control), 53 pp.

Lindström, L., Kadefors, R. and Petersen, I., 1977, An electromyographic index for localized muscle fatigue. *Journal of Applied Physiology*, **43**, 750–754.

Ljunggren, G., 1986, Observer ratings of perceived exertion in relation to self-rating and heart rate. *Applied Ergonomics*, **17**(2), 117–125.

Loebl, W., 1967, Measurement of spinal posture and range of spinal movement. *Annals of Physical Medicine*, **9**, 103–110.

Long, C., Conrad, P., Hall, E. and Furler, S., 1970, Intrinsic–extrinsic muscle control of the hand in power grip and precision handling. *Journal of Bone and Joint Surgery*, A,

52, 853–867.

Longhurst, J. C. and Mitchell, J. H., 1979, Reflex control of the circulation by afferents from skeletal muscle. *Cardiovascular Physiology,* **18,** 125–148.

Lord, J. and Rosati, L., 1958, Neurovascular compression syndromes of the upper extremity. *Ciba Clinical Symposia,* **10,** 35–62.

Lowry, O. H. and Passonneau, J. V., 1972, *A Flexible System of Enzymatic Analysis* (New York: Academic Press).

Lueder, K., 1983, Seat comfort: A review of the construct in the office environment. *Human Factors,* **25,** 701–711.

Lundervold, A. J. S., 1951 a, Electromyographic investigations of position and manner of working in typing. *Acta Orthopaedica Scandinavica,* Suppl. 84, **24,** 1–71.

Lundervold, A. J. S., 1951 b, Electromyographic investigations during sedentary work, especially typing. *British Journal of Physiological Medicine,* **14,** 32–36.

Lundervold, A. J. S., 1958, Electromyographic investigations during typing. *Ergonomics,* **1,** 226–233.

Luttmann, A., Jäger, M., Schoo, K.-C., Laurig, W. and Puhlvers, E., 1985, Wirkung erhöhter Wirbelsäulenbelastung beim Lastentransport auf die Häufigkeit von Rückenbeschwerden, in *Untersuchungen zum Gesundheitsrisiko beim Heben und Umsetzen schwerer Lasten im Baugewerbe,* Laurig, W., Gerhard, L., Luttman, A., Jäger, M. and Nau, H. E. (eds.). Schriftenreihe Bundesanstalt für Arbeitsschutz, Forschung-Fb 409 (Bremerhaven: Wirtschaftsverlag NW), pp. 123–180.

MacConaill, M. A., 1967, The ergonomic aspects of articular mechanics, in *Studies on the Anatomy and Function of Bones and Joints,* Evans, F. G. (ed.) (Berlin: Springer-Verlag), pp. 69–80.

McFarland, R. A., 1971, Understanding fatigue in modern life. *Ergonomics,* **14,** 1–22.

McLaughlin, T. M. and Miller, N. R., 1980, Techniques for the evaluation of loads on the forearm prior to impact in tennis strokes. *Journal of Mechanical Design,* **102,** 701–710.

MacNicol, M., 1982, Extraneural pressures affecting the ulnar nerve at the elbow. *Hand,* **14,** 6–11.

Maeda, K., 1977, Occupational cervicobrachial disorders and its causative factors. *Journal of Human Ergology,* **6,** 193–202.

Magora, A., 1972, Investigation of the relation between low-back pain and occupation. 3. Physical requirements: sitting, standing and weight lifting. *Industrial Medicine,* **41,** 5–9.

Magoun, H. W. and Rhines, R., 1946, An inhibitory mechanism in the bulbar reticular formation. *Journal of Neurophysiology,* **9,** 165–171.

Mandal, A. C., 1975, Workchair with tilting seat. *Lancet,* **i,** 642–643.

Mandal, A. C., 1976, Workchair with tilting seat. *Ergonomics,* **19,** 157–164.

Mandal, A. C., 1981, The seated man (Homo Sedens). *Applied Ergonomics,* **12,** 19–26.

Mandal, A. C., 1982, The correct height of school furniture. *Human Factors,* **24,** 257–269.

Manenica, I., 1980, The relationship between posture, holding time and rest pauses, in *Papers II,* Modic, S. (ed.) (Ljubljana: Klicnicki).

Manenica, I., 1982, Neki pokusaji odredivanja velicine staticnog napora. *Radovi Filozofskog Fakulteta u Zadru,* **20,** 221–230.

Marcellin, J. and Valentin, M., 1971, Etude comparative d'ouvriers de 40 à 44 ans travaillant sur chaîne dans deux ateliers de l'industrie automobile. *Travail Humain,* **34,** 342–348.

Marek, T. and Noworol, Cz., 1984 a, Some remarks on a measure of computer operator workload: changes in pupil reflex. in *Proceedings of the First Seminar on Man–Computer*

Interaction Research, Klix, F. (ed.) (Amsterdam: North-Holland).

Marek, T. and Noworol, Cz., 1984 b, The sequential approach to analysis of variance. *Polish Psychological Bulletin*, **15**, 195–200.

Marholf, K. L., 1972, Deformation of the thoracolumbar intervertebral joints in response to external loads. *Journal of Bone and Joint Surgery*, A, **54**, 511–533.

Martin, J. B. and Chaffin, D. B., 1972, Biomechanical computerized simulation of human strength in sagittal plane activities. *American Institute of Industrial Engineers Transactions*, **4**, 19–28.

Matiegka, J., 1960 (Reference to his 1921 paper in *Am. J. Phys. Anthrop.*, **4**, 223–230, in *Introduction to Physical Anthropology*, 3rd ed., Ashley-Montagu, M.F. (ed.) (Springfield, Illinois: Charles C. Thomas).

Michon, J. A., 1966, Tapping regularity as a measure of perceptual motor load. *Ergonomics*, **9**, 401–412.

Miller, J., Haderspeck, K. and Schultz, A., 1983, Posterior element loads in lumbar motion segments. *Spine*, **8**, 331–337.

Mills, K. R. and Edwards, R. H. T., 1984, Muscle fatigue in myophosphorylase deficiency: power spectral analysis of the electromyogram. *Electroencephalography and Clinical Neurophysiology*, **57**, 300–335.

Milner, N. P., Corlett, E. N. and O'Brien, C., 1985, Problems of assessing static postures, in *Ergonomics International '85*. 9th Congress of the International Ergonomics Association, Bournemouth, 2–6 September (London: Taylor & Francis), pp. 964–966.

Monod, H., 1972, How muscles are used in the body, in *The Structure and Function of Muscle*, Vol. I, Bourne, G. H. (ed.) (New York: Academic Press), pp. 27–74.

Monod, H. and Scherrer, J., 1965, The work capacity of a synergic muscular group. *Ergonomics*, **8**, 329–338.

Morioka, M., 1964, Some physiological responses to static muscular exercise. *Proceedings of the IEA Conference*, Dortmund (London: Taylor & Francis), pp. 35–40.

Morris, J. M., Lucas, D. B. and Bresler, B., 1961, The role of the trunk in the stability of the spine. *Journal of Bone and Joint Surgery*, A, **43**, 327–351.

Mortimer, J. T., Magnusson, R. and Petersen, I., 1970, Conduction velocity in ischemic muscle effect on EMG frequency spectrum. *American Journal of Physiology*, **219**, 1324–1329.

Mossin, J., 1972, *Operasjonsanalytiske emner* (Oslo: Johan Grundt Tanum Forlag) (in Norwegian), Chapter 5.

Muckart, R., 1964, Stenosing tendovaginitis of abductor pollicis longus and extensor pollicis brevis at the radial styloid (DeQuervain's Disease). *Clinical Orthopaedics*, **33**, 201–208.

Murray, M. P., Gardner, G. M., Mollinger, L. A. and Sepec, S. B., 1980, Strength of isometric and isokinetic contractions. *Phys. Ther.*, **60**, 412–419.

Murrell, K. F. H., 1965, *Ergonomics, Man in His Working Environment* (London: Chapman and Hall), pp. 145, 254.

Myhre, K. and Andersen, L. K., 1971, Respiratory response to static muscular work. *Respiration Physiology*, **12**, 77–89.

Nachemson, A., 1966, The load of lumbar discs in different positions of the body. *Clinical Orthopaedics*, **45**, 107–122.

Nachemson, A., 1976, The lumbar spine, an orthopaedic challenge. *Spine*, **1**, 59–71.

Nachemson, A. and Elfström, G., 1970, Intravital dynamic pressure measurements in lumbar discs. *Scandinavian Journal of Rehabilitation Medicine*, Suppl. 1.

Nachemson, A. and Morris, J. M., 1964, *In vivo* measurements of intradiscal pressure. *Journal of Bone and Joint Surgery*, A, **46**, 1077–1092.

Nag, P. K., 1982, Influence of posture and speed of arm and leg work with reference to physiological responses, *Journal of Sports Medicine*, **22**, 426–433.

Nag, P. K., 1984, Circulo-respiratory responses to different muscular exercise. *European Journal of Applied Physiology*, **52**, 393–399.

Nag, P. K., Panikar, J. T., Malvankar, M. G., Pradhan, C. K. and Chatterjee, S. K., 1982, Performance evaluation of lower extremity disabled people with reference to hand-cranked tricycle propulsion. *Applied Ergonomics*, **13**, 171–176.

Napier, J., 1956, The prehensile movements of the human hand. *Journal of Bone and Joint Surgery*, B, **38**, 902–913.

Newham, D. J., McPhail, G., Mills, K. R. and Edwards, R. H. T., 1983, Ultrastructural changes after concentric and eccentric contractions of human muscle. *Journal of the Neurological Sciences*, **61**, 109–122.

Newton, L., 1984, An ergonomic analysis of supermarket checkout systems. B.Sc. dissertation, University of Surrey (unpublished).

Nichols, H., 1967, Anatomic structures of the thoracic. *Clinical Orthopaedics and Related Research*, **51**, 17–25.

Nie, N. H., Hull, C. H., Jenkins, J. G., Steinbrenner, K. and Bent, D. H., 1975, *SPSS: Statistical Package for the Social Sciences* (New York: McGraw-Hill).

NIOH (National Institute of Occupational Health), 1982, Development of a method in predicting occupational fatigue resulting from work component. Part III. Quantification of physiological responses in combined static and dynamic work. *Annual Report* (Ahmedabad: NIOH), pp. 148–157.

NIOSH (National Institute for Occupational Safety and Health), 1981, *Work Practices Guide for Manual Lifting*. DHHS (NIOSH) Publication No. 81-122 (Cincinnati: Government Printing Office).

Nordin, M., 1982, Methods for studying workload, with special reference to the lumbar spine. Thesis, Göteborg.

Nordin, M., Örtengren, R. and Andersson, G. B. J., 1984, Measurements of trunk movements during work. *Spine*, **9**, 465–469.

Norin, K. and Seijmer-Andersson, B., 1982, Kan armstöd på stödstol avlasta nacke-skuldra vid sittande monteringsarbeten? Report, Industrial Health Care Centre, Ericsson, Stockholm.

O'Brien, C. and Paradise, M. G. A., 1976, The development of a portable non-invasive system for analysing human movement. *Proceedings of the Sixth Congress of the International Ergonomics Association*, Maryland, July, pp. 390–392.

Occhipinti, E., Colombini, D., Frigo, C., Pedotti, A. and Grieco, A., 1985, Sitting posture: analysis of lumbar stresses with upper limbs supported. *Ergonomics*, **28**, 1333–1346.

Ochanine, D., 1978, Le rôle de l'image opératrice dans la régulation des activités de travail. *Psychologie de l'éducation*, **2**, 63–65.

Ohara, H., Aoyama, H. and Itani, T., 1976, Health hazard among cash-register operators and the effects of improved working conditions. *Journal of Human Ergology*, **5**, 31–40.

Ohara, H., Itani, T. and Aoyama, H., 1982, Prevalence of occupational cervicobrachial disorders among different occupationals groups in Japan. *Journal of Human Ergology*, **11**, 55–63.

Okushima, H., 1970, Study on hydrodynamic pressure of lumbar intervertebral disc. *Archives of Japanese Surgery*, **39**, 45–57.

Olsson, G. and Åhlen, D., 1979, Handlovsstöd. Examensarbete, Skyddsingenjörer, 1979.

National Board of Occupational Safety and Health, Umeå.

Onishi, N., Nomura, H., Sakai, K., Yamamoto, T., Hirayama, K. and Itani, T., 1976, Shoulder muscle tenderness and physical feature of female industrial workers. *Journal of Human Ergology*, **5**, 87–102.

Onishi, N., Sakai, K. and Kogi, K., 1982, Arm and shoulder muscle load in various keyboard operating jobs of women. *Journal of Human Ergology*, **11**, 89–97.

Oxford, H. W., 1969, Anthropometric data for educational chairs. *Ergonomics*, **12**, 140–161.

Paillard, J., 1974, Le traitement des informations spatiales, in *De l'espace corporel à l'espace écologique*. P.U.F. (ed.), pp. 7–54.

Palla, S. and Ash, M. M., 1981, Power spectral analysis of the surface electromyogram of human jaw muscles during fatigue. *Archives of Oral Biology*, **26**, 547–553.

Palmar, A., Werner, F., Murphy, D. and Glisson, R., 1985, Functional wrist motion: a biomechanical study. *Journal of Hand Surgery*, A, **10**, 39–46.

Panjabi, M., Takata, K. and Goel, V., 1983, Kinematics of lumbar intervertebral foramen. *Spine*, **8**, 348–357.

Paradise, M. G. A., 1982, Recording human posture. Ph.D. thesis, University of Nottingham.

Partridge, M. J. and Walters, C. E., 1959, Participation of the abdominal muscles in various movements of the trunk in man: an electromyographic study. *Physical Therapy Review*, **39**, 791–800.

Patriarco, A. G., Mann, R. W., Simon, S. R. and Mansour, J. M., 1981, An evaluation of the approaches of optimization models in the prediction of muscle forces during human gait. *Journal of Biomechanics*, **14**, 513–525.

Pedotti, A., 1982, Functional evaluation and recovery in patients with motor disabilities, in *Proceedings of the IFIP-IMIA Working Conference on Uses of Computers in Aiding the Disabled*, Haifa, 3–5 November 1981 (Amsterdam: North-Holland), pp. 53–71.

Pedotti, A., Krishnan, V. V. and Stark, L., 1978, Optimization of muscle force sequencing in human locomotion. *Mathematical Biosciences*, **38**, 57–76.

Penrod, D. D., Davy, D. T. and Singh, D. P., 1974, An optimization approach to tendon force analysis. *Journal of Biomechanics*, **7**, 123–129.

Perey, O., 1957, Fracture of the ventral endplates in the lumbar spine. An experimental biomechanical investigation. *Acta Orthopaedica Scandinavica*, Suppl. 25, 1–101.

Perry, J. and Bekey, G. A., 1981, EMG-force relationship in skeletal muscle. *CRC Critical Reviews in Biomedical Engineering*, **12**, 1–22.

Persson, J. and Kilbom, A., 1983, VIRA—en enkel videofilmteknik för registrering och analys av arbetsställningar och rörelser. Undersökningsrapport 1983: 10, 23 pp. (English summary).

Peters, T., 1976, Anthropometrische und physiologische Grundlagen zur Gestaltung von Arbeitssitzen. *Ergonomics*, **12**, 162–170.

Peterson, W. and Lindell, J., 1981, Sjukfrånvaro och förtidspensionering bland byggnadsarbetare. Bygghälsan, Box 94, 182 11 Danderid.

Petrofsky, J. S., 1979, Frequency and amplitude analysis of the EMG during exercise on the bicycle ergometer. *European Journal of Applied Physiology*, **41**, 1–15.

Petrofsky, J. S. and Lind, A. R., 1979, Isometric endurance in fast and slow muscles in cats. *American Journal of Physiology*, **41**, 1–15.

Petrofsky, J. S. and Lind, A. R., 1980, The influence of temperature on the amplitude and frequency components of the EMG during brief and sustained isometric contractions. *European Journal of Applied Physiology and Occupational Physiology*, **44**, 189–200.

Phalen, G., 1966, The carpal tunnel syndrome. *Journal of Bone and Joint Surgery*, A, **48**, 211–228.

Phalen, G., 1972, The carpal tunnel syndrome. Clinical evaluation of 598 hands. *Clinical Orthopaedics and Related Research*, **83**, 29–40.

Pinsky, L., Kandarou, R. and Lantin, G., 1979, Le travail de saisie chiffrement sur terminal d'ordinateur. Rapport No. 65, Laboratoire de physiologie du travail et d'ergonomie du C.N.A.M., Paris.

Poirier, F., Jobin, M. and Roy, B., 1984, The working strategies of hands in heavy work. *Proceedings of the 1984 International Conference on Occupational Ergonomics*, Toronto, Ontario, 7–9 May.

Pottier, M., 1969, The effects of sitting posture on the volume of the foot. *Ergonomics*, **12**, 753–758.

Priel, V. Z., 1974, A numerical definition of posture. *Human Factors*, **16**, 576–584.

Ramazinni, B., 1713, *The Diseases of Workers*, translated by W. Wright, 1940 (Chicago: University of Chicago Press), p. 15.

Ranu, H. S., Danton, R. A. and King, A. I., 1979, Pressure distribution under an intervertebral disc—an experimental study. *Journal of Biomechanics*, **12**, 807–812.

RCGP, OPCS and DHSS (Department of Health and Social Security), 1979, *Morbidity Statistics from General Practice, 1971–1972. Second National Study* (London: HMSO).

Reilly, T., Tindell, A. and Troup, J. D. G., 1984, Circadian variation in human stature. *Chronobiology International*, **1**, 121–126.

Rhines, R. and Magoun, H. W., 1946, Brainstem facilitation of cortical motor responses. *Journal of Neurophysiology*, **9**, 219–240.

Rizzi, M. A. and Corelli, B., 1975, Die Nackenmuskulatur biomechanisch berechnet. *Manuelle Medizin*, **13**, 101–106.

Robbins, H., 1963, Anatomical study of the median nerve in the carpal tunnel and aetiologies of the carpal tunnel syndrome. *Journal of Bone and Joint Surgery*, A, **45**, 953–966.

Rohmert, W., 1960 a, Ermittlung von Erholungspausen für statische Arbeit des Menschen. *Internationale Zeitschrift fuer Angewandte Physiologie einschliesslich Arbeitsphysiologie*, **18**, 123–164.

Rohmert, W., 1960 b, *Statische Haltearbeit des Menschen* (Reprint of Refa reports) (Darmstadt: Verband fur Arbeitsstudien Refa e.V.).

Rohmert, W., 1962, *Untersuchungen uber Muskelermudung und Arbeitsgestaltung* (Berlin: Beuth-Vertrieb).

Rohmert, W., 1973, Problems in determining rest allowances. *Applied Ergonomics*, **4**, 91–95 and 158–162.

Rohmert, W., 1975, Analytical problems of work design, in *Measurement of Human Resources*, Singleton, W. T. and Spurgeon, P. (eds.) (London: Taylor & Francis).

Rohmert, W., 1984, Das Belastungs-Beanspruchungs-Konzept. *Zeitschrift für Arbeitswissenschaft*, **38**, 193–200.

Rohmert, W. and Landau, K., 1979, *Das Arbeitswissenschaftliche Erhebungsverfahren zur Tätigkeitsanalyse (AET)* (Bern: Hans Huber Verlag).

Rohmert, W. and Luczak, H., 1973, Ergonomische Untersuchung von Teilzeit-Schicht Systemen und Pausen bei informatorischer Arbeit. *Internationales Archiv der Arbeitsmedizin*, **31**, 171–91.

Rohmert, W., Scheibe, W. and Mainzer, J., 1979, Ergonomische Beurteilung und Gestaltung von Montagearbeitssystemen in der elektrotechnischen Industrie.

Forschungsbericht des Instituts für Arbeitswissenschaft der Technischen Hochschule Darmstadt, Darmstadt.

Rohmert, W., Haider, E., Mainzer, J. and Zipp, P., 1984 a, Arbeitswissenschaftliche Anforderungsanalyse, Beurteilung und Gestaltung von Mikroskopiertätigkeiten. Interner Bericht des Instituts für Arbeitswissenschaft der Technischen Hochschule Darmstadt, Darmstadt.

Rohmert, W., Mainzer, J. and Zipp, P., 1984 b, Untersuchung der Ursachen und Auswirkungen von Körperhaltungen bei zahnärztlicher Tätigkeit. Interner Bericht des Instituts für Arbetiswissenschaft der Technischen Hochschule Darmstadt, Darmstadt.

Rohmert, W., Mainzer, J. and Hecker, C., 1985 a, Ergomische Gestaltung von Mikroskopierarbeitsplätzen. Interner Bericht des Instituts für Arbeitswissenschaft der Technischen Hochschule Darmstadt, Darmstadt.

Rohmert, W., Wangenheim, M., Mainzer, J. and Zipp, P., 1985 b, Maximum holding times in static muscular work in relationship to different types of body postures and load levels. Interner Bericht des Instituts für Arbeitswissenschaft der Technischen Hochschule Darmstadt, Darmstadt.

Rolett, E. L., Sjøgaard, G., Kiens, B., Strange, S., Rowell, B. and Saltin, B., 1985. Lack of effect of β_2 agonist on K^+ release from contracting skeletal muscle in man. *Acta Physiologica Scandinavica*, **123**, A26.

Roozbazar, A., 1975, Biomechanics of lifting, in *Biomechanics*, Vol. IV, International Series on Sport Science (University Park: University Park Press), pp. 37–43.

Rosemeyer, B., 1971, Electromyographische Untersuchunged der Rücken- und Schultermuskulatur im Stehen und Sitzen unter Berucksichtigung der Haltung des Autofahrers. *Archiv für Orthopaedische und Unfall-Chirurgie*, **71**, 59–70.

Rosemeyer, B., 1972, Eine Methode zur Beckenfixierung im Arbeitssitz. *Zeitschrift für Orthopaedie und Ihre Grenzgebiete*, **110**, 514–517.

Ruffs, S., 1950, Brief acceleration: less than one second, in *German Aviation Medicine, World War II*, Vol. 1, Ferman (ed.) (Washington, D.C.: Government Printing Office), pp. 584–597.

Sackett, G. P., 1978, *Observing Behaviour*, Vol. 2 (Baltimore: University Park Press).

Sällström, J. and Schmidt, H., 1984, Cerviobrachial disorders in certain occupations, with special reference in the thoracic outlet. *American Journal of Industrial Medicine*, **6**, 45–52.

Saltin, B., Sjøgaard, G., Gaffney, F. A. and Rowell, L. B., 1981, Potassium, lactate, and water fluxes in human quadriceps muscle during static contractions. *Circulation Research*, **48**, Suppl. 1, 18–24.

Sämann, W., 1970, *Charakteristische Merkmale und Auswirkungen ungünstiger Arbeitshaltungen*. Arbeitswissenschaft und Praxis, Vol. 17 (Berlin: Beuth-Vertrieb GmbH).

Sammie (System for Aiding Man–Machine Interaction Evaluation) 1981, *Information Booklet* (4th edn.). The Sammie Research Group, Department of Production Engineering and Production Management, University of Nottingham, University Park, Nottingham.

Sanchez, J. and Monod, H., 1979, Physiological effects of dynamic work on a bicycle ergometer combined with different types of static contraction. *European Journal of Applied Physiology*, **41**, 259–266.

Scheving, L. E. and Pauly, J. E., 1959, An electromyographic study of some muscles acting on the upper extremity of man. *Anatomical Record*, **135**, 239–246.

Schiro, S. G., 1985, Predicting the potential benefits of work design changes. Ph. D. thesis,

University of Birmingham.

Schlesinger, G., 1919, Der mechanische Aufbau der künstlichen Glieder, in *Ersatzglieder und Arbeitshilfen* (Berlin: Springer-Verlag).

Schneider, H. J. and Decker, K., 1961, Gedanken zur Gestaltung des Sitzes. *Deutsche Medizinische Wochenschrift*, **86**, 1816–1820.

Schoberth, H., 1962, *Sitzhaltung, Sitzschaden, Sitzmöbel* (Berlin: Springer-Verlag).

Schultz, A. B., Andersson, G. B. J., Haderspeck, K., Örtengren, R., Nordin, M. and Bjork, R., 1982 a, Analysis and measurement of lumbar trunk loads in tasks involving bends and twists. *Journal of Biomechanics*, **15**, 669–675.

Schultz, A. B., Andersson, G. B. J., Örtengren, R., Haderspek, K. and Nachemson, A., 1982 b, Load on the lumbar spine. Validation of biomechanical analysis by measurements of intradiscal pressures and electric signals. *Journal of Bone and Joint Surgery*, **64**, 713–720.

Schultz, A. B., Andersson, G. B. J., Örtengren, R., Bjork, R. and Nordin, M., 1982 c, Analysis and quantitative myoelectric measurements of loads on the lumbar spine when holding weights in standing postures. *Spine*, **7**, 390–397.

Seireg, A. and Arvikar, R. J., 1973, A mathematical model for evaluation of forces in the lower extremities of the musculo-skeletal system. *Journal of Biomechanics*, **6**, 313–326.

Shackel, B., Chidsey, K. D. and Shipley, P., 1969, The assessment of chair comfort. *Ergonomics*, **12**, 269–306.

Sharkey, B. J., 1966, A physiological comparison of static and phasic exercise. *Research Quarterly*, **37**, 520–526.

Sillanpaa, J., 1984, Improved work conditions of industrial seamstresses. *Proceedings of the XXI Congress on Occupational Health*.

Singleton, W. T., 1959, The training of shoe machinists. *Ergonomics*, **2**, 148–152.

Singleton, W. T., 1960, An experimental investigation of speed controls for sewing machines. *Ergonomics*, **3**, 365–375.

Sjøgaard, G., 1983, Electrolytes in slow and fast muscle fibres of humans at rest and with dynamic exercise. *American Journal of Physiology*, **245**, 25–31.

Sjøgaard, G. and Saltin, B., 1982, Extra- and intracellular water spaces in muscles of man at rest and with dynamic exercise. *American Journal of Physiology*, **243**, R271–R280.

Sjøgaard, G. and Saltin, B., 1983, Potassium and lactate release from human skeletal muscle in prolonged dynamic exercise. *Acta Physiologica Scandinavica*, **118**, A12.

Sjøgaard, G. and Saltin, B., 1985, Muscle energy metabolism and electrolyte shifts during low-level long-term static contractions. *American Journal of Physiology* (submitted for publication).

Sjøgaard, G., Adams, R. P. and Saltin, B., 1985 a, Water and ion shifts in skeletal muscle of man with intense dynamic knee extension. *American Journal of Physiology*, **248**, R190–R196.

Sjøgaard, G., Kiens, B., Jørgensen, K. and Saltin, B., 1985 b, Intramuscular pressure, EMG and blood flow during low-level, prolonged static contraction in man. *American Journal of Physiology* (in press).

Skarabis, H., 1970, *Mathematische Grundlagen und praktische Aspekte der Diskrimination und Klassifikation* (Würzburg: Physica-Verlag).

Smith, E., Sonstegard, D. and Anderson, W., 1977, Contribution of the flexor tendons to the carpal tunnel syndrome. *Archives of Physical Medicine and Rehabilitation*, **58**, 379–385.

Smith, J. L., Smith, L. A. and McLaughlin, T. M., 1982, A biomechanical analysis of industrial manual materials handlers. *Ergonomics*, **25**, 299–308.

Snijders, C. and Nordin, M., 1985, unpublished material.

Snijders, C. J. and Philippens, M. M., 1985, Ontwikkeling van een nieuwe methode voor houdingsregistratie. *Tijdschrift voor Sociale Gezondheidszorg.*(in press).

Snijders, C. J., Snijder, J. G. N. and Gaasterland, N., 1977, De ogen als schakels tussen lichaamshouding en werkvlak. *Tijdschrijft voor Ergonomie*, **2**(4), 1–4.

Snook, S. H., 1978, The design of manual handling tasks. *Ergonomics*, **21**, 963–985.

Snook, S. H., 1982, Workloads, in *Low Back Pain and Industrial and Social Disablement*, Nelson, M. (ed.) (London: Back Pain Association), pp.30–36.

Snorrason, E., 1968 a, Easy chair problems. *Tidsskrift for Danske Sygehuse*, **44** (22), 3–15 (in Danish).

Snorrason, E., 1968 b, Hvilestolsproblemer. *Tidsskrift for Danske Sygehuse*, **44**, (22), 339–351.

Sommer, H. and Miller, N., 1980, A technique for kinematic modeling of anatomical joints. *Journal of Biomechanical Engineering*, **102**, 311–317.
vertebral column. *Journal of Kyoto Prefectural University of Medicine*, **71**, 659.

Spanns, F., 1970, Occupational nerve lesions, in *Handbook of Clinical Neurology*, Vol. 7, Vinken, P. and Bruyn, G. (eds.) (New York: Elsevier), pp. 326–343.

Stammerjohn, L., Smith, M. J. and Cohen, B., 1981, Evaluation of workstation design factors in VDT operations. *Human Factors*, **23**, 401–412.

Steen, B., 1964, The function of certain neck muscles in different positions of the head with and without loading of the cervical spine. *Acta Morphologica Neerlando-Scandinavica*, **6**, 301–310.

Stein, A., Ramsey, R. and Key, J., 1951, Stenosing tendovaginitis at the radial styloid process (DeQuervain's Disease). *Archives of Surgery*, **63**, 216–228.

Strasser, H., 1913, *Lehrbuch der Muskel- und Gelenkmekanik* (Textbook of muscle and joint mechanics) (Cited in Åkerblom (1948).)

Stubbs, D. A., 1984, Back pain: methodology of field measurements and case study application, in *Proceedings of the Society of Occupational Medicine Symposium on Occupational Aspects of Back Disorders*, Brothwood, J. (ed.), pp. 34–39.

Stubbs, D. A. and Buckle, P. W., 1984, Epidemiology of back pain in nurses. *Nursing*, **32**, 935–938.

Stubbs, D. A., Buckle, P. W., Hudson, M. P., Rivers, P. M. and Worringham, C. J., 1983 a, Back pain in the nursing profession. I. Epidemiology and pilot methodology. *Ergonomics*, **26**, 755–765.

Stubbs, D. A., Buckle, P. W., Hudson, M. P. and Rivers, P. M., 1983 b, Back pain in the nursing profession. II. The effectiveness of training. *Ergonomics*, **26**, 767–779.

Stubbs, D. A., Buckle, P. W., Fernandes, A. F., Baty, D., Hudson, M. P. and Rivers, P. M., 1984, Final report on back pain in the nursing profession. DHSS Report No. JR 125/120.

Sury, R. J., 1968, A comparative study of performance rating systems. *International Journal of Production Research*, **1**, 23–38.

Susnik, J. and Gazvoda, T., 1983, Biomehanska metoda za izracunavanje navora pri dviganju v sagitalni ravnini (Biomechanical method for computation of torque at lifting in the sagittal plane. I.). *Zdrav. vestn.*, **52**, 249–253.

Susnik, J. and Gazvoda, T., 1984, Biomehanska metoda za izracunavanje navora pri dviganju v sagitalni ravnini (Biomechanical method for computation of torque at lifting in the sagittal plane. II.). *Zdrav. vestn.* **54**, 1–6.

Suzuki, R., 1956, Function of the leg and foot muscles from the viewpoint of the electromyogram. *Journal of the Japanese Orthopaedic Surgeons Society*, **30**, 775–786.

Suzuki, N. and Endo, S., 1983, A quantitative study of trunk muscle strength and fatiguability in the low-back pain syndrome. *Spine*, **8**, 69–74.

Swansen, A., Matev, I. and Groot, G., 1970, The strength of the hand. *Bulletin of Prosthetics Research*, **10–14**, 145–153.

Tanzer, R., 1959, The carpal tunnel syndrome. *Journal of Bone and Joint Surgery*, A, **41**, 626–634.

Taylor, C., 1954, The biomechanics of the normal and of the amputated upper extremity, in *Human Limbs and Their Substitutes*, Klopsteg, P. and Wilson, P. (eds.) (New York: McGraw-Hill), Chapter 7.

Teiger, C., Laville, A. and Duraffourg, J., 1973, Tâches répétitives sous contrainte de temps et charge de travail. Rapport No. 39, Laboratoire de physiologie du travail et d'ergonomie du C.N.A.M., Paris.

Thompson, A., Plewes, L. and Shaw, E., 1951, Peritendinitis crepitans and simple tenosynovitis: a clinical study of 544 cases in industry. *British Journal of Industrial Medicine*, **8**, 150–160.

Tichauer, E. R., 1966, Some aspects of stress on forearm and hand in industry. *Journal of Occupational Medicine*, **8**, 63–71.

Tichauer, E. R., 1975, Occupational Biomechanics. An Introduction to the Anatomy of Function of Man at Work. New York: Institute of Rehabilitation Medicine, New York University Medical Center.

Tichauer, E. R., 1976, Biomechanics sustains occupational safety and health. *Industrial Engineering*, **8**, 46–56.

Tichauer, E. R., 1978, *The Biomechanical Basis of Ergonomics* (New York: Wiley).

Townsend, M. A. and Dul, J., 1985, Optimal modes of muscle recruitment and load sharing.

Townsend, M. A., Lainhart, S. P., Shiavi, R. and Caylor, J., 1978, Variability and biomechanics of synergy patterns of some lower-limb muscles during ascending and descending stairs and level walking. *Medical and Biological Engineering and Computing*, **16**, 681–688.

Tranel, N., 1962, Effects of perceptual isolation in introverts and extroverts. *Journal of Psychiatric Research*, **1**, 128–139.

Troup, J. D. G. and Chapman, A. E., 1969, The strength of the flexor and extensor muscles of the trunk. *Journal of Biomechanics*, **2**, 49–62.

Troup, J. D. G., Reilly, T., Eklund, J. A. E. and Leatt, P., 1985, Changes in stature with spinal loading and their relation to the perception of exertion or discomfort. *Stress Medicine*, **1**, 303–307.

Tsivian, I. L., Rakhinshtein, V. K., Motov, V. P. and Ovseychik, F. F., 1971, Results of clinical study of pressure within the intervertebral lumbar discs. *Ortopediia Travmatologiia i Protezirovanie*, **33** (6), 31–35.

Tuttle, W. W. and Horvarth, S. M., 1957, Comparison of effects of static and dynamic work on blood pressure and heart rate. *Journal of Applied Physiology*, **10**, 294–296.

Tyrrell, A. R., Reilly, T. and Troup, J. D. G., 1985, Circadian variation in stature and the effects of spinal loading. *Spine*, **10**, 161–164.

Umezawa, F., 1971, The study of comfortable sitting postures. *Journal of the Japanese Orthopaedic Association*, **45**, 1015–1022.

Vanderhoof, E. F., Imig, C. J. and Hines, H. M., 1961, Effect of muscle strength and endurance development on blood flow. *Journal of Applied Physiology*, **16**, 873–880.

Van Wely, P., 1970, Design and disease. *Applied Ergonomics*, **1**, 262–269.

Vernon, H. M., 1924 (cited by Åkerblom (1948), pp. 46–47).

Vihma, T., Nurminen, M. and Mutanen, P., 1982, Sewing-machine operators' work and musculo-skeletal complaints. *Ergonomics*, **25**, 295–298.

Vokac, Z., Bell, H., Bautz-Hoiter, E. and Rodhal, K., 1975, Oxygen uptake/heart rate relationship in leg and arm exercise, sitting and standing. *Journal of Applied Physiology*, **39**, 54–59.

Wachs, T. D., 1977, The optimal stimulation hypothesis and early development, in *The Structuring of Experience*, Uzgiri, I. C. and Weizman, F. (eds.) (New York: Plenum).

Wachsler, R. and Learner, D., 1960, An analysis of some factors influencing seat comfort. *Ergonomics*, **3**, 315–320.

Wadsworth, T. and Williams, J., 1973, Cubital tunnel compression syndrome. *British Medical Journal*, 662–666.

Walmsley, B., Hodgson, J. A. and Burke, R. E., 1978, Forces produced by medial gastrocnemius and the soleus muscle during locomotion in freely moving cats. *Journal of Neurophysiology*, **41**, 1203–1216.

Walsh, E. G., 1964, *Physiology of the Nervous System* (London: Longmans).

Wangenheim, M. and Holzmann, P., 1984, Measurement of whole-body vibration by double-pulsed holography. *Journal of Biomechanics*, **17**, 449–456.

Ward, J. and Mabey, M., 1977, Ergonomic research into optimum dimensions of office furniture. Project RXBO74F, Department of Human Sciences, University of Technology, Loughborough.

Waris, P., Kuorinka, I., Kurppa, K., Loupajarvi, T., Virolaninen, M., Pesonen, K., Nummi, J. and Kukkonen, R., 1979, Epidemiologic screening of occupational neck and upper limb disorders. *Scandinavian Journal of Work, Environment and Health*, **5**, Suppl. 3, 25–38.

Warwick, D., Novak, G., Schultz, A. and Berkson, M., 1980, Maximum voluntary strength of male adults in some lifting, pushing and pulling activities. *Ergonomics*, **23**, 49–54.

Weber, J., 1984, Het bepalen van normen bij het gebruik van schaine werkvlakken. Thesis No. OC-D-76, Twente University of Technology, Enschede.

Wernersson, S., 1982, Metod för belastningsregistrering. Rapport från Ericsson AB, Företagshälsovården, 13 pp.

Westgaard, R. H. and Aarås, A., 1984, Postural muscle strain as a causal factor in the development of musculo-skeletal illnesses. *Applied Ergonomics*, **15**, 162–174.

Westgaard, R. H. and Aarås, A., 1985, The effect of improved workplace design on the development of work-related musculo-skeletal illnesses. *Applied Ergonomics*, **16**, 91–97.

Williamson, R., 1984, Evaluation of a new technique for measuring spinal shrinkage in relation to car seat effectiveness. Ph.D. thesis, University of Nottingham.

Winer, B. J., 1970, *Statistical Principles in Experimental Design* (New York: McGraw-Hill).

Wisner, A., 1971, A quel homme le travail doit-il être adapté? Doc. ron. Laboratoire de physiologie du travail et d'ergonomie du C.N.A.M., Paris.

Wisner, A., 1981, Eléments de méthodologie ergonomique, in *Précis de physiologie du travail*, Scherrer (ed.) (Masson), pp. 521–539.

Wright, I., 1945, The neurovascular syndrome produced by hyperabduction of the arms. *American Heart Journal*, **29**, 1–19.

Wyss, Th. and Ulrich, S. P., 1954, Festigkeitsuntersuchungen und gezielte Extensions-behandlung der Lendenwirbelsäule unter Berücksichtigung des Bandscheiben-Vorfalls. *Vierteljahresschrift der Naturforschenden Gesellschaft in Zürich*, **99**, 1–144.

Yamaguchi, Y., Umezawa, F. and Ishinada, Y., 1972, Sitting posture: an electromyographic study on healthy and notalgic people. *Journal of the Japanese Orthopaedic Association*, **46**, 277–282.

Yates, J. W., Kamon, E., Rodgers, S. H. and Champney, P. C., 1980, Static lifting strength and maximal voluntary contractions of back, arm and shoulder muscles. *Ergonomics*, **23**, 37–47.

Yeo, B. P., 1976, Investigations concerning the principle of minimal total muscle force. *Journal of Biomechanics*, **9**, 413–416.

Younghusband, O. and Black, J., 1963, DeQuervain's Disease: stenosing tenovaginitis at the radial styloid process. *Canadian Medical Association Journal*, **89**, 508–512.

Zuckerman, M., 1979, *Sensation Seeking: Beyond Optimal Level of Arousal* (Hillsdale, New Jersey: Lawrence Erlbaum)

Zuidema, H., 1976, Rugbelasting door industriële arbeid. *Tijdschrijft voor Sociale Gezondheidszorg.*, **54**, 571–574, 578.

Index